Surface Tailoring of Inorganic Materials for Biomedical Applications

Editor

Lia Rimondini

Dipartimento di Scienze della Salute
Università del Piemonte Orientale "Amedeo Avogadro"
Novara
Italy

Co-Editors

Claudia L. Bianchi

Dipartimento di Chimica Fisica ed Elettrochimica
Università degli Studi di Milano, Milano
Italy

&

Enrica Vernè

Dipartimento di Scienza Applicata e Tecnologia
Politecnico di Torino, Torino
Italy

CONTENTS

FOREWORD

Biomaterials, the functional components of medical devices, are used extensively in the treatment of disease, trauma and disability. The history of biomaterials began in the pre-Christian era, with the use copper, bronze, precious stones or shells and natural fiber to substitute parts of the human body. Since then, many improvements have been implemented in medical device manufacturing, through the advent of new biomaterials and innovative design and characterization techniques.

Some of the most significant advances have taken place in the last 50 years, a period which has witnessed the introduction of pioneering joint and heart valve replacements and, more recently, the development of so-called bioactive materials which interact with the host tissues to assist healing. While these advances have been arguably sufficient to meet the medical challenges of the last century, society now faces even greater problems. These are caused mainly by an increase in lifespan which is frequently not matched by maintenance of health and quality of life.

The ability to tailor the surface of inorganic biomaterials introduces the possibility to broaden their area of application to include, for example, bioactive prostheses, biomimetic scaffolds for tissue engineering, and systems for drug delivery.

The implementation of nanotechnology has offered a completely new perspective to material scientists aiming to mimic the different types of extracellular matrices present in tissues. Techniques are now available which can produce structures on the nanometer scale which exhibit finely controlled atomic composition and architecture.

This eBook provides an interdisciplinary overview of the surface treatment of a wide range of nano- and micro-structured inorganic materials, both degradable and non-degradable, for a wide range of clinical and biomedical applications.

Luigi Ambrosio

Institute of Composite & Biomedical Materials (IMCB-CNR)
National Research Council of Italy

Piazzale F. Tecchio 80
80125 Naples, Italy
Tel.: +39-0812425929
Fax: +39-0812425932
Email: ambrosio@unina.it

Director of IMCB-CNR
President of the European Society for Biomaterials
Adjunct Professor of Artificial Organs at University of Naples Federico II

PREFACE

Biomaterial surfaces play a crucial role in medicine and biology, since most of the biological reactions occurring *in vivo* take place at surfaces and interfaces. As an example, when biomaterials are implanted, protein adsorption on the foreign surface occurs immediately. In protein-surface interactions a variety of governing factors have to be considered in defining the role of the solid-solution interface. Among them, surface charge, surface roughness, absorbed ions, surface composition, surface energy *etc.*, are determined by the physical state of the material.

In the last decades, the ability of scientists to investigate the composition and molecular structure of biomaterials surface has significantly increased, due to the evolution of the characterization techniques. Similar advancements have occurred in molecular biology and material science.

These advancements allowed the development of a biological model for surface science, aiming to understand how the biological behavior of a material can be controlled by its surface properties.

In this eBook, a number of examples of surface tailoring studies on different class of biomaterials are presented, offering a wide range state-of-the-art of biomedical surface science taking into account also the clinical applications.

The chapters of this eBook written by eminent scientists, everyone specialized in their own field, will serve as an important and useful reference source for biomedical engineers and biologists.

Lia Rimondini
Università del Piemonte Orientale "Amedeo Avogadro", Novara
Italy

Claudia L. Bianchi
Università degli Studi di Milano, Milano
Italy

Enrica Vernè
Politecnico di Torino, Torino
Italy

List of Contributors

Valentina Aina - Dipartimento di Chimica IFM & Interdipartimental Centre of Excellence “Nanostructured Interfaces and Surfaces - NIS”, Universita’ di Torino, Torino - Italy

Silvia Ardizzone - Dipartimento di Chimica Fisica ed Elettrochimica, Università di Milano, Milano, Italy

Muthupandian Ashokkumar - School of Chemistry, University of Melbourne, Parkville, VIC 3010, Australia

Nicolò Nicoli Aldini - Preclinical and Surgical Studies Laboratory - Rizzoli Orthopaedic Institute, Codivilla Putti Research Institute, Bologna, Italy

Luca Bertinetti - Dipartimento di Chimica IFM & Interdipartimental Centre of Excellence “Nanostructured Interfaces and Surfaces - NIS”, Università di Torino, Torino - Italy

Borvornwut Buranawat - King's College London, Biomaterials, Biomimetics & Biophotonics Group, Dental Insitute, at Guy's, King's College and St Thomas' Hospitals, UK

Fatih Büyükserin - *Gazi University* NanoMedicine & Advance Technology Research Center, Ankara - Turkey

Francesco Baino - Dipartimento di Dipartimento di Scienza Applicata e Tecnologia, Politecnico di Torino, Torino, Italy

Gürer G. Budak - *Gazi University* NanoMedicine & Advance Technology Research Center, Ankara - Turkey

Iolanda Biraghi - Dipartimento di Chimica Fisica ed Elettrochimica, Università di Milano, Milano, Italy

Chiara Vitale Brovarone - Dipartimento di Scienza Applicata e Tecnologia, Politecnico di Torino, Torino, Italy

Giuseppe Cappelletti - Dipartimento di Chimica Fisica ed Elettrochimica, Università di Milano, Milano, Italy

Tolga Çamli - *Gazi University* NanoMedicine & Advance Technology Research Center, Ankara - Turkey

Christophe Drouet - Université de Toulouse, Institut Carnot CIRIMAT, UPS - INPT - CNRS, ENSIACET, Toulouse, France

José Manuel Delgado - López - LEC, IACT (CSIC - UGR). Ed. Instituto López Neyra. University ogf Granada, Spain

Lucy Di Silvio - King's College London, Biomaterials, Biomimetics & Biophotonics Group, Dental Insitute, at Guy's, King's College and St Thomas' Hospitals, UK

Sara Ferraris - Dipartimento di Scienza Applicata e Tecnologia, Politecnico di Torino, Torino, Italy

Milena Fini - Preclinical and Surgical Studies Laboratory - Rizzoli Orthopaedic Institute, Codivilla Putti Research Institute, Bologna, Italy

Ismaela Foltran - Dipartimento di Chimica ''G. Ciamician'', Alma Mater Studiorum, Università di Bologna Bologna; Dipartimento di Scienze della Salute, Università del Piemonte Orientale Novara, Italy

Roberto Giardino - Preclinical and Surgical Studies Laboratory - Rizzoli Orthopaedic Institute, Codivilla Putti Research Institute, and University of Bologna Medical School, Bologna, Italy

Jaime Gomez - Morales - Laboratorio de Estudios Cristalograficos, IACT (CSIC - UGR) University of Granada, Spain

Neelam Gurav - King's College London, Biomaterials, Biomimetics & Biophotonics Group, Dental Insitute, at Guy's, King's College and St Thomas' Hospitals, UK

Michele Iafisco - Dipartimento di Chimica ''G. Ciamician'', Alma Mater Studiorum, Università di Bologna, Bologna, Italy; Dipartimento di Scienze della Salute, Università del Piemonte Orientale, Novara, Italy

Boguslaw Major - Institute of Metallurgy and Materials Science Polish Academy of Sciences, Krakow, Poland

Roman Major - Institute of Metallurgy and Materials Science Polish Academy of Sciences, Krakow, Poland

Gianmario Martra - Dipartimento di Chimica IFM & Interdipartimental Centre of Excellence "Nanostructured Interfaces and Surfaces - NIS", Università di Torino, Torino - Italy

Daniela Meroni - Dipartimento di Chimica Fisica ed Elettrochimica, Università di Milano, Milano, Italy

Marta Miola - Dipartimento di Scienza Applicata e Tecnologia, Politecnico di Torino, Torino, Italy, Politecnico di Torino, Torino, Italy.

Renato Mortera - Dipartimento di Scienza Applicata e Tecnologia, Politecnico di Torino, Torino, Italy

Barbara Onida - Dipartimento di Scienza Applicata e Tecnologia, Politecnico di Torino, Torino, Italy

Barbara Palazzo - Ghimas S.p.A., Cittadella della Ricerca Brindisi and Dipartimento di Scienze della Salute, Università del Pimeonte Orientale, Novara, Italy

Anna Paola Parrilli - Preclinical and Surgical Studies Laboratory - Rizzoli Orthopaedic Institute, Codivilla Putti Research Institute, Bologna, Italy

Yuriy Sakhno - Dipartimento di Chimica IFM & Interdipartimental Centre of Excellence "Nanostructured Interfaces and Surfaces - NIS", University of Torino, Torino - Italy

Antonio J. Salinas - Departamento de Química Inorgánica y Bioinorgánica, Facultad Farmacia; Universidad Complutense; Madrid; Spain, and Centro de Investigación Biomédica en Red: Bioingeniería, Biomateriales y Nanomedicina, CIBER - BBN, Spain

Stephanie Sarda - Institut Carnot CIRIMAT, UPS - INPT - CNRS, ENSIACET, Université de Toulouse, France

Francesca Spadavecchia - Dipartimento di Chimica Fisica ed Elettrochimica, Università di Milano, Milano, Italy

Silvia Spriano - Dipartimento di Scienza Applicata e Tecnologia, Politecnico di Torino, Torino, Italy, Politecnico di Torino, Torino, Italy

Simone Sprio - Italian National Research Council, Institute for Science and Technology for Ceramics, ISTEC - CNR, Faenza, Italy

Anna Tampieri - Italian National Research Council, Institute for Science and Technology for Ceramics, ISTEC - CNR, Faenza, Italy

Selman Yavuz - *Gazi University* NanoMedicine & Advance Technology Research Center, Ankara - Turkey

María Vallet-Regí - Departamento de Química Inorgánica y Bioinorgánica, Facultad Farmacia; Universidad Complutense; Madrid; Spain, and Centro de Investigación Biomédica en Red: Bioingeniería, Biomateriales y Nanomedicina, CIBER - BBN, Spain

Dominic Walsh - School of Chemistry, University of Bristol, Bristol, UK

CHAPTER 1

Physico-Chemical Tailoring of Material Surface Properties

S. Ardizzone*, I. Biraghi, G. Cappelletti, D. Meroni and F. Spadavecchia

Dipartimento di Chimica Fisica ed Elettrochimica, Università di Milano, Via Golgi 19, 20133 Milano, Italy

Abstract: Biomaterials' applications and performances are strongly related to their surface properties, such as surface chemistry, surface energy, roughness, surface charge, topography, hydrophobicity or hydrophilicity. In this chapter, the fundamental aspects of some of these properties – in particular wetting, adhesion, roughness and topography – will be elucidated. The relation between such surface features and the biocompatibility issues will also be discussed briefly. The three main classes of biomaterials – polymers, metals and ceramics – will be separately analyzed, listing a series of examples of the most commonly employed surface modifying techniques for that class.

Keywords: Wetting, adhesion, roughness, topography, hydrophobicity, hydrophilicity.

1. INTRODUCTION

Materials selection for most biomedical applications is based upon their bulk properties, however surface characteristics play a pivotal role in determining the success or failure of devices and implants. In fact, the biochemical reactions that determine the biocompatibility and useful life of devices and implants occur at the interface between the biomaterial and body fluids.

Biocompatibility is strongly related to the interactions between the implanted material and proteins and cells. Depending on the specific application, such interactions may be desirable or not. For example, in the case of orthopedic and dental implants, the occurrence of interactions with proteins that promote the function of osteoblasts for the new bone tissue formation is needed in order to obtain the implant integration in bone [1]. On the contrary, in the case of cardiovascular implants, the adsorption of fibrinogen induces activation of

***Address correspondence to Silvia Ardizzone:** Dipartimento di Chimica Fisica ed Elettrochimica, Università di Milano, *Via* Golgi 19, 20133 Milano, Italy; Tel: +390250314255; Fax: +390250314300; E-mail: silvia.ardizzone@unimi.it

Lia Rimondini, Claudia Bianchi and Enrica Vernè (Eds)

platelets and of the blood coagulation cascade; these events lead to blood coagulation, a major complication of vascular prostheses (Fig. **1**) [2].

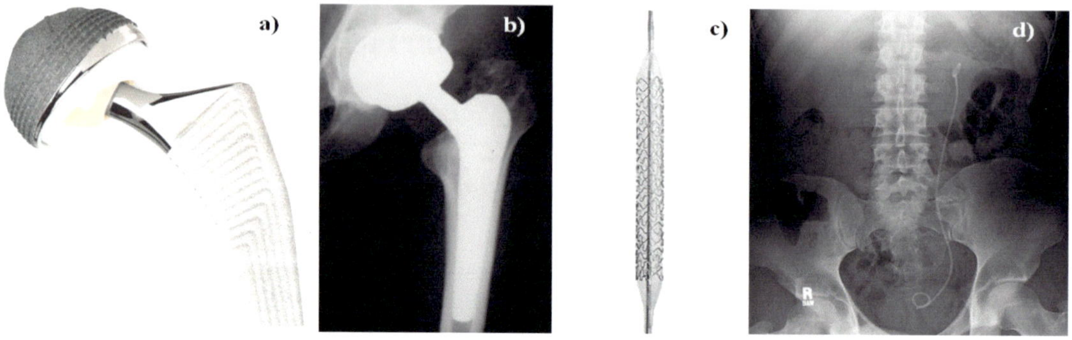

Figure 1: Examples of biomaterials applications: A total hip implant replacement (**a, b**), a coronary stent (**c**), and an urethral stent (**d**).

The surface properties play also a fundamental role in biomaterial associated infections, as bacterial contamination is related to physico-chemical interactions (*e.g.*, adhesion) between bacteria and the inert surface.

Furthermore, encrustation phenomena that take place on urethral stents and catheters are strictly related to the adhesive properties of the adopted biomaterial.

The capability of synthesizing and processing materials with tailored structures and topographies (through, for example, specific grain and pore size) provides the possibility of designing the novel proactive biomaterials, that elicit specific, timely and desirable responses from surrounding cells and tissues. Synthetic routes able of modulating the material properties as a function of the desired application must be adopted.

A great deal of efforts has been recently devoted to understanding how the surface properties of a material – such as surface chemistry and surface energy, roughness, surface charge [3], topography [4], hydrophobicity or hydrophilicity [5] – control the biological reactivity of a cell interacting with that surface. Numerous studies have shown that the surface properties of a material are directly related to *in vitro* biological performances such as protein adsorption and cell growth. The future challenge will be to fully develop a biological model for the surface state of the material in the highly complex and interactive *in vivo* biological environment.

Other examples of surface technology applied to biological problems are blood oxygenators, hemodialysis, affinity chromatography, surface diagnostics, cell culture surfaces and biosensors.

Although no material has been found to be truly biocompatible, many devices have low or acceptable risks of complications [6]. For these reasons, research efforts have focused on elucidating the mechanisms that govern protein interactions with various biomaterials including polymers [7, 8], metals [9-11], and ceramics [12, 13]. A recent development of great promise proved to be nanostructured formulations which, compared to their respective conventional counterparts (that is, micrometer-size surface features), promoted functions (such as adhesion [14], proliferation [15], *etc.*) of select cells (for example, osteoblasts *versus* fibroblasts).

In order to interact with precision with such complex biological systems, surfaces should be complex–multicomponent, multilayer, orientated, patterned. A relevant advantage is that surfaces provide high accessibility for reactions. The low energy barrier to mobility in the plane of the surface can be used to facilitate complex reactions (clustering, conformational changes, exposure and burial in membranes). Epitaxy-like phenomena can be readily exploited at surfaces. High surface area geometries can be created to increase reaction turnover rates. Unique microenvironments can enhance specific affinities and reactions. Much of the specificity of a material, particularly of a solid but of all the interphases too, is given by the surface, which is full of flaws, by its nature: God made solids, but surfaces were the work of the devil, as W. Pauli said.

The state of the surface can be modified both chemically and physically by procedures which depend on the properties of the material. In the following, before presenting examples specifically related to the main classes of biomaterials, a short introduction to the surface phenomena which mainly affect the interactions with a bio-environment will be given.

2. SURFACE WETTING, ROUGHNESS AND TOPOGRAPHY

2.1. Surface Wetting and Adhesion

The wettability of solid surfaces is an important property widely employed in several practical applications and, as well, is used as an analytical technique to characterize surfaces in fundamental materials research, mainly through contact angle measurements. The possible modulation of the wettability of a bio-material by the *in vivo* liquid is one of the main features which control a biomaterial performance and life-time. For example, wettability is important for designing membrane-type biosensors or filtration devices. Enhanced surface wettability shortens the analysis time, improves biosensor accuracy and promotes cell viability; further, hydrophylic surfaces may discourage bacterial adherence.

The wettability of a solid surface by a given liquid can be assessed experimentally by measuring the contact angle.

When a drop of liquid is placed on a solid surface, the triple interface formed between solid, liquid and gas will move in response to the forces arising from the three interfacial tensions until an equilibrium position is established. The situation is illustrated in Fig. **2a**, which shows a drop of liquid on a flat horizontal solid surface with air as the third phase. The angle, θ, between the solid surface and the tangent to the liquid surface at the contact with the solid is known as the **contact angle** (by convention, it is measured in the liquid phase).

At the equilibrium, the net force per unit length acting along the boundary line between the three phases must be zero, and thus:

$$\gamma_{GS} = \gamma_{LS} + \gamma_{GL} \cos\theta \qquad \textbf{(1)}$$

where γ_{GS}, γ_{LS}, γ_{GL} are the surface tensions between gaseous and solid phase, liquid and solid phase, gaseous and liquid phase respectively. The above equation is called **Young equation**. Although it describes the equilibrium contact angle in terms of the interfacial tension involved, an understanding of the origin of contact angle values for different liquids requires knowledge of the balance of forces between molecules in the liquid drop (**cohesive forces**) and those between the liquid molecules and the surface (**adhesive forces**). A surface that has polar groups, such as hydroxyl groups, will have a good affinity for water and therefore strong adhesive forces and a low contact angle (Fig. **2b**). Such a surface is called hydrophilic. On the contrary, we say that the surface is hydrophobic if it is made up of non-polar groups (as for polymer surfaces or surfaces covered by different kinds of organic layers), and it has a large contact angle (Fig. **2c**).

Wetting is determined by the equilibrium contact angle, θ. If the droplet will completely spread out on the solid surface, the contact angle will be close to 0°: The liquid is said to wet the solid and is very strongly attracted to the solid surface, as for water on a strongly hydrophilic solid (if θ = 0, there is complete or perfect wetting). Contact angles up to 90° characterize less strongly hydrophilic solids. If the solid surface is hydrophobic, the contact angle will be larger than 90°, being as high as ~120° for highly hydrophobic surfaces (as for water on fluorinated materials) and even greater than 150° for those kinds of materials with highly rough surfaces (superhydrophobic surfaces).

Fig. **2d** shows the so called Lotus effect, which refers to the very high water repellency (superhydrophobicity) exhibited by the leaves of the lotus flower (*Nelumbo*), while Fig. **2e** represents a SEM image of the same leaf. Due to their high surface tension, water droplets tend to minimize their surface trying to achieve a spherical shape. This effect can easily be demonstrated in many other plants, such as tropaeolum, cane and columbine, and on the wings of certain insects.

Figure 2: The surface tensions involved in the triple interface for a drop of liquid on a flat solid surface (**a**); contact angles of highly hydrophilic and highly hydrophobic surfaces, respectively (**b, c**); macroscopic and microscopic view of a lotus leaf: (**d**) The Lotus effect and **e**) micromorphological characteristics of *Nelumbo nucifera* leaf surfaces (C. Neinhuis *et al.*, Characterization and Distribution of Water-repellent, Self-cleaning Plant Surfaces, Annals of Botany, 1997, 79, 6, 667-677, by permission of Oxford University).

The contact angle is specific for any given system and not limited to a liquid/vapour interface; it is equally applicable to the interface between two liquids.

It is a frequent observation that the advancing angle (whether the measurement is performed on a drop of increasing size) and the receding angle (if the drop has a diminishing size), that represent the so called **dynamic contact angles**, may be very different. This apparent hysteresis is normally due to the roughness of the surface, and the difference between advancing and receding contact angles can be another useful and quick method of gaining qualitative information about the nature of a surface. Other causes of the hysteresis can be inhomogeneities in the distribution of adsorbates, contamination of either the liquid or the solid surface, and time-dependent surface rearrangements (for example, when a liquid contains a surfactant, low mobility can cause hysteresis of the contact angle).

The simplest way to perform **contact angle measurements** is that of placing a drop of liquid on a flat surface of the solid and viewing it with a magnifying lens. Then θ can be measured optically, usually by taking a digital image and using a software to elaborate the profile of the figure obtained.

Hydrophilicity/hydrophobicity may be the initial parameter affecting protein adsorption. It is well known that hydrophobic surfaces favor the adsorption of proteins from aqueous solution thermodynamically, but this may induce strong irreversible adsorption and promote denaturation of the protein's native conformation and bioactivity (a natural conformation of a protein is a prerequisite for its bioactivity). On the other hand, a highly hydrophilic surface may expel any protein molecule and inhibit protein adsorption.

Wetting and adhesion, as already seen, are strictly related. The physico chemical interactions involved in adhesion processes to inert surfaces (polymers, metals, glass, Teflon) are mainly of the acid–base, electrostatic and van der Waals type. They are directly related to the physico chemical properties of the interacting surface, such as the hydrophobic/hydrophilic, the surface charge and the acid–base properties, and the characteristics of the suspending medium. A knowledge of the physico chemical properties of both microbial cell and solid surface and of the characteristics of the suspending medium is important to understand and to predict the adhesion process.

It has been shown that osteoblastic cell adhesion, growth and differentiation are related to surface energy and roughness [16, 17].

2.2. Surface Roughness and Topography

The roughness of a biomaterial surface and its topography, by nanometer and micrometer scales, are found to deeply influence its performance, especially when, as in the case of orthopaedic implants, a vital role is played by integration of the material with the living tissue. Alteration in surface topography by physical placement of grooves and depressions changes the cell orientation and promotes attachment.

A solid surface is, as a rule, rich in defects, kinks, steps, ad-atoms, *etc.* but in order to obtain a reproducible roughness pattern either mechanical or chemical treatments are generally applied.The roughness of a given surface can be assessed by several methods. Common is the use of white light profilometry.

The intrinsic topography of a material strongly depends on the material composition and preparation procedure.

On rough, hydrophobic surfaces, the liquid can either follow the surface topography and show strong pinning or it can bridge from asperity to asperity while enclosing air beneath and showing almost no hysteresis in the contact angle. For the first case, the Wenzel model (see Fig. **3a**) introduces a roughness factor, r, which is defined by the following equation for the contact angle on a rough surface:

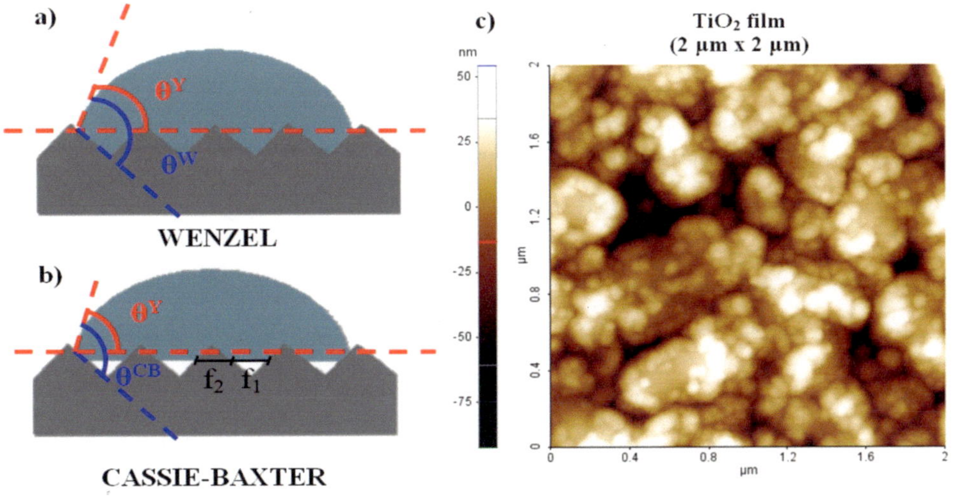

Figure 3: Drop edge on a rough surface: Wenzel-type wetting (**a**) and Cassie-type wetting (**b**). **c)** AFM image of the surface of a TiO_2 film prepared nanometric particles.

$$\cos\theta^{W} = r \cos\theta^{Y} \quad \textbf{(2)}$$

where θ^{W} is the apparent contact angle which corresponds to the stable equilibrium state (*i.e.* minimum free energy state for the system). The roughness ratio, r, is a measure of how surface roughness affects a homogeneous surface. It is defined as the ratio between the real (true, actual) surface area and the geometric surface area. θ^{Y} is the contact angle defined for an ideal surface, also called Young contact angle.

Although Wenzel's equation demonstrates that the contact angle of a rough surface is different from the intrinsic contact angle, when dealing with a heterogeneous surface, the Wenzel model is not sufficient. A more complex model is needed to measure how the apparent contact angle changes when various materials are involved: Cassie and Baxter modified Wenzel's equation by introducing the fractions f_1 and f_2, as seen in Fig. **3b**, where f_1 corresponds to the surface in contact with the liquid divided by the projected area, and f_2 is the area in contact with the air trapped beneath the drop, also divided by the projected area:

$$\cos\theta^{CB} = f_1 \cos\theta^{Y} - f_2 \quad \textbf{(3)}$$

Experimental results regarding the surface properties of Wenzel *versus* Cassie-Baxter systems showed the effect of pinning for a Young angle of 180° to 90°, a region classified under the Cassie-Baxter model. This liquid-air composite system is largely hydrophobic. After that point a sharp transition to the Wenzel regime is found where the drop wets the surface but no further than the edges of the drop.

The surface roughness is found to influence the cell morphology and growth. Alteration in surface topography by physical placement of grooves and depressions changes the cell orientation and attachment [18, 19].

It is understood that the microstructured rough surface enhances the effect of surface chemistry into superhydrophobicity, water repellency and reduced particle adhesion.

Topographical and chemical surface patterning, on different length scales generates different responses of the biological system. In order to obtain a positive and selective response, the molecular architecture should in general match some recognition site of biomolecules on the biological side of the hybrid interface.

3. POLYMERIC MATERIALS

Polymeric materials have wide applications in biomedical implants and devices, finding uses in areas such as the modulation of wound healing, implantable medical devices and artificial organs, prostheses, ophthalmology, dentistry, bone repair, and drug delivery systems [20]. Both synthetic and natural polymeric materials are used for manufacturing biomedical implants: The most commonly utilized include cellulose, chitosan, polytetrafluroethylene (PTFE), polyvinylchloride (PVC), segmented polyether urethane (SPU), polyethylene (PE), silicone rubber (SR), nylon and polysulfone (PSF) [21].

Polymeric biomaterials offer indeed a series of relevant advantages: They are relatively easy to manufacture into products with various shapes, at reasonable costs, and with desirable mechanical and physical properties. However, their main drawback is that they do not possess biocompatibility and antimicrobial activity.

The desired physicochemical and mechanical characteristics of polymeric materials, such as toughness, tensile strength, flexibility and wear resistance, depend on their bulk properties. In order to enhance the biocompatibility and antimicrobial activity of polymeric materials without altering their bulk properties, surface modification techniques are usually applied. They are aimed at modifying the most relevant surface properties of polymers, such as their roughness, conductivity, polarity, refractive index and wettability.

The various polymer treatment processes have advantages and disadvantages and a few techniques have been proved to be both highly effective and suitable for various *in vivo* environments.

3.1. Physical Treatments

When polymer surfaces are exposed to UV light or gamma rays, the radiation breaks the bond within the polymer chain thereby creating sites which can become either functional groups upon exposure to gas (oxygen containing species) [22] or reactive sites to which active molecules or ligands can be attached [23]. In this way, a range of surface chemical modifications can be produced, including etching, deposition and polymerization. Surface etching, for example, can lead to retention of adsorbed proteins through increased ionic, polar or hydrophobic interactions.

These techniques generally improve surface wettability and change surface morphology, enhancing cell proliferation and growth, while deposition of new reactive groups can help in subsequent covalent immobilization of the proteins [24].

High-energy electrons, gamma radiation, UV and visible light are also used to initiate grafting of polymers in a process known as surface photografting, which is mainly employed to improve polymer haemocompatibility and to generate surfaces that exhibit nonfouling properties. UV-induced graft polymerization can create micropatterned surfaces by using lasers and masks [25] and has the ability to tailor the depth of surface reactivity by varying wavelength and thus absorption coefficient [26]. A "photo-oxidization" method, by which peroxide groups can be universally introduced onto polymer surfaces [27-29], does not need special instruments and can be performed easily: The reaction proceeds by immersing the materials in hydrogen peroxide solution under UV irradiation, and by changing reaction conditions, such as the photo-oxidation time and grafting time, it is possible to modulate the density of the hydrophilic groups at the surface [30, 31].

However, UV irradiation also results in polymer degradation, which leads to reduction in molecular weight and loss of mechanical strength. Moreover, UV treatment can affect the optical properties of the polymer, and UV light can be blocked by particles, which may affect treatment consistency.

Another important category of surface modifications is constituted by ionized gas treatments, among which gas discharge, flame treatment and plasma techniques are frequently used.

As for gas discharge, corona discharge is a simple, low cost, continuous process in which an electrically induced stream of ionized air bombards the polymer surface [32]. However, this results in a broad range of oxygenated groups and therefore may be less useful when specific functionalities are to be introduced. Since it does not operate under vacuum, contamination may be an issue, and variations in local temperature and humidity can affect the consistency of the treatment. Finally, it has been reported that surface polar groups on corona treated polyolefins are particularly unstable. Materials should therefore be used shortly after treatment.

Like corona discharge, flame treatment is a non-specific surface functionalization method that bombards the polymer surface with ionized air, generating a broad spectrum of surface oxidation products to the top for several monolayers [33]. In this method, the reactive oxygen is generated by burning an oxygen rich gas mixture. Here, the enhanced adhesion and wettability can be achieved by the introduction of hydroxyl, aldehyde, and carboxylic acid functionalities, as already applied to polyethylene [32].

Plasma-based treatments enable adhesive metallic, ceramic and polymeric coatings to be applied to a range of substrates, including polymers. These techniques are used to etch, cross-link, oxidize and introduce specific chemical functionality at the material interface. Plasmas consist of highly excited species that can be obtained using gaseous (*e.g.*, radio frequency, microwave, *etc.*), metallic and laser-based sources. The resulting gaseous environment is highly reactive and capable of eliciting a range of plasma-surface interactions, even on inert materials such as ceramics. Plasma can provide modification of the top nanometers of a polymer surface without using solvents or generating chemical waste and with less degradation and roughening of the material than many wet chemical treatments [34, 35]. It can easily introduce polarized groups such as hydroxyl, carboxyl, amino and sulfate groups on polymer surfaces, depending on the selection of plasma gas (Ar, N_2, O_2, H_2O, CO_2, NH_3) and operating parameters (pressure, power, time, gas flow rate) [33]. Two categories of plasma surface modification techniques can be individuated: Plasma ion implantation and plasma modification/polymerization methods. The first one is used to introduce elements into the surface without thermodynamic constraints. In the latter, the polymer generally does not reproduce the chemistry of the monomer, instead the process tends to form complex, cross-linked polymers with random structures that incorporate various structural fragments of the monomer. In a process called radio frequency glow discharge (RFGD) plasma can be used to initiate surface graft polymerization of vaporized monomers [36]. Plasma treatments can be used as precursors to other surface modification techniques, for example, plasma activation followed by ultraviolet (UV) graft polymerization [37] or plasma activation followed by silanization [38, 39].

Ion beam techniques are commonly applied to polymers too. The ion implantation technique provides the implantation of ions of a desired material into another solid. Ions of the desired element are produced by an ion source, electrostatically accelerated to high energy (in the range of 10 to 500 keV) and then transferred to the target. Ion implantation has been used to increase surface hydrophilicity of silicone catheters to reduce the influence of thrombosis and to improve cell adhesion, proliferation and spreading on polyurethane and polyethersulphone. The incorporation of Ag ions has been used to impart antimicrobial properties. Nitrogen ion implantation of polycarbonate causes chain cross-linking at the polymer surface and thus improves wear and friction characteristics. Cross-linking results in surfaces with higher wear resistance, but polymer degradation can occur if processing parameters are not well-controlled, as chain cross linking and scission compete during the ion implantation process.

Ion beam assisted deposition (IBAD) is a thin film deposition process that combines physical vapor deposition with ion beam implantation. The coating material is vaporized using a high power electron beam or a PVD. The substrate is placed in the vapor, and individual coating atoms or molecules condense and stick on the surface of the substrate to form the coating. Simultaneously, highly energetic ions (100-2000 eV) are produced and directed at the substrate surface. The concurrent ion bombardment intermixes coating and substrate atoms and eliminates the columnar microstructure often observed in conventional, low temperature physical vapor deposition to create very dense, adherent film structures. This technique has been used to create alumina sealant coatings to prevent oxygen and water vapor permeation into polymers. Other applications are the deposition of antimicrobial Ag coatings and diamond like carbon coatings that are chemically inert, optically transparent, extremely hard and have low friction coefficient.

3.2. Chemical Treatments

A special category of polymers, water-soluble polymers, such as polyethylene glycol or polyethylene oxide (PEO), can prevent plasma protein adsorption, platelet adhesion and thrombus formation by decreasing the interfacial free energy and the repulsion force between the PEO chains and proteins. The addition of a low percentage of surface-modifying additives to the bulk polymer significantly changes its surface properties. For example, blending polysiloxane-containing copolymer with the base polymer before processing leads to the migration of the former to the surface during the fabrication, creating regions of alternating hydrophilic and hydrophobic domains [40]. Insertion of carboxylic groups in polyurethane using thioglycolic acid (TGA) increases surface hydrophilicity, limits water uptake, and reduces platelet adhesion and fibrinogen adsorption [41, 42].

Over the past forty years, a variety of techniques to graft water-soluble polymers onto bulk polymers have been employed; some of the techniques produce surfaces that are more biologically inert than others. Chemical polymer grafting is a technique used to chemically bond two or more types of resin molecules, and able to generate nonfouling surfaces. Moreover, grafting of polyfunctional agents can be utilized to functionalize the polymer surface in order to attach bioactive compounds.

Chemical polymer grafting can be achieved by graft polymerization of monomers or by direct coupling of existing polymer molecules to surfaces containing functional groups. An alternative way to the chemical graft polymerization uses high-energy radiation (both ionizing and non-ionizing) or oxidizing agents such as ozone or acid treatments.

Chemical polymer grafting can be classified among the so called "wet methods", which include also SAMs, biomolecule attachment and biomimetic surfaces. These methods are usually called "biomimetic processing", "biogenic processing", or "bio-inspired processing".

SAM stands for Self Assembled Monolayer, an organized layer of amphiphilic molecules in which one end of the molecule, the "head group", shows a special affinity for a substrate; the other end of the molecule, the tail, has a functional group at the terminal. Although research on surface modification using silane monolayers continues to focus primarily on inorganic substrates, this is certain to be an emerging field for polymer surface modification. SAMs are typically formed by adsorption of alkanethiols onto gold, silver, or copper surface, but also by chlorosilanes on OH-terminated oxide surfaces and by various acids on metal oxides [43, 44]. SAMs provides well defined surfaces that can be controlled at a molecular-level and that can also be patterned using microcontact printing and photolithography. They have been widely used as model surfaces for protein adsorption and cell adhesion, in order to determine the characteristics of functional groups that give the surfaces the ability to resist the nonspecific adsorption of proteins from solution [45]. Because of their nearly crystalline organization, SAMs offer the potential for more defined surface functionalization than typical wet chemical or ionized gas functionalization techniques [46]. However, the siloxane linkage can be hydrolyzed at high temperatures or alkaline pH [47, 48]. Silane coupling agents have been utilized to condition PET, ethylene vinyl alcohol (EVOH), and Nylon 6 surfaces with titanium oxide for the development of a bone tissue scaffold [49].

At variance with self-assembled monolayers, the Langmuir-Blodgett films do not form chemical bonds with the substrate, but the process is based on a physical adhesion. Thus, the application of LB films is critically related to their endurance. However, the stability of LB films can be improved by cross-linking the molecules after formation [50-55]. The Langmuir–Blodgett (LB) technique allows the deposition of highly ordered monomolecular films with densely packed structure and precisely-controlled thickness. The film formation is achieved by spreading amphiphilic molecules on a water surface; then the film is compressed by a movable barrier. The monolayer is subsequently transferred to a vertical plate at constant surface pressure, as can be seen in Fig. **4**: The LB film is formed on a solid substrate by vertical passage through a floating monolayer. A multilayer with controlled molecule orientation can be formed by depositing other monolayers onto the substrate; for example, once the first monolayer has been deposited on the substrate, subsequent movements in and out the surface often lead to the deposition of an additional layer on each passage. There are different types of deposition that only refer to the behavior during the deposition, not to the structure of the film that is formed.

Fig. **4** also shows experimental isotherms of surface pressure against area per molecule, from the shape of which it is possible to recognize the main monolayer phases: Gaseous (G), liquid-expanded (L_1), liquid-condensed (L_2) and solid (S) phases. In order to be transferred as a Langmuir-Blodgett film, the monolayer should show a liquid-condensed (L_2) state, with a rather compact but elastic structure.

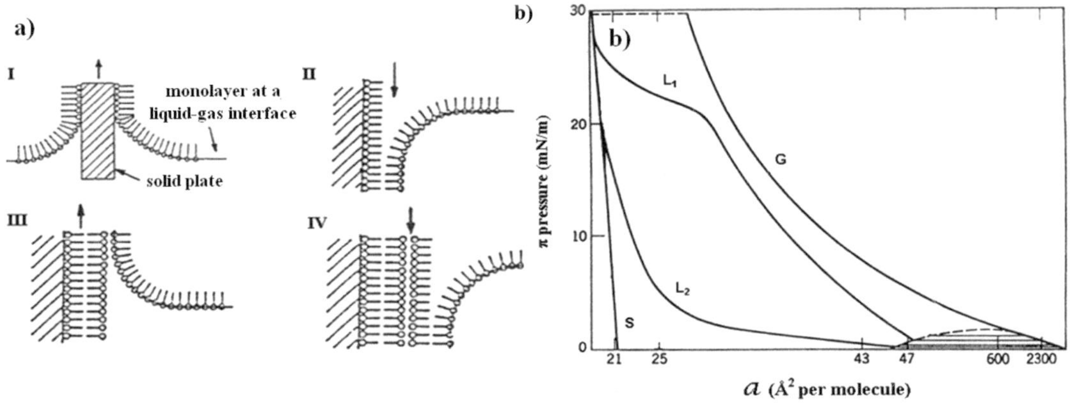

Figure 4: **a**) The Langmuir-Blodgett deposition process, with the formation of one (I) or more layers (II, III, IV), and **b**) Π (surface pressure) - *a* (area per molecule) isotherms indicating the main phases: G = gaseous, L_1 = liquid-expanded, L_2 = liquid-condensed with chains titled towards next nearest neighbor, S = solid.

Recent information, mostly derived from grazing-incidence X-ray diffraction measurements, shows that there is also a rich polymorphism within the condensed phases (S, I, L_2); similarities between these polymorphic phases and certain crystalline and smectic phases - that together with nematic phases forms the group of materials known as liquid crystals - have been identified [56].

As for biomolecule attachment, biomolecules such as proteins, peptides, polysaccharides, lipids and oligonucleotides can be immobilized onto biomaterial surfaces in order to produce surfaces able of eliciting specific, predictable and controlled biological responses (biomimetic surfaces). In order to obtain such a desired outcome, parameters like conformation, average density and spatial distribution of the immobilized molecules, must be controlled.

The major methods of immobilizing a bioactive compound to a polymeric surface are adsorption *via* electrostatic interactions, ligand–receptor pairing, and covalent attachment. Non-covalent adsorption is sometimes desirable, as in certain drug delivery applications or in the case of regenerable antimicrobial textiles. A technique called "entrapment" is widely used and consists in swelling the polymer surface using a mixture of a solvent and a non-solvent containing target biomacromolecules. then the biomacromolecules are "entrapped" on material surface by immersing the material into the pure non-solvent [49, 56]. However, non-specific adsorption results in a surface with randomly oriented molecules deposited in a non-controlled manner. This results in a short-term functionality with an efficiency that is diminished by the denaturation and biomolecule desorption.

On the contrary, covalent immobilization offers several advantages by providing the most stable bond between the compound and the functionalized polymer surface. In the biomedical field, a covalent immobilization can be used to extend the half-life of a biomolecule, prevent its metabolism or allow for continued bioactivity of in-dwelling devices (as in vascular devices, shunts, or catheters). Among polymer surface modifications for the attachment of bioactive compounds, covalent modification can be achieved using several techniques that have already been presented, such as plasma polymerization and photochemical grafting. Nevertheless, non specific covalent immobilization gives little control over the orientation and conformation of the introduced molecule and therefore over its biological activity. Thus, tethering a bioactive compound to a solid substrate *via* a spacer molecule, such as polyfunctional agents, can also improve bioactivity by reducing steric constraints and shielding the compound from hydrophobic surface induced denaturation.

4. METALLIC MATERIALS

Metallic biomaterials have been mainly employed in the fabrication of medical devices for the replacement of hard tissue – such as artificial hip joints, bone plates, and dental implants – because of their reliable mechanical performance.

Moreover, inorganic materials represent a valid alternative to polymers in some applications, since polymeric approach is facing recurrent problems [58, 59]. All polymers induce inflammatory reactions, some polymer formulations exhibit fracture during stent expansion and delayed endothelium growth, which can lead to acute stent thrombosis. In polymer based drug delivery, a common solvent is needed to dissolve drugs and polymers, thus restricting the number of drugs available – specifically polymers are unable to carry proteins or genes as antirestenoic agents. Furthermore, the polymeric coating may still contain small amounts of residual solvent that can be released into the blood together with the real drug. On this basis, inorganic biomaterials can represent an excellent attempt to overcome these problems and substitute polymers in specific applications.

Nowadays biodegradable metals are breaking the current paradigm in biomaterial science to develop only corrosion resistant metals. In particular, metals which consist of trace elements naturally present in the human body are promising candidates for temporary implant materials. Magnesium and its alloys have been investigated recently by many authors as a suitable biodegradable biomaterial. However, these materials undergo severe pitting corrosion *in vivo*, as the chloride content of the body fluids is about 150 mmol/L [60-62]. Therefore, the major recent advances in magnesium alloys as temporary biomaterials have been in understanding the interface and the interactions between magnesium alloys and their biological environment. Nowadays, a great deal of efforts are devoted to understanding the influence of the alloying elements on the formation of the corrosion protective layers and on the surrounding biological environment *in vitro* and *in vivo*.

Another class of promising metallic biomaterials are nearly equiatomic Ni-Ti alloys (Nitinol). Their use in medical devices is rapidly increasing, thanks to two generic properties of Nitinol alloys: Thermally induced shape recovery and superelasticity. However, several aspects of the alloys' short and long term efficacy have yet to be comprehensively determined, especially parameters relating to surface and corrosion performance. The surface conditions of alloys are modified using various advanced techniques, exploiting oxidization in air-water-steam mixtures, and electrochemical passivation procedures.

More ordinary materials for surgical implants include stainless steel (316LSS), cobalt chromium (Co–Cr) and titanium alloys. Stainless steel and Co-Cr alloys present several disadvantages compared to titanium alloys. They are found to release toxic elements such as Ni, Cr and Co, due to the corrosion in the body environment [63, 64]. In addition, both 316L SS and Cr–Co alloys possess much higher elastic modulus than bone: This mismatch in the elastic modulus prevents an efficient stress transfer from the implant to adjacent bone, resulting in bone resorption and loosening of the implant after some years of implantation [65]. Instead, titanium-based materials combine a series of outstanding characteristics such as high strength, low density (high specific strength), high immunity to corrosion, complete inertness to body environment, enhanced biocompatibility, low modulus and high capacity to join with bone and other tissues [66].

The superior biocompatibility of titanium is provided by a very stable passive layer of TiO_2 on its surface. Even if the passive layer is damaged, the layer is immediately rebuilt. Any metal, or metal alloy, is covered by a native oxide layer formed by interactions of the bare surface with oxygen/moisture, whose composition may vary, even largely, from that of the underlying material. In the case of titanium, the nature of the oxide film that protects the metal substrate from corrosion is of particular importance and its physicochemical properties such as crystallinity, impurity segregation *etc*., have been found to be quite relevant [67].

Table 1: Surface modification techniques that are utilized to improve several properties of titanium and its alloys.

Methods	**Modified properties**	**Refs.**
Ion implantation	Hardness	[68]
Plasma spray coating	Hardness	[69]
High-energy electron-beam irradiation	Hardness, wear resistance	[70]
Diamond-like carbon coating	Wear resistance, corrosion resistance, biocompatibility	[71, 72]
Mechanical treatments (grooving, sand blasting, acid etching, grit blasting, polishing)	Improved cell attachment and proliferation, biocompatibility	[73-77]
Heat treatment	Improved cell attachment	[78]
Laser Engineering Net Shaping	Bone ingrowth, modulus (biomechanical compatibility)	[79-81]
Deposition of calcium phosphate (dip coating, spin coating, electron-beam deposition, pulsed laser deposition, plasma spraying)	Biocompatibility, osseointegration	[82]
Thermo chemical surface treatments (nitriding, carburization, boriding)	Hardness, corrosion resistance (oxynitrides), biocompatibility	[83-87]
Oxygen diffusion hardening	Hardness, wear resistance	[88, 89]
Laser annealing	Hardness, corrosion resistance	[90]
Nanopatterning (scanning probe, electron beam, colloidal, imprint lithographies)	Improved cell attachment and proliferation, higher osseointegration	[91]
Surface coating with nanoparticles	Improved cell attachment and proliferation, higher osseointegration	[92-94]

Thus, titanium and its alloys are fast emerging as the first choice materials for bone replacement implants. A wide range of surface modification techniques have been developed to tailor their properties (Table **1**).

The mechanical, wear and corrosion properties of a material are largely determined by its microstructure. Another advantage of titanium alloys is linked to the wide spectrum of their microstructures: Titanium alloys properties can be tailored for specific requirements by varying the alloy chemistry and thermomechanical processing. In addition, various surface treatments have been exploited for improving the tribological properties of titanium and its alloys. Surface modification techniques such as physical deposition methods (*e.g.*, ion implantation) [68], plasma spray coating [69], thermo chemical surface treatments (*e.g.*, nitriding, carburization and boriding) have been used to improve the surface hardness of titanium alloys. Moreover, the enrichment of nitrogen in the passive film and formation of oxynitrides in the implanted and passivated layers enhances the corrosion resistance of these alloys. Though the corrosion resistance of ion-implanted surface is very high, the ion implant layer is often found to wear off with time. To overcome some of the problems associated with nitriding, high-energy electron-beam irradiation can be employed to produce Ti-based surface composites with improved hardness and enhanced wear resistance [70]. Another technique that has been studied with considerable interest to improve the abrasive wear of titanium alloys is oxygen diffusion hardening (ODH). Its enhanced performance is due to the adherent surface modification by oxygen diffusion, which does not spall or delaminate like the overlay coatings. Laser annealing is another innovative technique to harden the surface of the titanium alloys. Diamond-like carbon (DLC) coating is emerging as a promising technique for orthopedic implants, because it enhances the material tribological properties such as friction and wear [71, 72]. In addition, DLC coating possesses superior corrosion resistance, superior mechanical properties, and higher biocompatibility and haemocompatibility.

The physiological processes that occur during the process of ossointegration depend upon the surface properties such as surface chemistry, surface topography, surface roughness and mainly the surface energy which changes with all the discussed properties [95]. Nanometer surface roughness has proved to be benefical for osteoblast adhesion as the nano surfaces mimic the human bone. Therefore, the development of nanosurface topography is being studied with considerable interest. Thomas *et al.* observed increased osteoblast adhesion on novel surface topography created by carbon fibers with nanometer dimensions [96]. Experimental work of Jayaraman *et al.* on the behavior of grooved and sand blasted, acid etched titanium surfaces revealed that they offer better cell attachment and proliferation than rough surfaces [73]. Surface grit blasting and polishing methods enhance cell growth, improve fixation through increase in interlocking surface area [74-77]. These surface treatments also change the oxide thickness and hence the biocompatibility of titanium implants. The biocompatibility can also be improved by means of an appropriate heat treatment in oxygen or in air. MacDonald *et al.* revealed that heat-treating at low temperature enriches the surface with Ti and Al and promotes the cell attachment [78].

Waviness and porosity of the implant also play a vital role in bone integration. Bone ingrowth into porous surfaces can cause strong interlocking of surrounding bone tissue with the implant, resulting in improved biomechanical compatibility and high resistance to fatigue loading [79-81]. Remodeling of bone on porous surfaces has also been investigated in great detail. On the grounds of detailed studies made by several authors [97-100], we could conclude that the size of the pores should be in the range of 100–200 μm for better osseointegration. Porous biomaterials are also expected to lead to a good interface between the bone and the implant, since the elastic and mechanic modulus of such porous biomaterial is very low. By using a technique called Laser Engineering Net Shaping (LENS^{TM}), Vamsi *et al.* demonstrated that the modulus of a material can be tailored to greater extent by varying its porosity. They showed that the modulus of porous titanium implants can be varied from 1.7 to 47.7 GPa by choosing appropriate laser processing parameters [101].

The deposition of films of highly biocompatible calcium phosphate on the alloy surface has been studied in order to ensure a chemical bond with living bone. The methods more commonly used for coating alloys with calcium and phosphorous include dip coating, electron-beam deposition, pulsed laser deposition and plasma spraying [82]. Besides the physical deposition methods, several chemical methods have recently been reported. These methods are more economical and are able to produce uniform coatings on complex shaped implants. Hydroxylapatite (HA) or mixture of HA and TiO_2 are prepared using sol–gel coating technique [102, 103]. Generally, the ceramics are either deposited by dip or spin coating. In dip coating, the material is immersed in the sol and then withdrawn with a well defined speed under controlled conditions (temperature and atmosphere). The coating thickness in this method depends on the withdrawal speed, the solid content and the viscosity of the liquid. The atmosphere finally controls the evaporation of the solvent when pulled out and the subsequent destabilization of the sols by solvent evaporation, which leads to gelation processes and to the formation of thin film of the ceramic. On the other hand, in spin coating, the coating layer is obtained by spinning it around an axis that is perpendicular to the coating area. Centripetal acceleration causes most of the dispersion to spread out to, and eventually off, the edge of the substrate, leaving a thin film on the surface. Final film thickness and other properties depend on the nature of the dispersion (viscosity, drying rate, surface tension, *etc.*) and the parameters chosen for the spin process, such as final rotational speed, acceleration, and fume exhaust. The samples after coating are usually dried for long time and heated at low temperatures.

5. CERAMIC MATERIALS

Ceramic biomaterials are essentially inert within the body. This is the result of improved resistance to further oxidation. They offer temperature stability, low thermal conductivity and high cohesive strength. Although exhibiting more brittle behavior than metals, ceramics of small grain size have increased their use in orthopedics. There are several approaches now available to make ceramics more flexible, including nanophase technology [104]. If the brittle nature of ceramics is the only drawback in designing a new product, these technologies could be used to make tougher, more flexible implants.

There are several kinds of ceramic materials depending on the desired function or application: Biostable, bioactive (*e.g.*, hydroxyapatite, bioactive glasses, and glass-ceramics), flexible, bioresorbable (*e.g.*, tricalcium phosphate), *etc.*

In the case of biostable ceramics, they neither resorb nor induce osteoblastic apposition on their surfaces within the body. The major examples of this category are alumina (Al_2O_3) and zirconia (ZrO_2). Both oxides are inert, strong and stable; as a consequence, there is no need to follow degradation products. Disadvantages are weak interface with bone or tissue and low shock resistance. Many other oxides have been used experimentally as bioceramics, particularly in combination: $CaO \cdot Al_2O_3$, $CaO \cdot TiO_2$, $CaO \cdot ZrO_2$, as non-porous and porous implants in muscles and connective tissues. Alumina has been used clinically for many years as matched pairs of femoral head and acetabular components in total hip arthroplasty. The material porosity plays a vital role in bone integration. Mineralized bone growth into porous calcium aluminate skeletal implants requires a minimum interconnective pore size of 100 μm; the ingrowth of osteoid tissue is said to require a minimum pore size between 40 and 100 μm and the ingrowth of fibrous tissue requires a minimum pore size between 5 and 11 μm [105].

Zirconia has become a valid alternative as ceramic instead of alumina, due to its structural properties like good response to friction and high resistance to fracture (it possesses one of the highest fracture ratios among monolithic ceramics); for this reason zirconia femoral heads are particularly strong. The features of this material can be tailored by modulating the synthetic route of the ZrO_2 powder to give the concentrated suspension precursor of the high temperature treated ceramic material. In our laboratories, the synthesis of both pure zirconia and Y-doped zirconia was studied. The final features of the ceramic, calcined at 923 K, appeared to strongly depend on the adopted synthetic technique and type of dopant addition. Xerogels prepared by acrylamide gelification method (sample A), or by sol-gel and mechanical mixing (sample B) sol-gel synthesis (sample C) sol-gel and impregnation (sample D) show different morphological features (Table **2**).

Table 2: Morphological features of Y-doped zirconia samples prepared by different routes at constant annealing temperature (923 K): Crystallites diameter of the anatase phase ($d_{Scherrer}$), specific surface area (S_{BET}), total pore volume (V_{tot} BJH) of samples prepared by acrilamide gelification method (A), sol-gel mechanical mixing (B), sol-gel synthesis (C), and sol-gel impregnation (D).

Sample	$d_{Scherrer}$ (nm)	S_{BET} (m^2/g)	V_{tot} BJH (cm^3/g)
A	70	19	0.063
B	12	46	0.075
C	18	19	0.017
D	14	49	0.071

Samples A and C have similar surface area, but different total pore volume and pore size distribution; samples B and D exhibit similar surface area, similar total pore volume and pore size distribution. Therefore, when the guest cation is added in the precipitation reaction mixture, the surface area is drastically reduced, while surface deposition techniques seem to maintain both the porosity and surface area.

Bioactive ceramics provide ostheophilic surfaces so that bone-forming osteoblasts lay down bone in direct apposition on these surfaces; these bioactive glass and glass-ceramics well bind to bone and soft tissue. A wide variety of bioresorbable ceramics have been proposed and developed with differing rates of degradation and dissolution. One remarkable success of bioactive ceramics is the clinical use of sintered hydroxyapatite (HA) for implants, due to its bioactivity and osteoconductivity [106]. Synthetic hydroxyapatite exhibits good properties as biomaterial, such as biocompatibility, bioactivity, *etc.* and has interesting applications in the fields of bone tissue engineering and orthopaedic therapies. Biological HA is not stoichiometric according to the ideal formula $Ca_{10}(PO_4)_6(OH)_2$, but Ca^{2+} is replaced – even if at low extent – with other ions such as Na^+, K^+, Mg^{2+}, Sr^{2+}, while PO_4^{3-} and OH^- can be partially substituted from other anions such as $(CO_3)^{2-}$, $(HPO_4)^{2-}$, $(P_2O_7)^{4-}$, SiO_{4-}.

There exist numerous synthetic strategies for producing a substituted hydroxyapatite, including wet producing, hydrothermal and ultrasonic mobilization methods, electrodeposition technique, sol-gel and solid state synthesis. Size dimensions, porosity, morphology and surface properties represent the physicalchemical features that should be tailored in synthetic HA crystals for optimizing their specific biomedical applications.

Porous HA has been prepared using various technologies to control pores dimension, shape, distribution and interconnections. New formation methods at lower temperatures have been developed, allowing one to obtain porous bioceramics with a low degree of crystallinity. Colloidal processing, starch consolidation, gel casting and foams have yielded excellent results, producing bioceramics with a bimodal distribution of the pores size that can be modified as a function of the sintering conditions.

The interconnected network of pores promotes bone in-growth, but also allows bioceramics to be utilized as drug delivery agents, by inserting different bioactive molecules. Many studies have demonstrated that hydroxyapatite ceramics can be used to deliver steroids, antibiotics, proteins, hormones, anticancer drugs. The adsorption and release of bioactive molecules are strongly affected not only by the chemical properties of the drug molecule but also by the chemical and structural characteristics of the HA substrates. Hydroxyapatite nanocrystal materials can be synthesized with either plate-shaped or needle-shaped morphologies and with different physicochemical surface properties. The HA surface area and surface charge (Ca/P ratio) influence the adsorption and release kinetics of drugs.

Two groups of hydroxyapatite and biphasic calcium phosphate (BCP) ceramic powders with distinct microstructures can be fabricated by different drying processes [107]. The phase composition, microporosity and micropore size distribution of these ceramic powders have a notable impact on protein adsorption. The effect of their microporous structure on protein adsorption is related to the structures and properties of the proteins themselves: HA and BCP powders with more microporosities adsorb more fibrinogen or insulin, of smaller molecular size.

However, the low fracture toughness of HA ceramic limits the scope of clinical applications [108]. It is reported that compounds including SiO_2, *i.e.*, bioactive glasses [109], $CaSiO_3$ ceramics [110, 111], apatite/Wollastonite (A/W) glass-ceramics [112] and dicalcium silicate [113] show good biocompatibility [114]. Bioglass has been developed in the early '70s. It consists essentially of a glassy material designed to adhere directly to bone, composed of reactive silica, calcium oxide and phosphate groups in an alkaline pH. A typical composition is 45% by weight SiO_2, 24.5% CaO, 24.5% Na_2O, and 6% P_2O_5. While crystalline silicates have found few applications in this field, the most interesting applications are based on vitreous silica, in particular just in the field of bioglass. It has been shown that these materials can bind to bone tissues and in certain conditions also to soft tissues. Bioreactive materials change their composition and surface structure as a result of the graft, forming a layer of hydroxycarbonate apatite (HCA). This layer provides an interface able to bind tissues. The similar chemical composition of the mineral phase of bones and HCA explains the good adhesion which is created.

In recent years, some Si and Mg containing ceramics have drawn interests in the development of bone implant materials [115-118]; among the magnesia–silica systems, forsterite (Mg_2SiO_4) is considered to be a promising biomaterial [119]. *In vitro* studies showed significant osteoblast adhesion, spreading and growth on the surface of forsterite ceramic [120]. Webster *et al.* [121, 122] have designed and synthesized nanophase ceramic compacts with grain sizes less than 100 nm in diameter. Forsterite nanopowder, unlike micron-sized forsterite, possesses calcium phosphate layer formation ability and is bioactive. Similar results were obtained in the case of calcium phosphate ceramics [123, 124], which show that the *in vitro* and *in vivo* Ca^{2+} ion release from the synthetic HA nanopowders are similar to bone apatite and significantly faster than microscale conventional HA. The powders dissolution rate is largely dependent on the crystallinity degree, phase composition, microstructure, surface area and density. The nanometer-sized grains and the high-volume fraction of grain boundaries in nanostructured materials have been found to increase osteoblast adhesion, proliferation, osteointegration, and the deposition of calcium containing minerals on the surface of these materials [125].

Nanosized ceramic particles of silica, alumina, zirconia and titania are also widely investigated. Cell compatibility studies on these particles showed enhanced osteoblast function and hence large deposition of calcium minerals [16]. Further, wear particles generated from these nanophase ceramics are less detrimental on bone cells when compared to conventional ceramic wear particles [126]. The increase in cell adhesion is attributed to the increase in surface energy at the grain boundaries. Select enhanced osteoblast adhesion could be, in some cases, independent of surface chemistry and material phase but dependent on the surface topography (specifically on grain and pore size) of nanophase ceramics [122]. The above studies lead to conclude that a nanosurface seems to be advantageous from the biocompatibility and biomechanical compatibility points of view.

In order to study the interactions of a specific chemical functionality with biological environment and to improve the materials' biocompatibility, many researchers have applied the self-assembled monolayer (SAM) deposition technique to ceramic materials [127, 128]. This technique, which has already been presented in the section dedicated to polymeric materials, is just one of the different approaches of the self-assembly processes. They are considered effective and environmentally benign methods for surface chemical and biological functionalization of ceramic and composite materials.

6. NANOMATERIALS AND COMPOSITES

Biomaterials currently used in orthopedic implants do not replicate the surface as well as the replaced bone. This leads to failure of the implant due to insufficient bonding with juxtaposed bone, bone loss, implant loosening and fracture. Nanophase materials are presently considered a very promising technology for future orthopedic biomaterials, as they possess unique surface and mechanical properties similar to those of bones [129]. Nanosized materials are characterized by grain sizes in the nanometer range, whereas conventional materials have grain sizes in the micron range. Therefore, the ratio of surface area to volume is considerably higher in nanomaterials than in conventional materials, which results in larger surface energy. Thus nanomaterials exhibit entirely different behavior compared to the microsized grains of the same material.

The bone forming cells generally attach themselves to surfaces whose roughness are in the nanometer range. In fact, our bones consist of inorganic minerals of grain size varying from 20 to 80 nm long and from 2 to 3 nm in diameter [130]. Therefore, the presence of nanosurface roughness leads to desirable cellular responses, resulting in high osseointegration [131]. The cell adhesion behavior on submicron and nanometer structured titanium surface was compared with that on a flat smooth titanium surface by Dongwoo *et al.* [131]. Their results showed that both nanometer and submicron surfaces have very high surface energy, resulting in very high adhesion of bone cells on these surfaces at variance with the smooth Ti surface. Increased cellular adhesion has been observed in both nanometric metals and alloys (commercial pure Ti, Ti–6Al, 4V and CoCr), and nanoceramics such as alumina, titania, hydroxyapatite [16, 132]. For all materials, when the grain size is decreased from 167 to 24 nm, osteoblast adhesion is increased by 51% and the fibroblast adhesion responsible for encapsulation is reduced by 235%. Moreover, the cell density is observed to be relatively higher for the nanomaterials when compared to the conventional counterpart. Choudhary *et al.* [92] reported that titanium and CoCrMo nanoparticles exhibit an increased endothelial and vascular smooth muscle cell adhesion compared with conventional titanium and CoCrMo surfaces. Nanostructured commercial pure Ti and CoCrMo were created by separately utilizing either constituent commercial pure Ti or CoCrMo nanoparticles as opposed to conventional microsized particles. In addition, these kinds of muscle cells have a better spread morphology on the nanostructured metals compared with conventional methods.

The pore size on the surface plays also an important role in influencing protein and cell adhesion. A protein, victronectin, generally is adsorbed on pores of smaller sizes (0.69, 0.95 and 0.66 nm of Al_2O_3, TiO_2 and hydroxyapatite); on the other hand, proteins that decrease cell adhesion, such as laminin, generally adsorb to pore size 2.54, 2.33 and 3.1 μm, corresponding to Al_2O_3, TiO_2 and hydroxyapatite bioceramics [133]. Thus, small pores enhance cell adhesion because of the particular protein that attaches to that surface.

Besides cell adhesion, nanophase materials also exhibit high density of osteoclast adhesion, indicating bone remodeling and new bone formation, improved osteoblast proliferation and APT synthesis [129]. Ergun *et al.* [93] observed increased osteoblast adhesion onto nanoparticulate calcium phosphate prepared by wet chemistry.

The above studies point out that, irrespective of the type of material used for an implant, the crucial factor that determines the cell response and osseointegration is the grain size of the biomaterial surface. Nevertheless, the mechanism by which nanosurface alters the cell adsorption behavior remains to be fully understood and investigated.

A detailed review of nanotechnology applications of biomaterials has been recently written by Khang *et al.* [134].

The techniques most commonly employed to nanopattern biomaterials surface include scanning probe, electron beam, colloidal, and imprint lithographies (for a review, see Schmidt *et al.* [91]).

An alternative to nanopatterning is represented by surface coating with nanoparticles. Sato *et al.* [94] reported that the coating of titanium surfaces with nanocrystalline titania powders or nanocrystalline hydroxyapatite powders strongly influences cell behavior: Osteoblasts adherent on single-phase nanotitania coatings are well spread, whereas they become rounder and extend distinct filopodia on the composite and single-phase hydroxyapatite coatings. In their study, nanocrystalline hydroxyapatite powders were synthesized through wet chemistry and hydrothermal treatments, while nanocrystalline titania powders commercially obtained were mixed with the nanocrystalline hydroxyapatite powders at various weight ratios. The mixed powders were deposited on titanium utilizing a room-temperature coating process called IonTite^{TM}. Their results showed that such coatings maintained the chemistry and crystallite size of the original hydroxyapatite and titania powders.

Wet methods, such as precipitation and sol-gel methods, represent valuable synthetic techniques to produce nanoparticles: They are a cheap alternative to nanolithographies and the product characteristics can be easily tailored by varying a series of experimental parameters.

As previously pointed out, one of the most important features of a biomaterial surface is its wettability. Hereafter, a series of examples of modulation of the hydrophilicity/hydrophobicity of titanium dioxide nanopowders will be presented. Titanium dioxide is a ceramic widely employed as biomaterial, but all the techniques presented below can be applied to other oxides as well.

Titanium dioxide is normally a highly hydrophilic material. Under UV illumination it enhances its hydrophilicity (a phenomenon called photoinduced superhydrophilicity). Titania hydrophilicity is linked to the presence of physisorbed and chemisorbed water and hydroxide groups at the material surface: The higher their content the more hydrophilic the titania surface. Water removal in the case of oxides and inorganic compounds starts at low temperatures with the removal of physisorbed water and continues up to 1100-1200 K with the removal of strongly chemisorbed water. It is obvious that by keeping low the temperature treatment, the hydrophilicity of the layer is maintained high.

Hydrothermal post-synthetic treatments can be used to increase titania hydrophilicity. It is also possible to synthesize superwetting TiO_2 films (namely, a non-UV superwetting TiO_2 film) by hydrothermal-dip-coating method: These films are found to preserve their superhydrophilicity up to 6 months. Hydrothermal treatments are found to increase particle size as a result of an Ostwald ripening growth mechanism. Fig. **5a** illustrates the comparison between TEM images of a sample prepared in our laboratory by hydrothermal growth (named T_Hydro) and a TiO_2 sol-gel precursor with no such a growth step, calcined at 573 K. An extrapolation of the particle dimensions from diffraction patterns shows a weighted average particle diameter of 6 nm for sample T_573 and 9 nm for sample T_Hydro.

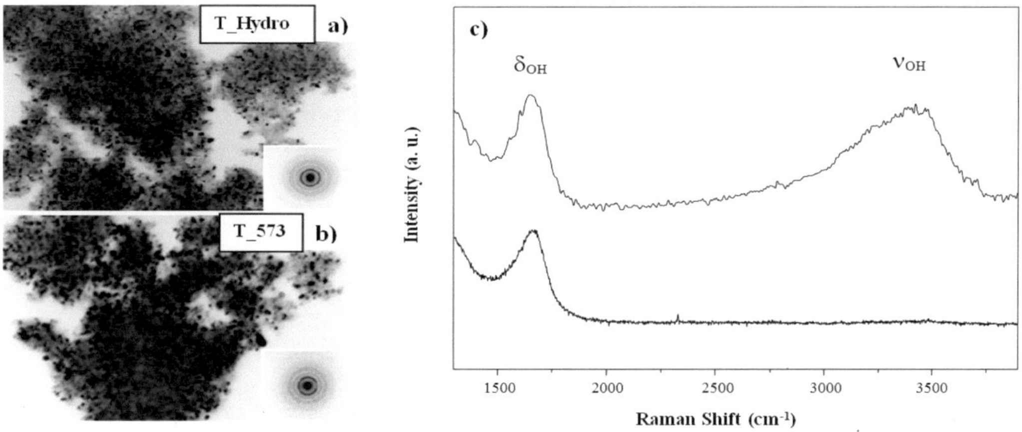

Figure 5: TEM images of TiO_2 precursor calcined at 573 K (T_573) (**a**) and hydrothermally treated for 552 h (T_hydro) (**b**); **c**) Raman spectra in the region 1500-4000 cm^{-1} of the same samples calcined at 873 K.

More important, Fig. **5c** shows Raman spectra in the region 1500-4000 cm^{-1} of the same samples named above, except for the difference that they have both undergone a subsequent step of calcination at 873 K [135] The analysis of the water oxide interactions by Raman scattering indicates that while the 873 K heating of the un-aged precursor provokes the total removal of free water, the hydrothermal sample has a larger amount of free liquid water, indicated by a broad peak in the OH stretching region (2900-3700 cm^{-1}). The residual presence of free water at the surface of the T_Hydro_873 is fully consistent with the higher thermal stability shown by the hydrothermal sample. Elaboration of Raman spectra in the low wavelengths region and XRD data suggests the formation of an ordered material in which long range and local structures are alike.

On the contrary, coating the TiO_2 surface with hydrophobic films is a technique widely exploited in order to make TiO_2 hydrophobic. This technique obviously requires the use of an appropriate biocompatible coating material, able of linking permanently to the titania surface. While many attempts have been done to coat microstructured TiO_2 surfaces, only a few examples are reported on the direct hydrophobization of nanostructured TiO_2. Several techniques are known for the microscale patternings of hydrophobic layers through the use of molded polymers [136] and waxes [137], by chemical methods such as plasma chemical roughening and by physical processing methods such as ion etching and compression of polymer beads [138], which can all result in ultrahydrophobic coatings. Risse *et al.* reported the preparation of mechanically stable mesoporous, rough TiO_2 coatings made by a mix of commercial 25P and acidic TiO_2 sols for technical substrates such as steel [139]. TiO_2 is by itself a coating on polystyrene beads [140] acquiring buoyancy properties. However, the hydrophobic character is often given by a matrix or support where the TiO_2 is dispersed.

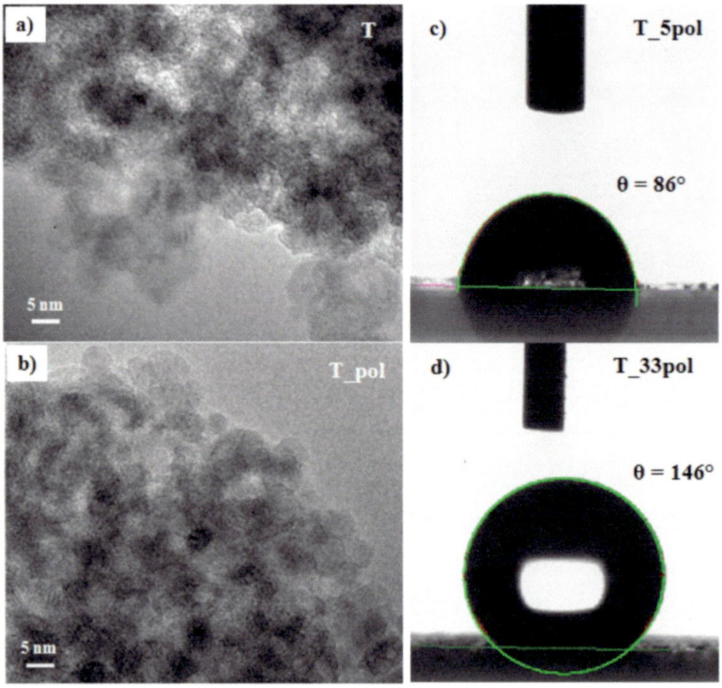

Figure 6: Wettability modulation of oxide surfaces by coating with hydrophobing agents: HRTEM images of **a**) nanometric bare TiO_2 and **b**) surface modified nano-TiO_2. Contact angles of nanometric TiO_2 coated with different amount of siloxane: **c**) 5% and **d**) 33 %.

In our laboratories, TiO_2 nanoparticles prepared by sol-gel method were clad using a trisubstituted siloxane [141]. By precisely dosing the amount of the coating hydrophobing agent, the final material achieves tunable hydrophobic properties (Fig. **6**). Contact angle measurements are the best to evaluate the hydrophobicity of the final material: A higher value of the contact angle is obtained for samples with higher amounts of the coating (Fig. **6 c**, **d**).

HRTEM images refer to the bare oxidic material (Fig. **6a**) and to the polymer-coated nano-TiO_2: The former material is made up of closely packed nanoparticles (average mean diameter: 5-8 nm), exhibiting smooth edges and a somewhat "amorphous" outer aspect, even if their inner aspect is highly crystalline (the presence of crystal planes is witnessed by the high incidence of fringe patterns). As for the coated material, a similar "crystalline" situation is observed: In fact, the same crystal distances are observed, but the particles clearly more separated exhibit better defined contours and sharper interference fringe patterns, indicating a less packed outer nature.

Nano-TiO_2 coated by functionalized siloxane shows high hydrophobicity properties, an excellent buoyancy on water and, at the same time, the hydrophobic coating does not modify the pristine physicochemical features of the oxide.

7. CONCLUSION

Surface properties are related critically to biomaterials performance and applications, since they determine material biocompatibility, bacterial contamination, occurrence of corrosion and encrustation phenomena, *etc.* Ultimately, surface characteristics such as surface energy, roughness, surface charge, topography and hydrophobicity/hydrophilicity, concur to determine the useful life of an implant.

In all the variety of possible situations, the interactions occurring at the interface between a solid material surface and its biological environment are a relevant aspect that, primarily, has to be understood, controlled and optimized.

The state of the surface can be modified both chemically and physically by procedures that depend on the properties of the material. Alongside the conventional classes of biomaterials, the emerging class of nanomaterials and composites has been reviewed briefly, since the interest for the application of nanosized and composite materials to the biomedical science is continuously growing up. Numerous studies provided evidence of improved biocompatibility and biomechanical compatibility of nanomaterials in comparison with their conventional counterpart.

The synthetic and surface modification techniques described so far will increase the ability of researchers to tailor biomaterials structures and topographies to the purpose of designing novel proactive biomaterials, able to elicit and control biological responses.

ACKNOWLEDGEMENTS

The authors acknowledge the Park company for providing the AFM image of TiO_2.

CONFLICT OF INTEREST

The author(s) confirm that this chapter content has no conflict of interest.

REFERENCES

[1] Puleo DA, Nanci A. Understanding and controlling the bone-implant interface. Biomaterials 1999; 20: 2311-21.

[2] Gorbet MB, Sefton MV. Biomaterial-associated thrombosis: roles of coagulation factors, complement, platelets and leukocytes. Biomaterials 2004; 25: 5681-703.

[3] Harkes G, Feijen J, Dankert J. Adhesion of *Escherichia coli* on to a series of poly(methacrylates) differing in charge and hydrophobicity. Biomaterials 1991; 12: 853-60.

[4] Gallardo-Moreno AM, González-Martín ML, Bruque JM, Pérez-Giraldo C. The adhesion strength of *Candida parapsilosis* to glass and silicone as a function of hydrophobicity, roughness and cell morphology. Colloids Surf A Physicochem Eng Asp 2004; 249: 99-03.

[5] Liu Y, Yang S-F, Li Y, Xu H, Qin L, Tay J-H. The influence of cell and substratum surface hydrophobicities on microbial attachment. J Biotechnol 2004; 101: 251-6.

[6] Walley VM, Stinson WA, Upton C, Santerre JP, Mussivand T, Masters RG, Ghadially FN. Device thrombosis and thromboembolism. Foreign materials found in the cardiovascular system after instrumentation or surgery (Including a guide to their light microscopic identification) Cardiovasc Pathol 1993; 2: 157–85.

[7] Werner C, Eichhorn KJ, Grundke K, Simon F, Grählert W, Jacobasch HJ. Insights on structural variations of protein adsorption layers on hydrophobic fluorohydrocarbon polymers gained by spectroscopic ellipsometry (part I). Colloids Surf. A 1999; 156: 3-17.

[8] Fang F, Satulovsky J, Szleifer I. Kinetics of Protein Adsorption and Desorption on Surfaces with Grafted Polymers. Biophys J 2005; 89: 1516-33.

[9] Rechendorff K, Hovgaard MB, Foss M, Zhdanov VP, Besenbacher F. Enhancement of Protein Adsorption Induced by Surface Roughness. Langmuir 2006; 22: 10885-8.

[10] Dolatshahi-Pirouz A, Rechendorff K, Hovgaard MB, Foss M, Chevallier J, Besenbacher F. Bovine serum albumin adsorption on nano-rough platinum surfaces studied by QCM-D. Colloids Surf. B 2008; 66: 53-9.

[11] Arvidsson S, Askendal A, Tengvall P. Blood plasma contact activation on silicon, titanium and aluminium. Biomaterials 2007; 28: 1346-54.

[12] Kurrat R, Prenosil JE, Ramsden JJ. Kinetics of Human and Bovine Serum Albumin Adsorption at Silica–Titania Surfaces. J Colloid Interface Sci 1997; 185: 1-8.

[13] Höök F, Vörös J, Rodahl M, Kurrat R, Böni P, Ramsden JJ, Textor M, Spencer ND, Tengvall P, Gold J, Kasemo B. A comparative study of protein adsorption on titanium oxide surfaces using *et al.* ellipsometry, optical waveguide lightmode spectroscopy, and quartz crystal microbalance/dissipation. Colloids Surf B 2002; 24: 155-70.

[14] Tsibouklis J, Stone M, Thorpe AA, Graham P, Peters V, Heerlien R, Smith JR, Green KL, Nevell TG. Preventing bacterial adhesion onto surfaces: the low-surface-energy approach. Biomaterials 1999; 20: 1222-35.

[15] Webster TJ, Ergun C, Doremus RH, Siegel RW, Bizios R. Enhanced functions of osteoblasts on nanophase ceramics. Biomaterials 2000; 21: 1803-10.

[16] Cooper LF, Masuda T, Yliheikkila PK, Felton DA. Generalizations regarding the process and phenomenon of osseointegration. Part II. *In vitro* studies. Int J Oral Maxillofac Impl 1998; 13: 163–74.

[17] Anselme K. Osteoblast adhesion on biomaterials. Biomaterials 2000; 21: 667–81.

[18] Brunette DM, Ed. The bone–biomaterial interface. Toronto: University of Toronto Press 1991; pp. 49–61.

[19] Boyan B, Hummert T, Kieswetter K, Schraub D, Dean D, Schwartz Z. Effect of titanium surface characteristics on chondrocytes and osteoblasts *in vitro*. Cells Mater 1995; 5: 323–35.

[20] Jagur-Grodzinski J. Biomedical application of functional polymers. React Funct Polym 1999; 39: 99–138.

[21] Mao C, Qiu Y, Sang H, Mei H, Zhu A, Shen J, Lin S. Various approaches to modify biomaterial surfaces for improving biocompatibility. Adv Colloid Interface Sci 2004; 110: 5–17.

[22] Benson RS. Use of radiation in biomaterial science. Nucl Instrum Methods B 2002; 191: 752-7.

[23] Zhang S, Du H, Ong S, Aung K, Too H, Miao X. Bonding structure and haemocompatibility of silicone-incorporated amorphous carbon. Thin Solid Film 2006; 515: 66–72.

[24] Hoffman AS, Kiaei D, Safranj A, Horbett TA, Hanson SR. Binding of proteins and platelets to gas discharge-deposited polymers. Clin Mater 1991; 8: 3–8.

[25] Welle A, Gottwald E. UV-based patterning of polymeric substrates for cell culture applications. Biomed Microdev 2002; 4: 33-41.

[26] Chan CM, Ko TM, Hiraoka H. Polymer surface modification by plasmas and photons. Surf Sci Rep 1996; 24: 3–54.

[27] Guan JJ, Gao CY, Feng LX, Shen JC. Functionalizing of polyurethane surfaces by photografting with hydrophilic monomers. J Appl Polym Sci 2000; 77: 2505–12.

[28] Ma ZW, Gao CY, Juan J, Ji J, Gong YH, Shen JC. Surface modification of poly-L-lactide by photografting of hydrophilic polymers towards improving its hydrophilicity. J Appl Polym Sci 2002; 85: 2163–71.
[29] Ma ZW, Gao CY, Shen JC. Surface modification of poly-l-lactic acid (PLLA) membrane by grafting acrylamide: an effective way to improve cytocompatibility for chondrocytes. J Biomater Scie – Polym. Ed. 2003; 14: 13–25.
[30] Zhu HG, Ji J, Lin RY, Gao CY, Feng LX, Shen JC. Surface engineering of poly(dl-lactic acid) by entrapment of alginate-amino acid derivatives for promotion of chondrogenesis. Biomaterials 2002; 23: 3141–8.
[31] Zhu YB, Gao CY, Jian J, Shen JC. Photo-induced grafting of poly(methacrylic acid) onto polycaprolactone towards promotion of cell compatibility. Chin J Mater Res 2002; 16: 233–7.
[32] Briggs D, Brewis DM, Dahm RH, Fletcher IW. Analysis of the surface chemistry of oxidized polyethylene: comparison of XPS and ToF-SIMS. Surf Interface Anal 2003; 35: 156–67.
[33] Lane JM, Hourston DJ. Surface treatments of polyolefins. Prog Org Coat 1993; 21: 269–84.
[34] Desai S, Singh RP. Surface modification of polyethylene. In: Albertsson AC, Ed. Long-term properties of polyolefins. New York, Springer, 2004; pp. 231–93.
[35] Ozdemir M, Yurteri CU, Sadikoglu H. Physical polymer surface modification methods and applications in food packaging polymers. Crit Rev Food Sci 1999; 39: 457–77.
[36] Biederman H, Boyaci IH, Bilkova P, Slavinska D, Mutlu S, Zemek J, Trchova M, Klimovic J, Mutlu M. Characterization of glow-discharge-treated cellulose acetate membrane surfaces for single-layer enzyme electrode studies. J Appl Polym Sci 2001; 81: 1341–52.
[37] Cen L, Neoh KG, Li YL, Kang ET. Assessment of *in vitro* bioactivity of hyaluronic acid and sulfated hyaluronic acid functionalized electroactive polymer. Biomacromolecules 2004; 5: 2238–46.
[33] Long TM, Prakash S, Shannon MA, Moore JS. Watervapor plasma-based surface activation for trichlorosilane modification of PMMA. Langmuir 2006; 22: 4104–9.
[39] Prissanaroon W, Brack N, Pigram PJ, Hale P, Kappen P, Liesegang J. Fabrication of patterned polypyrrole on fluoropolymers for pH sensing applications. Synth Met 2005; 154: 105–8.
[40] Kayirhan N, Denizli A, Hasirci N. Adsorption of blood proteins on glow-discharge-modified polyurethane membranes. J Appl Polym Sci 2001; 81: 1322–32.
[41] Francois P, Vandaux P, Nurdin N, Mathieu H J, Descouts P, Lew P. Physical and biological effects of a surface coating procedure on polyurethane catheters. Biomaterials 1996; 17: 667–78.
[42] Lee KY, Ha WS, Park WH. Bloodcompatibility and biodegradability of partially N acylated chitosan derivatives. Biomaterials 1995; 16: 1211–6.
[43] Filippini P, Rainoldi C, Ferrante A, Mecheri B, Gabrielli G, Bombace M, Indovina PL, Santini MT. Modulation of osteosarcom cell growth and differentiation by silane-modified surface. J Biomed Mater Res 2001; 55: 338-49.
[44] Textor M, Ruiz L, Hofer R, Rossi A, Feldman K, Hahner G, Spencer ND. Structural chemistry of self-assembled monolayers of octadecylphosphoric acid on tantalum oxide surfaces. Langmuir 2000; 16: 3257-71.
[45] Ostuni E, Chapman RG, Holmlin RE, Takayama S, Whitesides GM. A Survey of Structure-Property Relationships of Surfaces that Resist the Adsorption of Protein. Langmuir 2001; 17: 5605-20.

[46] Whitesides GM, Laibinis PE. Wet chemical approaches to the characterization of organic-surfaces—self-assembled monolayers, wetting, and the physical organic-chemistry of the solid liquid interface. Langmuir 1990; 6: 87–96.

[47] Xia Y, Zhao X-M, Whitesides GM. Pattern transfer: selfassembled monolayers as ultrathin resists. Microelectron Eng 1996; 32: 255–68.

[48] Wasserman SR, Tao YT, Whitesides GM. Structure and reactivity of alkylsiloxane monolayers formed by reaction of alkyltrichlorosilanes on silicon substrates. Langmuir 1989; 5: 1074–87.

[49] Balas F, Kawashita M, Nakamura T, Kokubo T. Formation of bone-like apatite on organic polymers treated with a silane-coupling agent and a titania solution. Biomaterials 2006; 27: 1704–10.

[50] Reinhardt-Schlegel H, Kawamura Y, Furuno T, Sasabe H. Microstructure of phospholipid monolayers studied by dark field electron and fluorescence microscopy. J Colloid Interface Sci 1991; 147: 295-06.

[51] Yang XM, Xiao D, Xiao SJ, Lu ZH, Wei Y. Observation of chiral domain morphology in a phospholipid Langmuir-Blodgett monolayer by atomic force microscopy. Phys Lett A 1994; 193: 195-8.

[52] Ekelund K, Sparr E, Engblom J, Wennerstroem H, Engstroem S. An AFM Study of Lipid Monolayers. 1. Pressure-Induced Phase Behavior of Single and Mixed Fatty Acids. Langmuir 1999; 15: 6946-9.

[53] Lee YL, Chen CY. Surface wettability and platelet adhesion studies on Langmuir-Blodgett films. Applied Surface Science 2003; 207: 51-62.

[54] Cho CS, Takayama T, Kunou M, Akaike T. Platelet adhesion onto the Langmuir-Blodgett film of poly(gamma-benzyl L-glutamate)-poly(ethylene oxide)-poly(gamma-benzyl L-glutamate) block copolymer. J Biomed Mater Res 1990; 24: 1369-75.

[55] Bird RL, Hall B, Hobbs KE, Chapman D. New haemocompatible polymers assessed by thrombelastography. J Biomed Eng 1989; 11: 231-4.

[56] Barnes GT, Gentle IR. Interfacial Science. An introduction. Oxford: Oxford University Press 2005; p. 98.

[57] Quirk RA, Davies MC, Tendler SJB, Chan WC, Shakesheff KM. Controlling biological interactions with poly(lactic acid) by surface entrapment modification. Langmuir 2001; 17: 2817–20.

[58] Nebeker JR, Virmani R, Bennett CL, Hoffman JM, Samore MH, Alvarez J, Davidson CJ, McKoy JM, Raisch DW, Whisenant BK, Yarnold PR, Belknap SM, West DP, Gage JE, Morse RE, Gligoric G, Davidson L, Feldman MD. Hypersensitivity cases associated with drug-eluting coronary stents. J Am Coll Cardiol 2006; 47: 175-81.

[59] Kaiser C, Brunner-La Rocca HP, Buser PT, Bonetti PO, Osswald S, Linka A, Bernheim A, Zutter A, Zellweger M, Grize L, Pfisterer ME. Incremental cost-effectiveness of drug-eluting stents compared with third generation bare metal stents in real-world setting: randomized Basel Stent Kosten Effktivitats Trail (BASKET). Lancet 2006; 56: 921-9.

[60] Xu LP, Yu GN, Zhang E, Pan F, Yang K. *In vivo* corrosion behavior of Mg–Mn–Zn alloy for bone implant: application. J Biomed Mater Res A 2007; 83A: 703–11.

[61] Witte F, Kaese V, Haferkamp H, Switzer E, Meyer-Lindenberg A, Wirth CJ, Windhagen H. *In vivo* corrosion of four magnesium alloys and the associated bone response. Biomaterials 2005; 26: 3557–63.

[62] Witte F, Fischer J, Nellesen J, Crostack HA, Kaese V, Pisch A, Beckmann F, Windhagen H. *In vitro* and *in vivo* corrosion measurements of magnesium alloys. Biomaterials 2006; 27: 1013–8.

[63] Okazaki Y, Gotoh E. Comparison of metal release from various metallic biomaterials *in vitro*. Biomaterials 2005; 26: 11–21.

[64] Wapner KL. Implications of metallic corrosion in total knee arthroplasty. Clin Orthop Relat Res 1991; 271: 12–20.

[65] Teoh SH. Fatigue of biomaterials: a review. Int J fatigue 2000; 22(10): 825–37.

[66] Niinomi M, Kuroda D, Fukunaga KI, Fukui H, Kato Y, Yashiro. Mechanical Performance of Newly Developed β-Type Titanium Alloys for Biomedical Use. St. Petersburg, Russia 2000; pp. 1195-1201.

[67] Geetha M, Singh AK, Asokamani R, Gogia AK. Ti based biomaterials, the ultimate choice for orthopaedic implants – A review. Prog Mater Sci 2009; 54: 397–425.

[68] Fukumoto S, Tsubakino H, Inoue S, Liu L, Terasawa M, Mitamura T. Surface modification of titanium by nitrogen ion implantation. Mater Sci Eng 1999; 263A: 205–9.

[69] Liu X, Chu PK, Ding C. Surface modification of titanium, titanium alloys, and related materials for biomedical applications. Mater Sci Eng 2004; R47: 49–121.

[70] Oh JC, Yun E, Lee S. Correlation of Microstructure with the Hardness and Wear Resistance of (TiC, SiC)/Ti-6AI-4V Surface Composites Fabricated by High-Energy Electron-Beam Irradiation. Metall Mater Trans A 2004; 35: 525–34.

[71] Ma WJ, Ruys AJ, Mason RS, Martin PJ, Bendavid A, Liu Z, Ionescu M, Zreiqat H. DLC coatings: Effects of physical and chemical properties on biological response. Biomaterials 2007; 28: 1620–8.

[72] Kim S, Lee JB, Koak JY, Heo SJ, Lee KR, Cho LR, Lee SS. An abutment screw loosening study of a Diamond Like Carbon-coated CP titanium implant. J Oral Rehab 2005; 32: 346–50.

[73] Jayaraman M, Meyer U, Buhner M, Joos U, Wiesmann HP. Influence of titanium surfaces on attachment of osteoblast-like cells *in vitro*. Biomaterials 2004; 25: 625–31.

[74] Pilliar RM. Overview of surface variability of metallic endosseous dental implants: textured and porous surface-structured designs. Implant Dent 1998; 7: 305–14.

[75] Wennerberg A, Hallgren C, Johansson C, Danielli S. A histomorphometric evaluation of screw-shaped implants each prepared with two surface roughnesses. Clin Oral Implant Res 1998; 9: 11–9.

[76] Martin JY, Schwartz Z, Hummert TW, Schraub DM, Simpson J, Lankford J Jr, Dean DD, Cochran DL, Boyan BD. Effect of titanium surface roughness on proliferation, differentiation, and protein synthesis of human osteoblast-like cells (MG63). J Biomed Mater Res 1995; 29: 389-401.

[77] Buser D, Schenk RK, Stienemann S, Fiorelini JP, Fox CH, Stich H. Influence of surface characteristics on bone integration of titanium implants. A histomorphometric study in miniature pigs. J Biomed Mater Res 1991; 25: 889–902.

[78] Mac Donald DE, Rapuano BE, Deo N, Stranick M, Somasundaram P, Boskey AL. Thermal and chemical modification of titanium–aluminum–vanadium implant materials: effects on surface properties, glycoprotein adsorption, and MG63 cell attachment. Biomaterials 2004; 25: 3135–46.

[79] Hulbert SF, Cooke FW, Klawitter JJ, Leonard RB, Sauer BW, Moyle DD, Skinner HB. Attachment of prostheses to the musculoskeletal system by tissue ingrowth and mechanical interlocking. J Biomed Mater Res 1973; 7: 1–23.

[80] Sauer BW, Weinstein AM, Klawitter JJ, Hulbert SF, Leonard RB, Bagwell JG. The role of porous polymeric materials in prosthesis attachment. J Biomed Mater Res 1974; 8: 145–53.

[81] Roderic L. Composite biomaterials. In: Bronzino JD, Ed. The biomedical engineering handbook. Boca Raton, CRC Press, 1995; pp. 309–14.

[82] Li SJ, Yang R, Ninomi M, Hao YL, Cui YY. Formation and growth of calcium phosphate on the surface of oxidized Ti-29Nb-13Ta-4.6Zr alloy. Biomaterials 2004; 25: 2525–32.

[83] Alves Jr C, Guerra Neto CLB, Morais GHS, da Silva CF, Hajek V. Nitriding of titanium disks and industrial dental implants using hollow cathode discharge. Surf Coat Technol 2005; 194: 196–204.

[84] Sundararajan T, Kamachi Mudali U, Rajeswari S, Subbaiyan M. Electrochemical studies on nitrogen ion implanted Ti6Al4V alloy. Anti Corros Meth Mater 1998; 45: 162–6.

[85] Hintermann H, Joint implant. US 4687487. 1987.

[86] Thair L, Kamachi Mudali U, Bhuvaneswaran N, Nair KGM, Asokamani R, Raj Baldev. Nitrogen ion implantation and *in vitro* corrosion behavior of as-cast Ti–6Al–7Nb alloy. Corros Sci 2002; 44: 2439–57.

[87] Thair L, Kamachi Mudali U, Rajagopalan S, Asokamani R, Raj Baldev. Surface characterization of passive film formed on nitrogen ion implanted Ti–6Al–4V and Ti–6Al–7Nb alloys using SIMS. Corros Sci 2003; 45: 1951–67.

[88] Robert SM, Heinz W, Rolf S, Manfred S. New surface modification for Ti-6Al-7Nb alloy: oxygen diffusion hardening (0DH). Biomaterials 1991; 12: 125-9.

[89] Poggie RA, Kovacs P, Davidson JA. Oxygen Diffusion Hardening of Ti-Nb-Zr Alloys. Materials and Manufacturing Processes, 1996; 11: 185–97.

[90] Geetha M, Kamachi Mudali U, Asokamani R, Raj Baldev. Microstructural and Corrosion Evaluation of Laser Surface Nitrided Ti-13Nb-13Zr Alloy. Surf Eng 2004; 20: 68–74.

[91] Schmidt RC, Healy KE. Controlling biological interfaces on the nanometer length scale. J Biomed Mater Res A 2009; 90: 1252–61.

[92] Choudhary S, Berhe M, Haberstroh KM, Webster TJ. Increased endothelial and vascular smooth muscle cell adhesion on nanostructured titanium and CoCrMo. Int J Nanomedicine 2006; 1: 41–9.

[93] Ergun C, Liu H, Webster TJ. Osteoblast adhesion on novel machinable calcium phosphate/lanthanum phosphate composites for orthopedic applications. J Biomed Mater Res A 2009; 89: 727–33.

[94] Sato M, Aslani A, Sambito MA, Kalkhoran NM, Slamovich EB, Webster TJ. Nanocrystalline hydroxyapatite/titania coatings on titanium improves osteoblast adhesion. J Biomed Mater Res A 2008; 84: 265–72.

[95] Zinger O, Anselme K, Denzer A, Habersetzer P, Wieland M, Jeanfils J, Hardouin P, Landolt D. Time-dependent morphology and adhesion of osteoblastic cells on titanium model surfaces featuring scale-resolved topography. Biomaterials 2004; 25: 2695–711.

[96] Webster TJ, Ellison K, Price RL, Haberstroh KM. Increased osteoblast function on nanostructured materials due to novel surface roughness properties. Mater Sci Forum 2003; 426–432: 3127–32.

[97] Stangl R, Pries A, Loos B, Muller M, Erben RG. Influence of pores created by laser superfinishing on osseointegration of titanium alloy implants. J Biomed Mater Res 2004; 69A: 444–53.

[98] Hulbert SF, Young FA, Mathews RS, Klawitter JJ, Talbert CD, Stelling FH. Potential of ceramic materials as permanently implantable skeletal prostheses. J Biomed Mater Res 1970; 4: 433–56.

[99] Li J, Liao H, Fartash B, Hermansson L, Johnsson, T. Surface-dimpled commercially pure titanium implant and bone growth. Biomaterials 1997; 18: 691–6.
[100] Gotz HE, Muller M, Emmel A, Holzwarth U, Erben RG, Stangl R. Effect of surface finish on the osseointegration of laser-treated titanium alloy implants. Biomaterials 2005; 25: 4057–64.
[101] Krishna BV, Bose S, Bandoyopadhyay A. Low stiffness porous Ti structures for load-bearing implants. Acta Biomater 2007; 3: 997-06.
[102] Gupta R, Kumar A. Bioactive materials for biomedical applications using sol-gel technology. Biomed Mater 2008; 3: 034005.
[103] Boiadjieva T, Cappelletti G, Ardizzone S, Rondinini S, Vertova A. The role of surface electrification on the growth and structural features of titania nanoparticles. Phys Chem Chem Phys 2004; 6: 3535–39.
[104] Dhert WJA, Klein CPAT, Jansen JA, van der Velde EA, Vriesde RC, Rozing PM, de Groot K. A histological and histomorphometrical investigation of fluorapatite, magnesium whitlockite, and hydroxylapatite plasm-sprayed coatings in goats. J Biomed Mater Res 1993; 27: 127-38.
[105] Haher T. New and Transfer Technology Material; Spinal Instrumentation from Harrington to the Future, Continuing Education Course by AAOS and SRS, Kansas City: Kansas 1992.
[106] Ohgushi H, Goldberg VM, Caplan AI. Hetrotopic osteogenesis in porous ceramics induced by marrow cells. J Orthop Res 1989; 7: 568–78.
[107] Zhu XD, Zhang HJ, Fan HS, Li W, Zhang XD. Effect of phase composition and microstructure of calcium phosphate ceramic particles on protein adsorption. Acta Biomater 2010; 6: 1536–41.
[108] Suchanek W, Yoshimura M. Processing and properties of hydroxyapatite- based biomaterials for use as hard tissue replacement implants. J Mater Res 1998; 13: 94–117.
[109] Izquierdo-Barba I, Salinas AJ, Vallet-Reg M. Effect of the continuous solution exchange on the *in vitro* reactivity of a $CaO–SiO_2$ sol–gel glass. J Biomed Mater Res 2000; 51: 191–99.
[110] Lin K, Zhai W, Ni S, Chang J, Zeng Y, Qian W. Study of the mechanical property and *in vitro* biocompatibility of $CaSiO_3$ ceramics. Ceram Int 2005; 31: 323–26.
[111] Binnaz Yoruc Hazar A. Preparation and *in vitro* bioactivity of $CaSiO_3$ powders. Ceram Int 2007; 33: 687–92.
[112] Yoshii S, Kakutani Y, Nakamura T, Kitsugi T, Oka M, Kokubo T. Strength of bonding between A-W glass ceramic and the surface of bone cortex. J Biomed Mater Res 1988; 22: 327–32.
[113] Gou Z, Chang J, Zhai W. Preparation and characterization of novel bioactive dicalcium silicate ceramics. J Eur Ceram Soc 2005; 25: 1507-14.
[114] Muralithran G, Ramesh S. The effects of sintering temperature on the properties of hydroxyapatite. Ceram Int 2000; 26: 221–30.
[115] Mardare CC, Mardare AI, Fernandes JRF, Correia RN. Deposition of bioactive glass-ceramic thin-films by RF magnetron sputtering. J Eur Ceram Soc 2003; 23: 1027–30.
[116] Kim SR, Lee JH, Kim YT, Riu DH, Jung SJ, Lee YJ, Chung SC, Kim YH. Synthesis of Si, and Mg substituted hydroxyapatites and their sintering behaviors. Biomaterials 2003; 24: 1389–98.
[117] Webster TJ, Massa-Schlueter EA, Smith JL, Slamovich EB. Osteoblast response to hydroxyapatite doped with doped with divalent and trivalent cations. Biomaterials 2004; 25: 2111–21.

[118] Wu CT, Chang J. Synthesis and apatite-formation ability of akermanite. Mater Lett 2004; 58: 2415–17.

[119] Fathi MH, Kharaziha M. Mechanically activated crystallization of phase pure nanocrystalline forsterite powders. Mater Lett 2008; 62: 4306–09.

[120] Ni S, Chou L, Chang, J. Preparation and characterization of forsterite (Mg_2SiO_4) bioceramics. Ceram Int 2007; 33: 83–8.

[121] Webster TJ, Siegel RW, Bizios R. Design and evaluation of nanophase alumina for orthopedic/dental applications. Nanostruct Mater 1999; 12: 983–6.

[122] Webster TJ, Ergun C, Doremus RH, Siegel RW, Bizios R. Specific proteins mediate enhanced osteoblast adhesion on nanophase ceramics. J Biomed Mater Res 2000; 51: 475–83.

[123] Wise DL, Trantolo DJ, Kai-Uwe L, Gresser JD, Cattaneo MV, Yaszemski MJ. Biomaterials Engineering and Devices: Human Applications, Orthopedic, Dental and Bone Graft Applications vol. 2. Humana Press. Totowa, USA, 2000.

[124] Murugan R, Ramakrishna S. Coupling of therapeutic molecules onto surface modified coralline hydroxyapatite. Biomaterials 2004; 25: 3073–80.

[125] Webster TJ, Siegel RW, Bizios R. Osteoblast adhesion on nanophase ceramics. Biomaterials 1999; 20: 1221–27.

[126] Gutwein LG, Webster TJ. Increased viable osteoblast density in the presence of nanophase compared to conventional alumina and titania particles. Biomaterials 2004; 25: 4175–83.

[127] Deng L, Mrksich M, Whitesides GM. Self-assembled monolayers of alkanethiolates presenting tri(propylenesulfoxide) groups resist the adsorption of protein. J Am Chem Soc 1996; 188: 5136-7.

[128] J.C. Lin, W.H. Chuang. Synthesis, surface characterization, and platelet reactivity evaluation for the self-assembled monolayer of alkanethiol with sulfonic acid functionality. J Biomed Mater Res 2000; 51: 413-23.

[129] Webster TJ. Nanophase ceramics: the future orthopedic and dental implant material. In: Ying JY, Ed. Nanostructured materials. New York, Academic Press, 2001; pp. 125–67.

[130] Kaplan FS, Hayes WC, Keaveny TM. Form and function of bone. In: Simon S, Ed. Orthopaedic basic science. Rosemont, American Academy of orthopaedic Surgeons, 1994; pp. 127–85.

[131] Dongwoo K, Jing L, Chang Y, Haberstroh, KM, Webster, TJ. The role of nanometer and sub-micron surface features on vascular and bone cell adhesion on titanium. Biomaterials 2008; 29: 970–83.

[132] Webster TJ, Ejiofor JU. Increased osteoblast adhesion on nanophase metals: Ti, Ti6Al4V, and CoCrMo. Biomaterials 2004; 25: 4731–9.

[133] Ayad S, Boot-Handford R, Humphries MJ, Kadler KE, Shutleworth A. The extracellular matrix facts book. Academic Press. San Diego, US 1994.

[134] Khang D, Carpenter J, Chun YW, Pareta R, Webster TJ. Nanotechnology for regenerative medicine. Biomed Microdevices 2008; 19:19.

[135] Cappelletti G, Ricci C, Ardizzone S, Parola C, Anedda A. Aged titania nanoparticles: the simultaneous control of local and long-range properties. J Phys Chem B 2005; 109: 4448-54.

[136] Bico J, Marzolin C, Quere D. Pearl drops. Europhys Lett 1999; 47: 220-6.

[137] Shibuichi S, Onda T, Satoh N, Tsujii K. Super waterrepellent surfaces resulting from fractal structure. J Phys Chem 1996; 100: 19512-7.

[138] Chen W, Fadeev A, Hsieh MC, Oner D. Ultrahydrophobic and ultralyophobic surfaces: some comments and examples. Langmuir 1999; 15: 3395-9.

[139] Risse G, Matys S Bottcher H. Investigation into the photoinduced change in wettability of hydrophobized TiO_2 films. Appl Surf Sci 2008; 254: 5994-01.

[140] Fabiyi ME, Skelton RL. Photocatalytic mineralization of methylene blue using buoyant TiO_2-coated polystyrene beads. J Photochem Photobiol A 2000; 132: 121-8.

[141] Milanesi F, Cappelletti G, Annunziata R, Bianchi CL, Meroni D, Ardizzone S. Siloxane-TiO_2 Hybrid Nanocomposites. The Structure of the Hydrophobic Layer. J Catal C 2010; 114: 8287-93.

CHAPTER 2

Calcium Phosphate Surface Tailoring Technologies for Drug Delivering and Tissue Engineering

Christophe Drouet[1], Jaime Gomez-Morales[2], Michele Iafisco[3,4,*] and Stephanie Sarda[1]

[1]Université de Toulouse, Institut Carnot CIRIMAT, UPS-INPT-CNRS, ENSIACET, 31030 Toulouse Cedex 4, France; [2]Laboratorio de Estudios Cristalográficos (IACT-CSIC-UGR). Avda. Las Palmeras 4. E-18100 Armilla, Spain; [3]Dipartimento di Chimica "G. Ciamician", Alma Mater Studiorum, Università di Bologna via Selmi 2, 40126 Bologna, Italy and [4]Dipartimento di Scienze della Salute, Università del Piemonte Orientale via Solaroli 4, 28100 Novara, Italy

Abstract: Calcium phosphates with apatitic structure (or apatites) constitute the mineral part of hard tissues in vertebrates. The structure and properties of apatites open several possibilities in medical sciences and also play an important role in living organisms for biomineralization processes. One of the most interesting characteristics of apatitic calcium phosphate is its surface reactivity and capability to exchange mineral ions and small molecules upon interaction with surrounding fluids. Such surface properties can be exploited and tailored in materials science to obtain nanostructured and bioactive biomaterials, in particular in view of drug delivery and tissue engineering.

In this chapter, biogenic calcium phosphates will be first presented, giving details about the principal characteristics of bone, tooth and pathological calcifications where calcium phosphates are present. We will then expose a general presentation of calcium phosphates used as biomaterials, especially in terms of ceramics, cements and coatings. Then the surface chemistry of synthetic and biogenic calcium phosphates will be examined in detail, summarizing the principal surface characteristics of biomimetic nanocrystalline apatites and of some other calcium phosphates of biological interest. In a subsequent part of the chapter, we will discuss surface interaction processes between calcium phosphate and molecules of biological interest: on one hand in the case of biomolecules (including peptides and proteins) for biomaterials development and biomineralization aspects, and on the other hand in the case of bioactive molecules for specific applications in drug delivery. Finally, we will conclude this chapter by presenting some *in vitro* and *in vivo* results obtained on functionalized calcium phosphates with the aim to develop innovative drug delivery devices.

Keywords: Calcium phosphates, biomineralization, nanocrystalline apatites, bone tissue engineering, drug delivery.

***Address correspondence to Michele Iafisco:** Dipartimento di Chimica "G. Ciamician", Alma Mater Studiorum, Università di Bologna, *via* Selmi 2, 40126 Bologna, Italy; Tel: +390512099489; Fax: +390512099593; E-mail: michele.iafisco@unibo.it

Lia Rimondini, Claudia Bianchi and Enrica Vernè (Eds)

1. BIOGENIC CALCIUM PHOSPHATES

Calcium phosphates are very attractive compounds in many interdisciplinary fields of science, including: geology, chemistry, biology and medicine; in particular they are the most important inorganic constituents of hard tissues in vertebrates. The atomic arrangement of calcium phosphates is built up around a network of orthophosphate (PO_4) groups, which provides stability to the structure. The majority of calcium phosphates are scarcely soluble in water and insoluble in alkaline solutions, but all of them are easily soluble in acids. All chemically pure calcium orthophosphates are white coloured crystals of moderate hardness [1]. In biological systems, calcium phosphates are the principal inorganic constituent of normal (bones, teeth, fish enameloid, deer antlers and some species of shells) and pathological (dental and urinary calculus and stones, atherosclerotic lesions) calcifications [2]. Except for small portions of the inner ear, all hard tissue of the human body is formed of calcium phosphates. Structurally, they occur mainly in the form of poorly crystallized non-stoichiometric F, Na, Mg, and carbonate substituted (hydroxy)apatite [3]. We will thus give details in this section about the principal characteristics of bone, tooth and pathological calcifications where the calcium phosphates are present highlighting the fact that the living organisms can produce different amazing high-performance materials.

1.1. Bone

Bones are rigid organs that form part of the endoskeleton of vertebrates. Their function is to move, support, and protect the various organs of the body, produce red and white blood cells and store minerals [4, 5]. Bones appear in a variety of shapes and have a complex internal and external structure, they are lightweight, yet strong and hard, in addition to satisfying their many other functions. One of the types of tissues that constitutes bone is the mineralized osseous tissue, also called bone tissue, that gives it rigidity and honeycomb-like three-dimensional structure [6]. Other types of tissues found in bones include marrow, endosteum and periosteum, nerves, blood vessels and cartilage [7]. There are 206 bones in the adult body and about 300 bones in the infant body. Bone tissue consists of cells embedded in a fibrous, organic matrix, the osteoid, which is primarily constituted by type I collagen (90%) and 10% amorphous ground substance (primarily glycosaminoglycans and glycoproteins). Osteoid comprises approximately 50% of bone by volume and 25% by weight [8].

Osteoblasts are mono-nucleate cells that are responsible for the secretion of osteoid, and subsequent bone formation through the mineralization of the osteoid matrix. They strongly produce alkaline phosphatase, an enzyme that has a role in the mineralisation of bone, as well as many matrix proteins. When osteoblasts are trapped in the bone matrix produced by themselves they become star-shaped cells named osteocytes, the most abundant cell found in bone [9].

Osteocytes are mature bone cells, networked to each other *via* long processes that occupy tiny canals called canaliculi, which are used for exchange of nutrients and waste. They are actively involved in the maintenance of bony matrix, through various mechano-sensory mechanisms regulating bone response to stress. Bone is a dynamic tissue that is constantly being reshaped by osteoblasts, cells which build bone, and osteoclasts, which resorb bone [10].

Osteoclasts are multi-nucleated cells responsible for the resorption of bone through the removal of the bone mineralized matrix. The removal process begins with the attachment of theosteoclast to the osteon (predominant structures found in compact bone, see later); the osteoclast then induces an infolding of its cell membrane and secretes collagenase and other enzymes important in the resorption process, such as tartrate resistant acid phosphatase, secreted against the mineral substrate. During childhood, bone formation exceeds resorption, but as the aging process occurs, resorption exceeds formation [11].

The characteristic rigidity and strength of bone derives from the presence of mineral salts, that permeate the organic matrix, formed by the osteoid mineralization, due to the secretion of vesicles containing alkaline phosphatase, by the osteoblasts [12]. The mineral phase comprises approximately 50% of bone by volume and 75% by weight. The principal constituent of bone mineral is calcium phosphate, mainly in the form of nanocrystalline carbonated (hydroxy)apatite, with traces of sodium, magnesium, silicon and fluoride [13] (although this poorly-crystallized apatitic phase has long been confounded, especially in the earliest studies, with a mixture of amorphous calcium phosphate and calcium carbonate, the identification of bone mineral to nanocrystalline calcium-deficient apatites appears now well established).

The whole architecture of bone is very complex: starting from the smallest constituting elements to the largest scales, the bone is not only organized in an anisotropic manner, but the arrangement of its constituting elements is hierarchical; this means that its structural units are organized at increasing size levels, and this feature confers unique properties to the whole bone structure. To better understand the complex bone architecture, several hierarchical models have been proposed. Weiner and Wagner have identified seven discrete levels of hierarchical organization in bone (Fig. **1**), which we describe here [14]. In their model, bone is considered as a family of materials with the mineralized collagen fiber as the primary building block for subsequent higher order architectures.

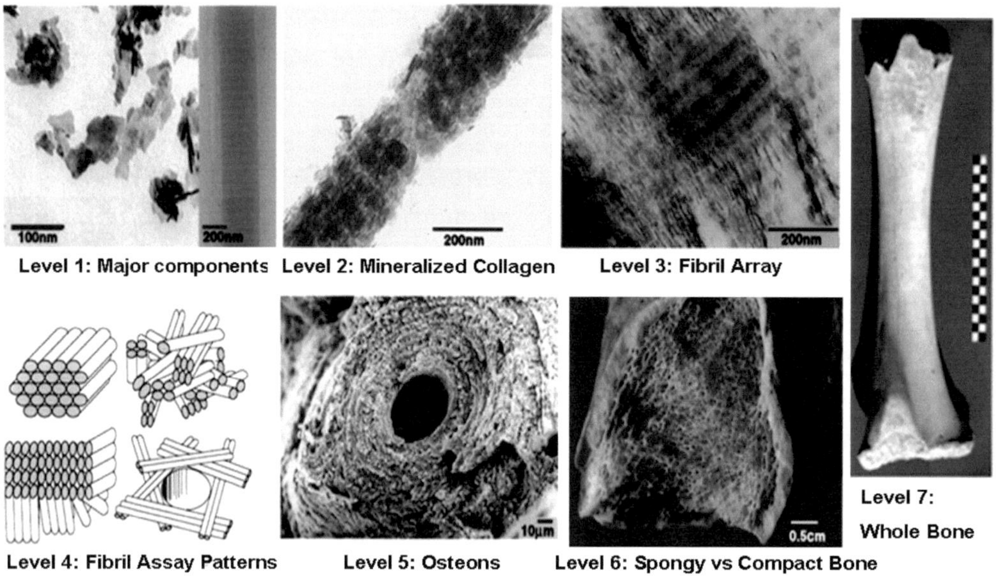

Figure 1: Seven hierarchical levels of organization of the bone family of materials as proposed by Weiner and Wagner.

The structure of bone differs greatly among different locations in the skeleton, but the basic nanoscale structure consisting of mineralized collagen remains the same throughout. The first level of hierarchy consists of the molecular/crystalline components: water, apatite, collagen, and other proteins. The apatite nanocrystals are plate-shaped and are among the smallest known biological crystals. Though the size of biological apatite crystals reported in the literature varies due to different treatment methods and analysis techniques, it is generally around the nanometric level, with values in the ranges of 30-50 nm (length), 15-30 nm (width) and 2-10 nm (thickness) [15]. In early studies, apatite needles were observed, but other studies suggest that platelets are the dominant morphology and that the apparent needles are most likely to be platelets viewed edge-on [16]. Further details of the structure of bone hydroxyapatite can be found in the third section. Non-collagenous proteins (NCPs) are also present but make up 10% or less of the total protein content in the bone matrix. The specific functions of the NCPs are still not completely understood. In addition to influencing crystal nucleation and growth, NCPs also play roles in cell signalling and ion homeostasis [17]. The second level is formed by the mineralization of collagen fibrils. Tiny plate-like crystals of biological apatite in bone occur within the discrete spaces within the collagen fibrils and grow with specific crystalline orientation along the *c*-axes, which are roughly parallel to the long axes of the collagen fibrils [18]. The hydroxyapatite growth is somehow limited by these small inter-collagenous spaces, which are approximately 50 nm in length, 25 nm in width and 2-3 nm thick [19]. The location of these crystals in the fibril was demonstrated in a study by Traub *et al.* that showed that mineralized collagen fibrils had the same banded pattern as negatively stained collagen fibrils [16]. This indicated that mineral is concentrated in the hole zones of the fibril. It was proposed that these mineral platelets were arranged in parallel like a stack of cards within the interstices of the fibril. Olszta *et al.* concluded from electron diffraction studies that the mineral plates are not quite as ordered as previously assumed [20]. This imperfect arrangement of nearly parallel crystals has been supported by recent small-angle X-ray scattering (SAXS) and transmission electron microscopy (TEM) data from Burger *et al.* [21]. The third level of hierarchy is composed of arrays of these mineralized collagen fibrils. These fibrils are rarely found isolated but rather almost always associated as bundles or other arrangements, often aligned along their long axis. Type I collagen molecules are self-assembled into fibrils with a periodicity of 67 nm and with 40 nm gaps between the ends of their molecules, into which the apatite crystals are placed. A composite of these two constituents forms mineralized fibrils [22]. The fourth level is the pattern of arrays that are formed. These include parallel arrays, woven arrangements, plywood like structures, and radial arrays like those found in dentin. Cylindrical structures called osteons make up the fifth level. Osteons are formed with significant cellular activity and remodelling; osteoclasts resorb bone and form a tunnel, and osteoblasts subsequently lay down lamellae in stacked layers until only a small channel (Haversian canal) is left behind. These channels serve as a conduit for nerves and blood supply to the bone cells. The sixth level of bone organization is the classification of osseous tissue as either spongy (trabecular or cancelous) or compact (cortical). Cancellous bone is extremely porous (75-95% porosity), providing space for marrow and blood vessels, but has much lower compressive strength. Cortical bone is the dense outer layer (5-10% porosity) that allows the many functions of bone. Therefore, the mechanical properties of cortical bone represent the benchmark for synthetic bone [23]. The seventh level is simply the whole bone on the macroscopic scale, incorporating all of the lower levels of hierarchy.

1.2. Tooth

The mammalian body contains numerous mineralized tissues as reported in the previous paragraph, but the tissue with the most robust mechanical properties is enamel. Enamel is the hardest material formed by vertebrates and is the most highly mineralized skeletal tissue present in the body [24]. Mature enamel is composed of 95-97% by weight of apatitic calcium phosphate (exhibiting a chemical composition close to that of stoichiometric hydroxyapatite, HA) with less than 1% organic material. The high degree of mineralization makes enamel a fascinating model for understanding fundamental mineralization processes and processes that occur within an extracellular matrix. It is distinct from bone in terms of architecture, pathology, and the biological mechanisms mediating its formation. Understanding the biological formation of different mineralized structures could lead to innovative approaches toward engineering novel scaffolds and providing new therapeutics. Additionally, unlike other biomineralized tissues, such as bone and dentin, mature enamel is acellular and does not resorb or remodel. As a result, enamel regeneration cannot occur *in vivo* and is therefore an attractive target for future biomimetic and therapeutic approaches.

The mammalian tooth is made up of four distinct structures: enamel, dentin, pulp, and cementum [25]. The pulp contains nerves, blood vessels, fibroblasts, and lymphocytes, while the mineralized organs of the tooth include enamel, dentin, and cementum. Enamel makes up the uppermost 1-2 mm of the tooth crown and contains a high mineral content, giving it a high modulus but also making it susceptible to cracking. Dentin lies below the enamel and is tougher, forming the bulk of the tooth and absorbing stresses from enamel, preventing its fracture [26]. The composition of dentin is similar to that of bone. The cementum is the mineralized layer that surrounds the root of the tooth covering the dentin layer and some of the enamel layer. The cementum allows for the anchoring of the tooth to the alveolar bone (jawbone) through the periodontal ligament. The primary function of the tooth is for mastication of food; however, some species use them for attacking prey and for defence. It also faces the lifelong challenge of maintaining robust mechanical properties in a bacteria-filled, often acidic environment. It is interesting to recall here that calcium phosphates are ionic compounds, for which the three-dimensional cohesion is mostly ensured by electrostatic bonding. In this context, it is interesting to notice then that Nature has learned to adapt the physico-chemical features of biological apatites to the functions that they have to play *in vivo*. In particular, the chemical composition of enamel is found to be close to the stoichiometry of HA, thus leading to a high thermodynamic stability and greater resistance to dissolution (*e.g.* in acidic conditions present in the dental environment), which is well adapted to the function of tooth enamel. In contrast, bone mineral is composed of non-stoichiometric nanocrystalline apatite which is less thermodynamically stable and thus enables the bone remodelling process (through the osteoclast/osteoblast cycle).

The enamel and dentin tissues give rise to a tough, crack-tolerant, and abrasion-resistant tissue through their unique architectures and mineral compositions. Enamel is highly patterned and consists of organized interweaving bundles of crystallites (called rods or prisms). It has a higher reported toughness than that of well-crystallized HA, indicating that the organization of the crystallites is essential for enamel function [27]. Because of the high mineral content and minimal organic, enamel is brittle. Interestingly, the architecture of the enamel crystallites can deflect a propagating crack preventing it from reaching the dentin-enamel junction (DEJ), which also has been shown to resist delamination of the tissues despite their differences in composition [28]. The mechanical properties of enamel, dentin, and the DEJ are not completely understood and are a significant area of research. Understanding the properties of these tissues could serve to motivate further engineering of more robust dental materials as well as to inspire fabrication of nonbiologically-oriented materials.

Like for bone, enamel possesses a complex architecture, which can be broken up into several hierarchical levels from the nanoscale to the macroscale [29]. On the nanoscale, the protein-protein and protein-mineral interactions in the presence of supersaturated fluids create a highly organized array of HA crystallites that grow preferentially along the *c*-axis [30]. The sizes of these crystallites depend on the stage of the mineralization. The crystallites grow primarily in length during the secretory stage and continue to grow in width and thickness during the maturation stage. The assembly of amelogenin has been shown to be crucial for the proper development of enamel crystallites. Disruption of the assembly alters formation on the nanoscale, subsequently affecting larger length scales and giving rise to a diseased or malformed enamel phenotype.

On the mesoscale level, there are three main structural components: the rod, the inter rod, and the aprismatic enamel. The main component of enamel on the mesoscale includes rods, which are bundles of aligned crystallites that are "woven" into intricate architectures that are approximately 3-5 μm in diameter [31].

The second structural component of the enamel matrix is the interrod (or interprismatic) enamel, which surrounds and packs between the rods. The difference between the rod and the interrod is the orientation of HA crystals; the rod contains aligned crystallites, whereas the mineral in the interrod is less ordered. These structures coalesce to form the tough tissue of enamel, which can withstand high forces and resist damage by crack deflection.

The third structure, aprismatic enamel, refers to the structures containing HA crystals that show no mesoscale or macroscale alignment. The macroscale architecture includes specific zones of enamel that have unique characteristics, which contribute to the whole tissue. The enamel adjacent to the DEJ exhibits a gradual transition from dentin to enamel. Aprismatic regions of enamel have been proposed to be primitive areas of the tooth serving as a toughening mechanism due to their flexible nature [32]. Several authors have identified these aprismatic areas to be located adjacent to the DEJ and at the incisal surface of both deciduous and permanent human enamel [33]. Tomes' process, a unique structure present at the secretory pole of an enamel-forming cell, is responsible for aligned mineral formation in the prismatic enamel. The absence of this process may give rise to the aprismatic zone in the tooth.

1.3. Pathological Calsifications

In the body of mammals, osteoblasts and odontoblasts fix ions of calcium and orthophosphate andthen precipitate biological apatite onto an organic matrix. This is the process of physiological biomineralization that is restricted to the specific sites in skeletal tissues, including growth plate cartilage, bones and teeth. Unfortunately, owing to ageing, various diseases and under certain pathological conditions blood vessels and some internal organs are calcifiedas well [34]. This process is called pathological calcification or ectopic mineralization and leads to increased mortality. In general, any type of abnormal accumulation of calcium phosphates in wrong places is accounted for by a disruption of systemic defence mechanism against calcification [35].

Unwanted depositions always lead to various diseases, such as: soft tissue calcification (in damaged joints, blood vessels, dysfunctional areas in the brain, diseased organs, scleroderma, prostate stones) [36], kidney and urinary stones [37], dental pulp stones and dental calculus [38], salivary stones, gall stones, pineal gland calcification, atherosclerotic arteries and veins [39], coronary calcification, cardiac skeleton, damaged cardiac valves (Fig. **2**) [40], calcification on artificial heart valves, carpal tunnel, cataracts, malacoplakia, calcified menisci, dermatomyositis and other [41]. All these cases are examples of a calcinosis, which might be described as a formation of calcium phosphate deposits in any soft tissue.

Contrary to the mineral phases of the normal calcifications (bone, dentine, enamel, cementum, antlers), which consist of only one type of calcium phosphate (namely, biological apatite), the mineral phases of abnormal and/or pathological calcifications are found to occur as single or mixed phases of other types of calcium phosphates and/or other phosphatic and non-phosphatic compounds in addition to or in place of biological apatite [42]. This happens because in the places of pathological calcifications the solution pH is often relatively low.

However in some cases, the chemical composition of an unwanted inorganic phase might depend on the age of the pathological calcification and its location [43]. It is interesting to note that the mineral phases of animal calculus (*e.g.,* from dog) were found to consist of calcium carbonate and biological apatite, while human calculi do not contain calcium carbonate. Some findings suggested that the mechanisms and factors regulating the physiological biomineralization might be similar to those influencing the ectopic mineralization: both were initiated by various organics (*i.e.,* membrane-enclosed particles released from the plasma membrane of mineralization-competent cells), that were present [44].

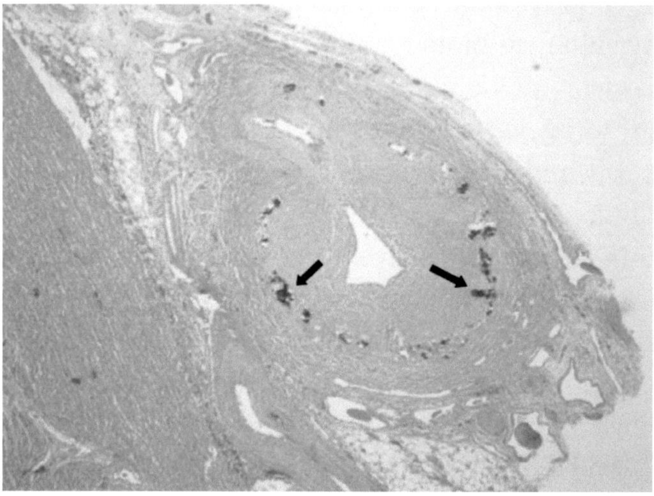

Figure 2: Coronary artery calcification in human nucleotide pyrophosphatase/phosphodiesterase 1 deficiency. Hydroxyapatite crystal depositions (arrows) predominantly at the level of the internal elastic lamina in a patient with generalized arterial calcification of infancy, who died of myocardial infarctus at the age of 4 weeks. Von Kossa, original magnification ×100. (from Ref. [40] with permission).

2. SYNTHETIC CALCIUM PHOSPHATES BIOMATERIALS

One of the most exciting and economical rewarding research areas of materials science involves applications of materials to health care, especially to reconstructive surgery. In the past, many implantations failed due to infections ora lack of knowledge about toxicity of the selected materials. In this case, the use of calcium phosphates is valid due to their chemical similarity to the mineral component of mammalian bones and teeth as previously exemplified [45]. Thus, calcium phosphate-based biomaterials are now used in a number of different applications throughout the body, covering all areas of the skeleton. Applications include dental implants, percutaneous devices and use in periodontal treatment, treatment of bone defects, fracture treatment, total joint replacement (bone augmentation), orthopaedics, cranio-maxillofacial reconstruction, otolaryngology and spinal surgery [45-47]. Depending upon the desired characteristics of the material (resorbable or not, highly-bioactive or not...), different calcium phosphates might be used [48]. In this section we will give some general knowledge about three different methods to use calcium phosphates as biomaterials: ceramics, cements and coatings.

2.1. Ceramics

Currently, calcium phosphate ceramics can be prepared from various sources [49-51]. Unfortunately, up to now, all attempts to synthesize bone replacement materials for clinical applications featuring the physiological tolerance, biocompatibility and a long-term stability have had only a relative success, underlining the superiority and complexity of the natural structures [52]. Chemically, the vast majority of calcium phosphate ceramics is based on HA, β-tricalcium phosphates (β-TCP), α-tricalcium phosphates (α-TCP) and/or biphasic calcium phosphate (BCP), which is an intimate mixture of either β-TCP+HA [53-52] or α-TCP+HA [56-58]. The preparation techniques of these calcium phosphates have been extensively reviewed in literature [46, 48]. When compared to both α- and β-TCP, HA is a more stable phase under the physiological conditions, as it has a lower solubility and, thus, a slower resorption kinetics [58]. Therefore, the BCP concept is determined by the balance of a stable phase of HA and a more soluble TCP. Due to a higher biodegradability of the α- or β-TCP component, the reactivity of BCP increases with the TCP/HA ratio increasing. Thus, *in vivo* bioresorbability of BCP can be controlled through the phase composition [55]. However, it should be noted here that this association of phases is only a way to increase artificially the resorption rate of the bioceramics, since implants containing calcined HA can be detected in the initial bone defects even many years after implantation [59].

A sintering stage appears to be of great importance to produce ceramics with the required properties. Several processes occur during the sintering of calcium orthophosphates. Firstly, moisture, carbonates and all volatile chemicals remaining from the synthesis stage, such as ammonia, nitrates and any organic compounds, are removed as gaseous products. Secondly, the removal of these gases facilitates production of dense ceramics with subsequent shrinkage of the samples. Thirdly, these chemical changes are accompanied by a concurrent increase in crystal size and a decrease in the specific surface area. Fourthly, a chemical decomposition of all acidic orthophosphates and their transformation into other phosphates takes place. Besides, sintering causes toughening and increases the mechanical strength [60, 61].

HA powders can be pressureless sintered up to theoretical density at 1000-1200 °C. Processing at higher temperatures may lead to exaggerated grain growth and decomposition because HA becomes unstable at temperatures exceeding ~1250-1300 °C [62]. The decomposition temperature of HA ceramics is a function of the partial pressure of water vapor. Processing under vacuum leads to an earlier decomposition, while processing under high partial pressure of water prevents the decomposition. On the other hand, the presence of water in the sintering atmosphere is reported to inhibit densification of HA and accelerate grain growth [63]. A definite correlation between hardness and a grain size in sintered HA ceramics was found: the hardness started to decrease at a certain critical grain size limit despite exhibiting high bulk density [64].

Ceramics made of calcium phosphates often possess limited mechanical properties that do not allow them to be used in load-bearing areas, such as artificial teeth or bones [65]. For example, fracture toughness of HA ceramics does not exceed the value of ~1.0 $MPam^{1/2}$ (human bone: 2-12 $MPam^{1/2}$). It decreases almost linearly with porosity increasing [63]. Generally, fracture toughness increases with decreasing grain size. However, in some materials, especially non-cubic ceramics, the fracture toughness reaches a maximum and rapidly drops with decreasing grain size. Due to high brittleness (associated to low crack resistance), the biomedical application of calcium phosphate ceramics is focused on production of non-load-bearing implants, such as pieces for middle ear surgery, filling of bone defects in oral or orthopaedic surgery, as well as coating of dental implants and metallic prosthesis. In order to improve the reliability of calcium phosphate ceramics, various reinforcements (ceramics, metals or polymers) have been used to manufacture biocomposites and hybrid biomaterials [66].

In spite of the mechanical limitations (limiting the use in weight-bearing sites), ceramics of calcium phosphates are available in various physical forms: powders, particles, granules (or granulates), dense blocks, porous scaffolds, injectable formulations, self-setting cements, implant coatings and composite component of different origin (natural, biological or synthetic) [67]. Furthermore, custom designed shapes like wedges for tibial opening osteotomy, cones for spine and knee and inserts for vertebral cage fusion are also available. Surface area of porous bodies and scaffolds is much higher, which guarantees good mechanical fixation in addition to providing sites on the surface that allow chemical bonding between the ceramics and bones. Furthermore, pore sizes and pore morphology are directly related to bone formation, since they provide surface and space for cell adhesion and bone in growth. On the other hand, pore interconnection provides the way for cell distribution and migration, as well as it allows an efficient *in vivo* blood vessel formation suitable for sustaining bone tissue neo-formation and possibly remodelling. Porous apatite-based ceramics can be colonized by bone tissues [68, 69]. Therefore, interconnecting macroporosity (pore size $>$ 100 mm), which is defined by its capacity to be colonized by cells, is intentionally introduced in solid ceramics [70]. Macroporosity is usually formed due to release of various volatile or highly-soluble materials and, for that reason, incorporation of pore creating additives (porogens) is the most popular technique to create macroporosity. Among porogens additives we can mention sodium chloride, sugars (*e.g.* glucose), paraffin, naphthalene, fluoride salts, hydrogen peroxide, polyvinyl butyral, *etc.* [71-73]. These additives are mixed to apatite powders or slurries. After moulding, the organics burn away from the moulding body during sintering. This approach allows direct control of the pore characteristics, which are a function of the amount and properties of the volatile/soluble phase. Several other techniques, such as replication of polymer foams by impregnation, dual-phase mixing technique, freeze casting, strereolithography and foaming of gel-casting suspensions and transformation of porous calcium carbonate from biological origin (marine corals or echinoderms skeletons) into porous calcium phosphates by treatment with phosphate solutions under hydrothermal conditions have been applied to fabricate porous calcium phosphate ceramics [74-77]. By using the latter technique the interconnected porosity of the aragonitic or calcitic skeletons of corals or echinoderms is maintained after transformation to apatite or mixtures of apatite and other calcium phosphates [78]. The preparation methods of porous bioceramics has been recently reviewed by Sopyan *et al.* [79].

Calcium phosphates ceramics in a wide number of forms and compositions are currently either in use or under a consideration in many areas of dentistry and orthopedics, with even more in development. For example, bulk materials, available in dense and porous forms, are used for alveolar ridge augmentation, immediate tooth replacement and maxillofacial reconstruction. Other examples include orbital implants increment of the hearing ossicles, spine fusion and repair of bone defects. Today, a large number of different calcium orthophosphate ceramics for the treatment of various defects is available on the market.

2.2. Cements

The possibility to obtain a monolithic calcium phosphate ceramics at ambient or body temperature *via* a cementation reaction was put forward by Le Geros *et al.* and Brown and Chow in the early 1980s [80, 81]. Currently this type of materials is known as calcium phosphate cements (commonly referred to as CPC), and, due to their suitability for repair, augmentation and regeneration of bones, they might be named as calcium phosphate bone cements (occasionally referred to as CPBC). Due to a good bioresorbability, calcium phosphate cements belong to the second generation of bone substituting biomaterials. These cements are blends of amorphous and/or crystalline calcium phosphate powder(s) with an aqueous solution, which might be distilled water, phosphate- buffered saline (PBS), an aqueous solution of sodium phosphate (0.25 M), orthophosphoric acid, 0.5 M aqueous solution of citric acid [82], or even revised simulated body fluid (SBF) [83]. After the powder and the solution are mixed together, a viscous and modulable paste is formed that sets to a firm mass within a few minutes. When the paste becomes sufficiently stiff, it can be placed into a defect as a substitute for the damaged part of bone, where it hardens *in situ* within the operating room. The proportion of solid-to-liquid or the powder-to-liquid (P/L) ratio is a very important characteristic because it determines bioresorbability and injectability. As the paste is set and hardened at room or body temperature, direct application in healing of bone defects became a new and innovative treatment modality in the end of the twentieth century. Moreover, calcium phosphate cements can be injected directly into fractures and bone defects, where they intimately adapt to the bone cavity regardless of its shape

[46]. Besides, they were found to promote the development of osteoconductive pathways, possess sufficient compressive strengths, be noncytotoxic, create chemical bonds to the host bones, restore contour and have both the chemical composition and X-ray diffraction patterns similar to those of bone. Finally, but yet importantly, they are osteotransductive, *i.e.,* after implantation calcium phosphate cements are replaced by a new bone tissue [84].

The first calcium orthophosphate cement formulation consists of the equimolar mixture of tetracalcium phosphate (TTCP) and dicalcium phosphate (DCPA or DCPD), which is mixed with water at a P/L ratio of 4:1; the paste hardened in about 30 min and formed apatite [80]. This highly viscous, non-injectable paste can be molded and is therefore used mainly as a contouring material in craniofacial surgery. In 1990s, it was established that there were about 15 different binary combinations of calcium orthophosphates, which gave pastes upon mixing with water or aqueous solutions, so that the pastes set at room or body temperature into a solid cement. The list of these combinations is available in the literature [85]. From these basic systems, secondary formulations could be derived containing additional or even non-reactive compounds but still setting like cements [85]. However, depending upon the pH value of a cement paste, after setting all calcium phosphate cements can only form two major end-products: a precipitated poorly-crystallized apatite (for $pH > 4.2$) and DCPD (under $pH < 4.2$), the later situation being however transient *in vivo* (and DCPD being then hydrolyzed into apatite for greater pH values) [86]. Therefore, all existing formulations of calcium orthophosphate cements have been divided into two major groups: apatite cements and brushite cements. The final sitting product of the cements is of the paramount importance because it determines the solubility and, therefore, *in vivo* bioresorbability. Since the chemical composition of mammalian bones is similar to ion-substituted non-stoichiometric nanocrystalline apatite, apatite cements have been more extensively investigated. However, many research papers on brushite cements have been published as well [46].

2.3. Coatings

Commercially pure titanium (cpTi) and Ti-6Al-4V alloys are the most commonly used metallic implant materials, as they are highly biocompatible materials with excellent mechanical properties and corrosion resistance [87, 88]. The biocompatibility of titanium implants is attributed to the stable oxide layer (with a thickness of 3-10 nm) that spontaneously forms when titanium is exposed to oxygen [89]. This reaction prevents the formation of fibrous tissue around the implant, and creates direct contact to osseous tissue. Nevertheless, when applying Ti (O_2) as implant material, a non-physiological surface is exposed to a physiological environment. However, by generating a coating onto a titanium surface that mimics the organic and inorganic components of living bone tissue, a physiological transition between the non-physiological titanium surface and surrounding bone tissue can be established. In this way, the coated titanium implant functions as scaffold for improved bone cell attachment, proliferation and differentiation [90, 91]. Such a coating is supposed to further enhance early and strong fixation of a bone-substituting implant by stimulating bone formation starting from the implant surface. As such, a continuous transition from tissue to implant surface can be induced. Consequently, research efforts have focused on modifying the surface properties of titanium to control the interaction between the implant and its biological surrounding [92].

Calcium phosphates ceramics are too brittle for use as bulk material under loaded conditions, but such ceramics are frequently applied as coatings onto the surface of metallic implant materials in order to combine the mechanical strength of metals with the excellent biological properties of calcium phosphates ceramics [93]. Calcium phosphates coatings for orthopaedic and dental implants were introduced by de Groot and Geesink [94, 95]. Since them numerous reports have been published about the osteoconductive properties of calcium phosphates-coated implants (osteo conduction refers to the ability of a biomaterial to support the growth of bone over its surface). These coatings are described to induce an increased bone-to-implant contact, to improve the implant fixation, and to facilitate the bridging of small gaps between implant and surrounding bone. The calcium phosphates-layer guides bone growth along the implant surface, and as a result bone formation now occurs from both the surrounding tissue and the implant surface, in which calcium phosphates functions as a physiological-like transition between the non-physiological titanium surface and surrounding bone [96].

From a commercial point of view, the most successful method to apply calcium phosphates coatings to implants has been the plasma-spraying technique, due to its high deposition rate and the ability to coat large areas [97, 98]. The osteoconductive and bone-bonding behaviour of plasma-sprayed coatings was assessed by numerous studies. However, it should be noted that, despite recent advances especially with low-energy plasmas, the chemical composition of plasma coatings generally involve a mixture of several phases (including amorphous calcium phosphate in particular), not always in a reproducible manner. Therefore, researchers have been continuously inspired in the past two decades to explore alternative or complementary techniques for deposition of coatings onto an implant surface. Various deposition methods have been proposed, including magnetron sputtering [99, 100], electrophoretic deposition [101], hot isostatic pressing, sol-gel deposition [102], pulsed laser deposition [103], biomimetic deposition [104], ion beam dynamic mixing deposition [105], electrospray deposition [106], electrolytic deposition induction and heating deposition from solution [97]. Generally, the properties of the produced coatings differ considerably in terms of chemical structure, composition, thickness, mechanical properties, *etc.* (see Fig. **3** for an illustration of the large variation in surface morphology of three common calcium phosphates coating techniques) [107].

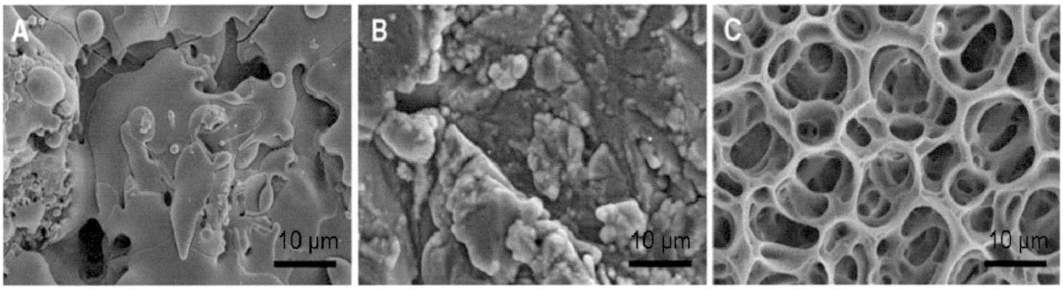

Figure 3: Scanning electron micrographs of CaP coating morphologies of (A) plasma-spray coating, (B) RF magnetron sputter coating, and (C) electrospray deposition coating.

Therefore, caution should always be taken when directly comparing the success rates of these coating techniques without a proper understanding of the physicochemical nature of the specific calcium phosphates coatings. Generally, it should be kept in mind that conclusions about the biological/clinical performance of coatings cannot be made without a complete set of characterizations that enable correlation of material properties to biological response [108]. Clinically, each application demands specific requirements, and in that respect the wide range of available coating techniques offers the possibility to select the most appropriate deposition method for each specific implant application.

3. SURFACE CHEMISTRY OF SYNTHETIC AND BIOGENIC CALCIUM PHOSPHATES

As was said previously, among the calcium phosphate compounds widely used nowadays in bone tissue engineering, stoichiometric hydroxyapatite (HA) is undoubtedly the most commonly found. However this compound is thermodynamically stable and microcrystalline, which explains the rather low intrinsic bioactivity and poor resorption rate observed *in vivo*. The design of future-oriented innovative biomaterials for bone-tissue engineering will, in contrast, require to consider not only the physical filling of bone defects but also the possibility to trigger simultaneously cellular functions and activate bone regeneration and healing processes. With this view in mind, it is then probably essential to consider developing biomaterials exhibiting physico-chemical features that are closer to the minerals actually observed in biological systems, which noticeably depart from stoichiometric HA. We will thus summarize in the following section the principal characteristics of biomimetic nanocrystalline apatites (either natural or synthetic), with a special focus on their surface characteristics. A few additional comments have been added at the end of this section concerning some noticeable surface features of other calcium phosphates of interest in bio-medical approaches.

3.1. Nanocrystalline Apatites

As previously mentioned, nanocrystalline calcium phosphate apatites represent the main inorganic constituent of hard tissues in vertebrates [109, 110]. Therefore, much attention has been lately focused on the study of their peculiar physico-chemical characteristics and, thereafter, on the possibility to imitate bone mineral for the development of new "advanced" synthetic biomaterials. Moreover, the possibility to synthesize in the lab synthetic biomimetic compounds makes it possible to consider these systems as a good bone mineral "model", enabling then to investigate on the one hand the interaction between bone-like apatite nanocrystals and the components of surrounding fluids (ions, proteins...), and on the other hand to follow surface interactions with drugs aimed at being delivered *in vivo*.

Among the peculiarities of these compounds, their plate-like morphology (elongated towards the c-axis) is one obvious specificity as compared to regular HA. Another difference is of course the mean crystallite size, often of the order of 15-30 nm in length and ca. 6-9 nm in width [111]. From a chemical point of view, the composition of nanocrystalline apatites also strongly differs from that of HA. Although it has been the object of much controversy during several decades in the middle of the 20th Century, the global chemical composition of biological apatites (or their synthetic analogues) can generally be described as:

$$Ca_{10-x}\,(PO_4)_{6-x}\,(HPO_4 \text{ or } CO_3)_x\,(OH \text{ or } \tfrac{1}{2}\,CO_3)_{2-x} \text{ with } 0 \leq x \leq 2$$

(except maybe for very immature nanocrystals which may depart from this generic formula). In particular, this expression underlines the presence of vacancies in both Ca and OH sites. For example, Legros *et al.* [112] analyzed various cortical bone specimens, suggesting the following relatively homogeneous composition, unveiling high vacancy contents:

$$Ca_{8.3}\,(PO_4)_{4.3}\,(HPO_4 \text{ or } CO_3)_{1.7}\,(OH \text{ or } \tfrac{1}{2}\,CO_3)_{0.3}$$

Minor substitutions are also found in biological apatites involving for example monovalent cations (especially Na^+) in cationic sites. In this case, charge compensation mechanisms have to be taken into account.

It should however be kept in mind that such chemical formulas only enable to have a "global" insight on the nature and amount of ions present in the compound, but it does not reflect possible local variations that may be observed on the nanocrystals or *in vivo* within the osteons [113], linked to the local apatite crystal formation and, *in vivo*, to continuous bone remodelling.

The characterization of nanocrystalline apatites is a rather arduous task due to their poor crystallinity and metastability. But despite the complexity of these systems, recent in-depth investigations have succeeded to unveil their particular surface features (and related reactivity in wet media) that may be exploited for the setup of new biomaterials, or for a better understanding of natural or pathological biomineralization phenomena. Recent advances in the characterization of apatite nanocrystals were obtained thanks to the use of spectroscopic techniques, and in particular Fourier Transform Infrared (FTIR) spectroscopy and solid-state Nuclear Magnetic Resonance (NMR). FTIR and Raman methods are especially useful for drawing conclusions on the local chemical environment of phosphate, carbonate and hydroxide ions as well as water molecules in such systems. Detailed analyses of the phosphate groups by FTIR have enabled to distinguish (Fig. **4**) in nanocrystalline apatites the presence of additional bands that cannot be attributed to phosphate groups in a regular apatitic environment [114, 115]. These chemical environments have been referred to by Rey as "non-apatitic" environments.

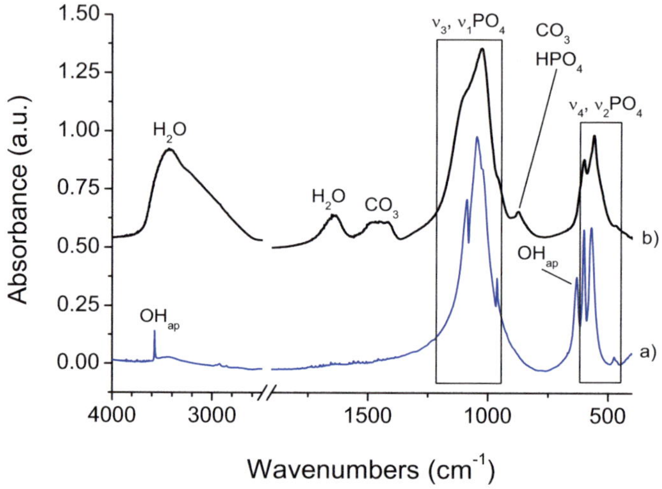

Figure 4: FTIR spectrum of a) stoichiometric hydroxyapatite and b) a carbonated nanocrystalline apatite.

These additional bands are best observed in the ν_4PO_4 vibration domain (Fig. **5**), where two shoulders can be seen beside apatitic bands, respectively assigned to non-apatitic PO_4^{3-} ions (around 617 cm^{-1}) and non-apatitic HPO_4^{2-} ions (around 535 cm^{-1}). Interestingly, this last assignment was also confirmed by chemical analysis [116]. The presence of such non-apatitic environments is thought to be related to the formation conditions of the apatite nanocrystals. The analysis of the ν_3PO_4 vibration domain is also of interest, in particular in the case of apatites studied in the wet state (*e.g.* freshly precipitated) where a fine structuration of the ν_3PO_4 band can be witnessed by FTIR (Fig. **6**). Indeed, in these conditions, the presence of several fine details can be evidenced in the form of successive shoulders (which is not the case for stoichiometric HA).

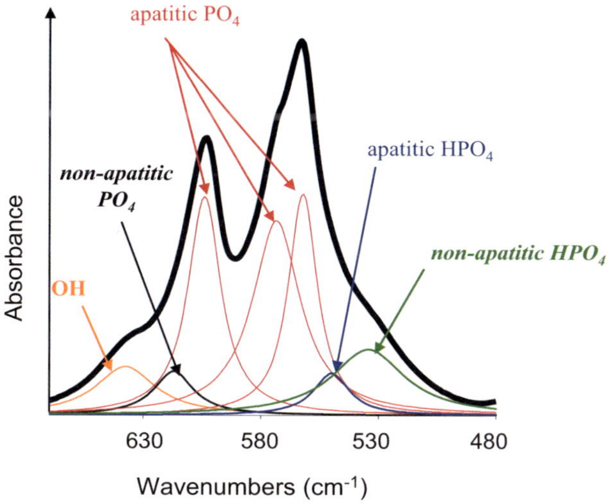

Figure 5: Example of ν_4PO_4 FTIR band decomposition (curve-fitting).

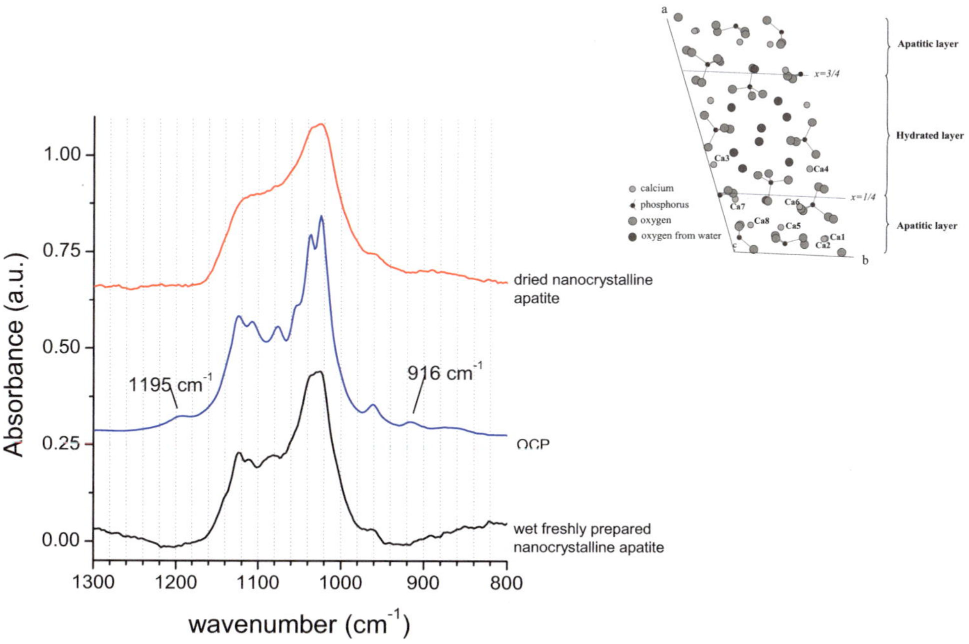

Figure 6: Details of the ν_3PO_4 FTIR band for nanocrystalline apatites (wet and dried) and for OCPt. Figure inlet: OCPt unit cell.

It is interesting to signal that these bands somewhat resemble the fine details observed for triclinic octacalcium phosphate (OCPt) [117, 118], for which the structure can be described as a continuous superimposition of so-called "apatitic" and "hydrated" layers, where the hydrated layers contain calcium ions as well as all the HPO_4^{2-} content of the OCPt unit cell and several water molecules (see Fig. **6**).

However, it can be noted that some bands present on the spectrum of OCPt (*e.g.* around 916 and 1195 cm^{-1}) are not detected for nanocrystalline apatites (it should probably also be mentioned here that the typical (100) XRD peak observed at low 2θ angles for OCPt is not observed for such apatites and that it is not possible to incorporate carbonate ions in the OCPt structure). Nevertheless, this rather high resemblance between the FTIR spectral features of OCPt and biomimetic apatites may suggest some degree of similarity regarding local chemical environments and/or interactions with water molecules; however these aspects will require some additional structural studies.

Similarly, for carbonated nanocrystalline apatites, additional vibrational bands can bedetected in all biological or synthetic apatites synthesized at physiological pH, and this was linked to the occurrence of non-apatitic carbonate ions. A characteristic band around 866 $cm^{-1}$is in particular clearly visible by FTIR in the ν_2CO_3 vibration domain, and can be exploited for the quantitative assessment of the amount of non-apatitic carbonate ions in the sample. Interestingly, these ions have been shown (based on ion exchange experiments, *e.g.* [119] to share the same surface domains as the non-apatitic HPO_4^{2-} ions. Such ion exchange experiments were found to be fast phenomena and mostly reversible, therefore strongly suggesting surface processes rather than involving bulk ions. These findings thus led to the concept of rather labile surface ions localized in non-apatitic chemical environments. Most importantly, 2D ^{13}C-^{1}H solid-state NMR analyses on biomimetic apatites [120] recently showed the existence of a clear correlation between the non-apatitic carbonate species and water molecules [121]. Taking into account the circumstances of formation of apatite nanocrystals in biomimetic conditions, a strong interaction of the nanocrystals surface with water molecules is indeed quite probable. Some ^{1}H solid-state NMR peaks have indeed been related to water molecules associated with bone apatite nanocrystals or in synthetic analogues [122, 123]. This interaction may in fact play a role in the thermodynamic stabilization of the nanocrystals surface (lowering of surface energy). In this context, the study of the potential impact of drying procedures on the physico-chemical characteristics of nanocrystalline apatites was also investigated [121]. FTIR spectroscopy analyses pointed out a significant modification of vibration bands upon drying (see Fig. **3**): while the wet, freshly precipitated apatite nanocrystals exhibit (as stated previously) very thin bands, *e.g.* in the ν_3PO_4 domain, this fine band structure is lost by drying and a noticeable line broadening takes place.

The presence of non-apatitic environments in biomimetic apatites has also been pointed out from solid-state NMR. In stoichiometric HA, the ^{31}P chemical shift of all phosphate groups is the same and leads to a single peak at 2.3-2.9 ppm with respect to H_3PO_4 [124-126]. In contrast, in bone or bone-like apatite, NMR data suggest phosphate environments different from those of regular HA [127, 129]. In addition, an asymmetrical broad ^{31}P peak of the main phosphate is often observed [129, 130], and this was related to HPO_4^{2-}ions, that can be especially evidence during 2-D $^{1}H\leftrightarrow^{31}P$ cross-polarization solid-state NMR [129]. The presence of a partially disordered hydrated surface layer on biomimetic apatites was again confirmed by Kaflak *et al.* in 2008, based on additional NMR experiments [131]. These specificities are in good agreement with the above-mentioned results, especially those obtained by FTIR spectroscopy.

Taking into account all the above summarized data arising from complementary characterization techniques, nanocrystalline apatites (whether biological or their synthetic analogues prepared under close-to-physiological conditions) may thus most probably be described as the association of an *apatitic core* (often nonstoichiometric) and a structured by fragile *surface hydrated layer* containing water molecules and rather labile ions (*e.g.* Ca^{2+}, HPO_4^{2-}, CO_3^{2-}...) [132] occupying non-apatitic crystallographic sites, and some striking similarity can be seen with the hydrated layers observed in triclinic OCPt (although in the case of biomimetic apatites the layer is directly exposed on the surface and not included in a "sandwich-like" structure between two "apatitic" layers). A schematic model for such nanocrystalline apatites is given in Fig. **7**.

The presence of this hydrated surface layer is thought to be responsible for most of the properties of biomimetic apatites, and in particular their high surface reactivity in relation with surrounding fluids (which is probably directly linked to a high mobility of ionic species contained within this layer) and this reactivity may explain, from a physico-chemical viewpoint, the role of bone mineral in homeostasis *in vivo*. This layer indeed contains labile ions that can potentially be exchanged by other ions from the surrounding solution, or by small molecules, which may be exploited for couplings with proteins or drugs (as commented later in this chapter).

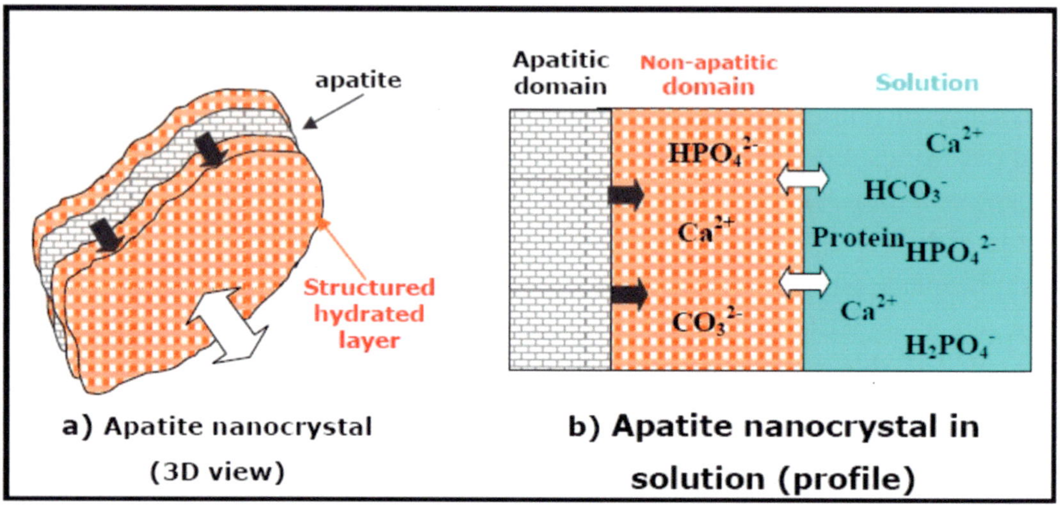

Figure 7: Schematic modelling of a biomimetic apatite nanocrystal (a) and interaction with surrounding fluids (b) (from Ref. [118], with permission)

It is interesting to remark that the typical non-apatitic features mentioned above tend to progressively disappear during the ageing of the nanocrystals in solution [133, 134]. This process is referred to as "maturation" and has been related to the progressive growth of apatite domains at the expense of the surface hydrated layer [111, 134].

This maturation process is thought to be linked to the metastability of such poorly-crystallized nonstoichiometric apatites, which steadily evolve in solution towards stoichiometry and better crystallinity [133, 135, 136]. This evolution can be for example witnessed by the decrease of the amount of non-apatitic HPO_4^{2-} ions upon ageing, or else by the decreased potentialities to undergo ion exchanges [134]. One illustration of such effects can be given for example by the decreased exchangeability of HPO_4^{2-} by CO_3^{2-} observed on carbonated apatites matured for incremented amounts of time. Additionally, beside this compositional evolution, some structural and microstructural features also tend to evolve such as the mean crystallite size which increases upon maturation. The control of synthesis parameters such as pH, temperature or maturation time can thus enable one to tailor the physico-chemical properties of biomimetic apatites [133], and in particular their surface reactivity, so as to mimic for example mature bone mineral or else newly-formed bone apatite. This control of surface reactivity can for instance be clearly evidenced by following the effect of maturation on ionic exchanges (whether cationic or anionic) (Fig. **8**) [11, 119, 134, 137] or on adsorption phenomena [138-140].

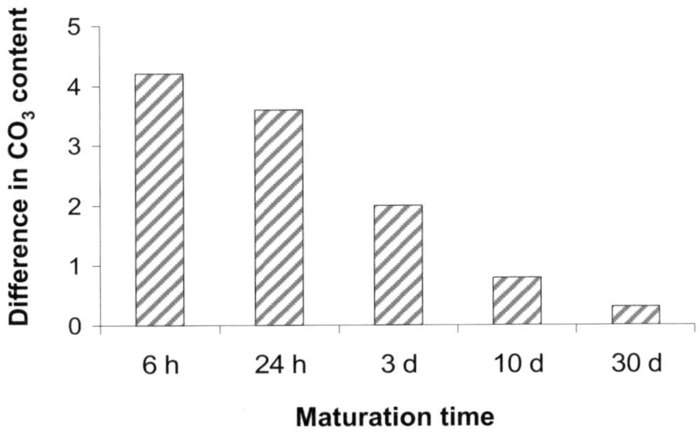

Figure 8: Evolution of CO_3^{2-}/HPO_4^{2-} exchange capabilities for carbonated biomimetic apatites *versus* maturation time (carbonated apatites with varying maturation states, subsequently subjected to a carbonated medium; form Ref. [119] with permission).

The possibility to tailor bulk and surface properties of biomimetic apatites can then be exploited for the preparation of advanced biomaterials with enhanced bioactivity and resorbability properties. This can for example be used for enriching biomimetic apatite-based materials with specific ions such as magnesium or strontium [117, 141, 142] which are known to exhibit advantageous biological properties (osteoblast cells activation and/or osteoclast cells inhibition), or with active molecules. The same idea can also be exploited to activate existing poorly-bioactive scaffolds (whether metallic or ceramic) by way of a coating with nanocrystalline apatites [143, 144]. Finally, the presence of the hydrated layer on apatite nanocrystals and the related ionic mobility has been shown recently to allow the preparation of a new generation of bioceramics by consolidating biomimetic apatites at "low" temperatures (*e.g.* at 150°C by spark plasma sintering, SPS), while preserving most of the characteristics of the initial powders, and in particular their nanometric size, hydrated characters and non stoichiometry [133, 145-147]. In such consolidation processes, the superficial hydrated layers of adjacent nanocrystals are thought to interact and form strong inter-crystal (or crystal-substrate) contacts, and the high ionic mobility is thought to substitute the thermally-activated ion diffusion mechanisms usually put into play during conventional sintering at high temperatures.

As mentioned previously, biomimetic apatites are nano-sized materials and the constitutive nanocrystals exhibit a highly reactive surface that may be exploited in view of biomedical applications. Although most applications of these compounds are envisioned in link with bone tissue regeneration, it should be noted that the exceptional reactivity of these materials have also lately attracted attention in view of the development of other biomedical applications not related anymore to bone tissue engineering, such as drug delivery, transfection or medical imaging. Indeed, a few recent studies revealed the possibility to prepare colloidal apatite nanoparticles through engineered protocols. These methods are based on the control of the agglomeration state of apatite by way of surface interaction, during precipitation, with biocompatible molecules presenting anionic functional groups such as amino acids [148], DNA [149] or phospholipid derivatives [150-152]. In a quite similar approach, apatite compounds have been synthesized in the presence of surfactants for controlling particle morphologies [153]. These methods exploit, for this agglomeration control, the competition for calcium ions between phosphate ions and the anionic organic molecules present in the medium, and the consequent possibility to tailor nanoparticle size then opens new attractive potential applications for example for direct interactions with cells (*e.g.* cancer cells), either in view of intracellular drug delivery or for diagnosis applications (luminescent nanoprobes).

In all cases, independently of the targeted applications, the surface properties of nanocrystalline apatites will play a key role in the resulting biological responses. The impact of surface energy has already been mentioned above. Another important parameter is also surface charge, which is intimately related to the concentration of specific ions such as calcium, phosphate and hydroxide (or proton) ions in the surrounding fluid. Then, a variety of concentration conditions can lead to a “zero surface charge” [154]. The point of zero charge for HA is generally found in the range 6.4-8.5 [155] (a negatively-charged surface is observed for HA beyond the pH and a positive charge is found below it). Ionic substitutions can however modify the surface charge of apatite compounds [156] and the point of zero charge generally increases with the solubility of the substituted apatite. Also, the presence of ionic organic molecules adsorbed on the surface of apatite compounds, including nonstoichiometric ones, can modify their surface charge [151, 157]. It must however be noted that for nonstoichiometric nanocrystalline apatites, the existence of the surface layer leads to a more complex situation adding some difficulty to evaluate the amounts of phosphate or calcium ions actually exposed on the surface.

3.2. Surface Features for Other Calcium Phosphates of Interest

Surface data concerning other calcium phosphates of interest for biomedical applications are rather scarce. Some reports are however worth mentioning in this section. In particular, zeta potential was measured on OCPt in solutions equilibrated with respect to that phase [158], indicating a positively-charged surface for a wide range of pH values (from 5 to 11). Also, the interfacial energy values 9.0, 18.5, 4.3 and 0. 4 mJ/m^2 (with respect to water) have been reported for HA, FA, OCPt and DCPD [159-160]. Interestingly, although very different structures are involved for these compounds, it can be noted that the lowest interfacial tensions are found for the hydrated phases (DCPD and OCPt) as compared to the anhydrous ones. It may be hypothesized, that the presence of water molecules within the structured hydrated layer on the surface of biomimetic apatite nanocrystals can originate from a need for energetic stabilization of the nanocrystals surface. The peculiarities observed for such water molecules as compared to loosely bound physisorbed ones, especially by solid-state NMR [131], seem to underline the particular role played by the former on biomimetic apatites.

4. SURFACE INTERACTION BETWEEN CALCIUM PHOSPHATES AND BIOMOLECULES - APPLICATION TO BIOMATERIAL AND BIOMINERALIZATION SCIENCE

Materials can interact with biomolecules as well as with living systems and these interactions can be used to develop new materials and technologies [161]. In particular, surface interaction of proteins and peptides with calcium phosphates based biomaterial represents the first and pivotal event when materials are implanted within an organism and must integrate in the host tissues, affecting the accompanying biological responses, namely cell attachment and activation. In some cases this interaction produces adverse effects, strongly determining the final success of the implanted device. For example, the adsorption of serum proteins, such as fibrinogen, fibronectin or vitronectin, can influence the adhesion of leukocytes, macrophages or platelets, and ultimately lead to fibrous encapsulation [162]. Therefore, a detailed understanding of the protein and/or peptides materials interactions is essential to tailor the surface structure of synthetic biomaterials, to optimize the interactions with biological tissues and this topic is acquiring importance in many fields of medicine [163]. Despite the many and deep investigations, the mechanisms involved in protein and peptides adsorption on inorganic substrate is still far from being fully understood. In fact, this is a complex process consisting of many events, including protein conformational changes and co-adsorption of different ions. The most important forces involved in the molecular adsorption are expected to be of electrostatic nature, associated with the charged groups exposed at the protein/biomaterial interface [164]. Also protein/protein binding and protein unfolding are expected and the protein conformational changes which result in entropic gain, contribute to the protein surface adhesion with hydrophobic/hydrophilic forces. However, the effective immobilization of proteins on the surface of inorganic nanocrystals remains a challenge to the life science industry, due to the requirement of: (i) a sufficiently strong and specific affinity of proteins with the surface; (ii) site-directed immobilization; and (iii) preservation of a native conformation and biological function of proteins in their adsorbed stages [165]. Most immobilization methods developed until now involve modification or coating of the surface with appropriate substances in order to immobilize the proteins by physical adsorption *via* van der Waals, hydrophobic or electrostatic forces, or chemical bonding [166].

In addition, in order to test and develop a biomaterial device, the knowledge of the interaction between proteins and (hydroxy)apatite is an important parameter for addressing biomineralization sciences. In fact, Nature has developed sophisticated strategies for engineering hard tissues through the interaction of proteins and, ultimately, cells with inorganic mineral phases [167, 168]. The remarkable material properties of bones and teeth thus result from the activities of proteins that function at the organic-inorganic interface. The underlying molecular mechanisms that control biomineralization are of significant interest to both medicine and dentistry, since disruption of biomineralization processes can lead to bone and tooth hypomineralization or hypermineralization, artherosclerotic plaque formation, artificial heart valve calcification, kidney and gall stone formation, dental calculus formation, and arthritis, among other pathological calcification processes. A better understanding of the biomolecular mechanisms used to promote or retard crystal growth could provide important design principles for the development of calcification inhibitors and promoters in orthopedics, cardiology, urology, and dentistry. At the level of fundamental science, it is important to note the paucity of molecular structure information available for biomineralization proteins in general, and in particular for mammalian proteins that directly control calcification processes in hard tissue. Even the most fundamental questions about how proteins interact at the biomineral surface (regarding for example their general structure and orientation on calcium phosphate surfaces), or whether the acidic residues are truly interacting directly with the crystal surface remain largely uncharacterized at the experimental level.

In this section we will give some examples on the study of the interaction and growth of calcium phosphates with proteins and small peptides highlighting the important role of the surface.

4.1. Peptides

Amino acids are biological molecules of particular interest, due to their great importance for living organisms. The effects of hydroxyproline, tyrosine, serine, glycine, cysteine, cystine, glutamine and lysine on HA crystallization have been studied by the "Chemo-stat method" by Koutsopoulos *et al.* [169-171]. Using this technique, the chemical potentials of the species in the working solution are kept constant during the crystallization process by the stoichiometric addition of reactants, and therefore the crystal growth reaction is performed under pseudo-steady-state conditions. Diverse inhibiting activities were observed as a function of the different amino acid side groups. All these studies have utilized biomolecules as simple growth inhibitors of HA crystallization, rather than considering their use as a strategy to fine-tune the bioactivity of the apatite crystals. In fact, considering that the presence of proteins in biological materials is intrinsic to the bioactivity of apatite, amino acids can be considered as agents that can increase the bioactivity of synthetic apatites. This idea was first developed by Gonzales *et al.* [172], who obtained, by hydrothermal crystallization, stable aqueous colloids of positively charged amino acid-functionalized apatite nanorods less than 80 nm in length and ca. 5 nm in width. Greatly elongated apatite nanorods, of ca. 150 × 3. 5 nm, were prepared with aspartic acid. The above work inspired further investigations into the nature and stability of the interaction of a series of amino acids with apatite [173-175]. Together with the "Chemo-stat method", these studies also suggested that crystallization kinetics is affected through the inhibition by amino acids of the active growth sites on the surface of apatite crystals. The active growth site on the apatite surface could be the calcium or phosphate sites (less probably hydroxyl groups). These experiments have been conducted either under supersaturation conditions or at high temperature. The complex behavior observed by these studies suggests that there may be a complex multi-site adsorption process, perhaps associated with different faces of the HA crystal. Changing pH may effectively "switch" certain sites for adsorption on or off, and it is further likely that the nature of the interaction between the amino acid and the HA (*e.g.* a surface complex between COO^- and Ca^{2+}) surface may be different at different sites.

HA nanocrystals with amino acid (alanine, arginine and aspartic acid) surface functionalities and different morphology in dependence on the amino acid used as co-reagent during the synthesis have been obtained by Palazzo *et al.* [176]. A self-assembly mechanism could be supposed, considering that the crystals domain size appeared slightly decreased by the amino acid presence, while the nano-crystal grows one-directionally. The ζ-potential measurements showed that the amino acid functionalized HA surface charge is inverted, being shifted towards neutrality in respect to the HA (Table **1**).

Table 1: ζ potential measurements for HA synthesized in the absence of aminoacids and amino acid-functionalized HA nanocrystals.

	pH of suspensions	ζ potential (mV)
HA	7. 5	-21. 2 ± 1. 4
Alanine-HA	7. 8	-5. 8 ± 1. 5
Aspartic acid-HA	7. 6	-6. 9 ± 1. 2
Arginine-HA	7. 6	-4. 1 ± 1. 0

This chemical surface modification should dramatically affect the hydroxyapatite biological properties and should offer the potential for the nanocrystals to be used as carriers of bioactive molecules linked to the amino acidic residue. Amino acid is the anchoring agent between apatite and bioactive molecules, dramatically affecting the adsorption and released kinetics. It was found that, while the introduction of the amino acids during the HA nanocrystal synthesis induces a modification in the primary crystallite domains along the c direction and along its transverse directions, only polar amino acids induce a morphological and dimensional variation in HA nanocrystals. Considering that the crystal domains sizes appear slightly decreased by the presence of amino acid while the nanocrystal grows unidirectionally, a self-assembly mechanism can be supposed. As was said earlier, the ζ-potential measurements show that the amino acid-functionalized HA surface charge is inverted compared to the unfunctionalized HA shifting towards neutrality. The role of the lateral residue in the binding mechanisms, involving either the carboxylate group (in the case of alanine and aspartic acid) or the amino lateral group in the case of arginine, has been underlined.

4.2. Proteins

Interactions of cytokines or proteins from blood with biomaterials have been studied to improve knowledge on osteconduction mechanism, trombogenesis or phagocytosis and have provided useful information to develop successful surgical devices or more efficient carriers for drugs. Recently, in evaluating the interaction of HA with serum and bone morphogenetic proteins, the interaction of HA with recombinant human mature bone morphogenetic protein 2 (rhBMP-2m) [177], bovine serum albumin (BSA), human serum albumin (HSA) [178] and myoglobin (Mb) [179] has been clearly elucidated.

RhBMP-2m has been expressed to study its adsorption onto HA, and the influence on the adsorption process of different parameters such as pH and concentrations of calcium and phosphate has been investigated [177]. The adsorption proceeds rapidly at the initial stages and the maximum adsorbed amount is reached after four hours. The adsorption data fit well to a Langmuir adsorption isotherm. The process is notably influenced by adding calcium or phosphate to the system but, while calcium ions increase the adsorption of rhBMP-2m onto HA, phosphate ions inhibit it. The influences of pH and NaCl concentration are notable but less important than those of calcium and phosphate. To explain this finding the surface of stoichiometric HA has been tentatively modeled by several authors. Based on the fact that HA contains calcium and phosphate groups, some authors have initiated some discussions on the presence of Ca and P sites on the surface of HA, related to emerging calcium and phosphate ions in the surface structure. Similarly, in aqueous solution, the formation of hydroxylated species on these Ca and P sites has been proposed. However, if this modeling may be assumed, in a first approximation, in the case of well-crystallized stoichiometric hydroxyapatite, the situation in the case of biomimetic nanocrystalline apatites (whether of biological nature of for synthetic analogues prepared in physiological-like conditions) deviates from this formalism due to the presence on the surface of the nanocrystals of a less-structured surface layer (so-called "hydrated surface layer", see preceding sections) which exposes labile, mostly bivalent, ions such as Ca^{2+}, HPO_4^{2-}, CO_3^{2-} as well as water molecules. In this context, the adsorption of organic molecules on the surface of biomimetic apatites can often be explained (on the basis of experimental data) by the replacement of some surface ions by some charged/polarized end-groups present in the organic molecule. For instance, molecules exposing negatively-charged functional groups may displace phosphate surface ions, while positively-charged groups may displace Ca^{2+} ions. Since the ionic species in solution such as H^+, OH^-, Na^+, Cl^-, Ca^{2+} and HPO_4^{2-}/PO_4^{3-} may influence such surface reactions, the net surface charge of the apatite phase will depend basically on the surface features, and on the pH and concentration of the mentioned ions in solution.

There exist a wide range of point of zero charge (zpc) values reported in the literature [180, 181]. Nevertheless, most of them are situated within the acidic and nearly neutral pH range. Yin *et al.* found that, for hydroxyapatite (HA) the ζ-potential, at whatever explored pH, always exhibited negative values, except when the authors added calcium ions to the solution [182]. Kawasaki *et al.* also reported a negative ζ-potential value for HA, even in the presence of adsorbed proteins [183]. These results can be interpreted by the predominance of phosphate species on the surface. To better explain the surface behavior of HA as a function of pH, Boix *et al.* suggested to draw the surface structure using a mixture of a surface complexation model and a surface hydroxylation model (Fig. **9**).

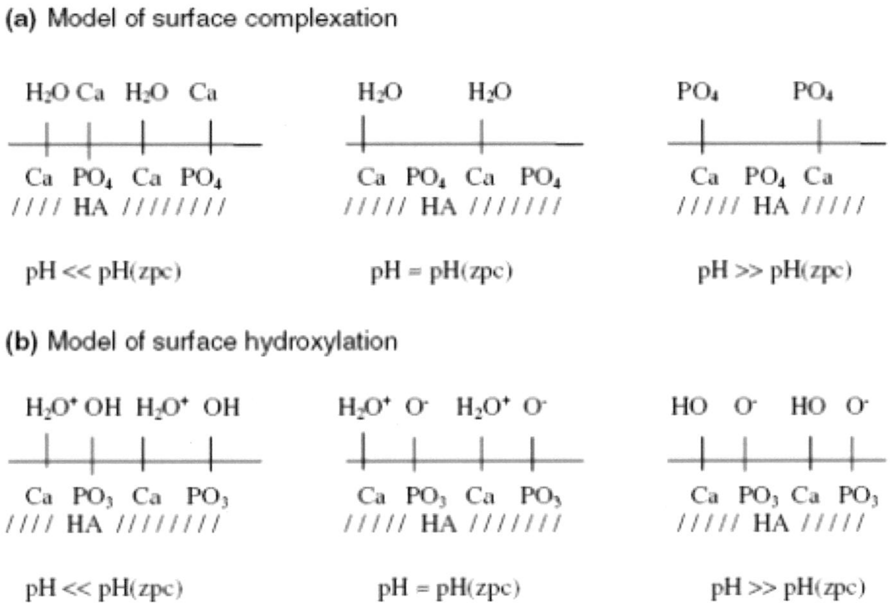

Figure 9: Models for HA surface proposed by Boix *et al.* : (a) model of surface complexation, (b) model of surface hydroxylation. (from Ref. [177] with permission).

Calculations using the values reported by Wu *et al.* [181] indicate that at the pHs used the work of Boix *et al.* a small predominance of negative surface phosphate over positive calcium species exist. Then, a small net negative surface charge is expected. From the point of view of the adsorbate, the rhBMP-2m presents acidic and basic residues, whose ionization depends on the solution pH. The isoelectric point (IEP) of rhBMP-2m is around pH 7.9. The fact that the working pH values are close to the IEP of the protein leads to small net charge. Phosphate anions in solution compete with carboxylate groups of the protein for adsorption on Ca sites. When phosphate is added, it interacts with Ca sites, and the surface charge becomes more negative, and though the global charge of the protein is slightly positive, the adsorption is inhibited. Smith *et al.* have observed by chemical force microscopy (CFM) that in the presence of phosphate ions the adsorption of BSA onto HA is inhibited [184] . However, they observed that some molecules remain adsorbed, although not detectable macroscopically, in specific sites. On the other hand, when Ca^{2+} ions are added in solution, they can interact with the phosphate sites of the HA surface allowing a more favorable protein adsorption through the carboxylate groups. Although the HA surface charge becomes positive and the global charge of the protein is slightly positive, so the global electrostatic interaction is unfavorable, an increase in adsorption is observed. Then, the adsorption results indicate that rhBMP-2m adsorbs or binds preferably on the Ca^{2+} sites, and through the carboxylate groups of the acidic residues. This type of interaction has also been proposed by other authors dealing with protein adsorption on HA [139].

The interaction mechanism of BSA and HSA with two kinds of biomimetic apatite nanocrystals with different dimensions, surface areas, crystallinity degree and surface properties has been elucidated by Iafisco *et al.* [178], pointing out some important discriminating factors. First, the two proteins reach a different maximum coverage especially towards the apatite nanocrystals with the higher surface area, the smaller size, the lower crystallinity degree and plate shape morphology. Second, the adsorption isotherms could be described by both the Langmuir and Freundlich models because of the matrix behaviour as a mixture of an energetically homogeneous and of an energetically heterogeneous support surface, with non-identical adsorption sites, in agreement with the typical surface of the biomimetic hydroxyapatite nanocrystals with low crystallinity degree and a non-stoichiometric Ca/P molar ratio. Although HSA adsorptions on apatite nanocrystals needle-shaped (HAns) and plate-shaped (HAps) appear quite similar, on the basis of both models adopted, Iafisco *et al.* conclude that HAps surface has a greater affinity towards both BSA and HSA. The different cross sectional surface areas of BSA on carbonated apatite (CHA) allowed to hypothesize that BSA interacts with HAps preferentially oriented in an end-on way, while it interacts with HAns with a side-on orientation. On the contrary, no conclusions can be drawn on the orientation of the adsorbed HSA molecules on the CHA nanoparticles due to the comparable values of cross sectional surface areas. The protein conformations were differently affected by the inorganic substrates showing a quite different structural stability. In particular, by using circular dichroism (CD) and FTIR spectroscopy, it was found that BSA and HSA behaviour after interaction with the surface of both the synthetic HAns and HAps nanocrystals were different due to the adsorption effects on both the adducts and the detached proteins, depending on the nature of the sorbent surface and, possibly, on the mode of interaction. From FTIR spectroscopy it can be generally appreciated that the protein adhesion on HA surface resulted in a significant reduction in helicity for both proteins on the two substrates, as a function of the coating extent as well as of the β-structure, which appears significantly modified. From CD spectroscopy on the protein solutions exchanged after adsorption, it can be concluded that the protein helical structure is partly regained, but not up to the extent of the native protein. This is particularly evident upon desorption from HAns for both BSA and HSA. In the tertiary structure region, a distinct behaviour is shown by HSA, in particular after desorption from HAps, as far as the environment at the Phe and Tyr residues, and possibly disulfide bridges, is involved.

Myoglobin (Mb) can be considered as a model protein because of its well-known structure and properties, commercial availability and relatively small sizes. A new trend of research is to compare the protein interaction not only with the "clean" substrate, but also with the functionalized materials. In particular, the Mb affinity *versus* the alendronate surface functionalised biomimetic HA with respect to unfunctionalised hydroxyapatite has been tested by Iafisco *et al.* [179]. The Mb recognising and assembling on drug-loaded apatite crystals not only allows the better understanding of the protein adhesion mechanism, but it could also aid the development of surface coatings to improve the biocompatibility of bone implantable biomaterials and for hard-tissue engineering and regeneration technologies. The HA-Mb interaction mechanism has been elucidated and, in particular, using UV/Vis and surface-enhanced Raman spectroscopy, it has been found that the spin state of Mb heme moiety changes from the six-coordination high spin native state to six-coordination low spin state as a consequence of the interaction with biomimetic hydroxyapatite nanocrystals. The spin state of myoglobin is relevant in relation to the catalytic activity of the protein. The surface electrostatic potential map of protein allows us to hypothesis a preferential interactive mechanism through one defined region of protein surface, in spite of random adhesion mechanisms towards other inorganic supports. Heme moiety is attracted towards the apatitic surface as a consequence of the surface structure on the nanocrystals (structured hydrated layer), which is connected to a negative surface charge in respect to the crystalline core. These results have highlighted that the immobilization of bisphosphonates onto the HA is an important strategy to set up a bone-specific drug delivery device. Considering that the interaction of protein with HA/biomolecules conjugates tailored for specific therapeutic applications plays a key role as biological probe, the Mb affinity towards the HA/alendronate bioconjugates has been tested. The alendronate avoids both the adsorption and the conformational changes of Mb heme moiety. Mb behaviour towards alendronate grafted HA crystals evidences that this functionalization imprints surface selectivity to HA and drives the biological environment response towards it.

A section about the interaction of statherin with HA also merits to be commented in this section. Statherin is a salivary protein, involved in several functions in the oral cavity but mostly associated with the regulation of HA crystal nucleation and growth in saliva [189]. Tooth enamel structural integrity is maintained through the super saturation of saliva with respect to calcium and phosphate salts and through the lubricative action of proteins in the pellicle coating the oral surfaces [186, 187]. The visco elastic properties of statherin, PRPs, and mucins in the salivary pellicle reduce the mastication load on the tooth surfaces 20-fold [188]. Statherin and PRPs inhibit formation of mineral precipitates termed accretions on the tooth surface by inhibiting both spontaneous calcium phosphate precipitation and HA secondary crystal growth [186]. Statherin can achieve these properties by binding to calcium ions in solution and by adsorbing to HA surfaces with a significant binding affinity and coverage [189]. The N-terminus of statherin contains a pSpSEE (where pS is phosphorylated serine) acidic motif that is also found in larger calcium-binding proteins such as osteopontin, as well as in biomineralization proteins that regulate calcium carbonate crystallization in lower organisms. The effects of statherin on HA crystal growth kinetics have been widely described before. Nancollas has demonstrated that derivative peptides from the N-terminus of statherin displays functional activities in controlling HA growth that are comparable to those of the full length protein [190]. This work has stimulated investigations of the proximity and dynamics of side chains in an N-terminal fragment of statherin to the surface of HA [191-193]. Isothermal titration micro calorimetry was successfully utilized to probe the thermodynamics of statherin adsorption onto HA crystals [94]. When analyzed in conjunction with results collected from adsorption isotherm, a clearer understanding of various adsorption mechanisms was achieved. At low coverage, statherin adsorbs onto HA with a significant favorable free energy change that is driven by both entropy and enthalpy. At higher coverage, the adsorption becomes enthalpically neutral and is driven solely by the positive entropy change associated with the release of bound water molecules. The higher affinity of the protein to the mineral at low coverage may be associated with adsorption to high energy surface sites like crystal growth sites and/or with a decreased affinity as a function of increased coverage due to unfavorable protein-protein interactions that lower the net free energy of adsorption. By blocking crystal growth sites first, statherin can inhibit HA crystal growth at relatively low coverage. Statherin is a good model system whose structural transition upon adsorption to HA crystallites provides leads for the structural basis for the function of biomineralization proteins. Many extracellular matrix proteins like osteopontin have segments in their primary sequence that are similar to statherin and have been found to be unstructured in solution [195]. In the case of statherin, the protein folding on the mineral surface may prevent further HA nucleation. Previous studies have underscored the role of the C-terminus in retardation of HA nucleation by statherin [194]. The tertiary folding of statherin is the basis for its inhibition of both primary and secondary crystallization of HA, and this folding is naturally induced by contact between the HA surface and the protein. Thus, a protein that is unstructured in solution may nevertheless be functional on a surface where a folding transition occurs. This remarkable feature of statherin provides with a unique case of surface-induced structuring, which reverses the common paradigm observed in many proteins where surface-induced unfolding is observed [185, 194].

5. SURFACE INTERACTION BETWEEN CALCIUM PHOSPHATES AND BIOACTIVE MOLECULES - APPLICATION TO DRUG DELIVERY

Numerous bioactive molecules can be associated to calcium phosphates, including antibiotics, growth factors, anticancer and anti osteoporosis drugs [52]. However only a very few studies have been made to understand adsorption behaviour of therapeutic molecules on apatitic calcium phosphates, especially nanocrystalline apatites. Yet these observations can have several consequences on the regulatory role of bone determining drugs uptake and release. Adsorption on biomimetic biomaterials appears as a powerful way to control the release and the activity range of therapeutic molecules.

5.1. Adsorption of Bioactive Molecules

Bioactive molecules often exhibit high binding affinity for calcium phosphate surfaces. Generally the adsorption equilibrium of such drugs strongly interacting with an apatitic surface is well described by a Langmuir isotherm, a behavior similar to that presented by proteins and aminoacids [139, 196-198], characterized by two adsorption parameters: the affinity constant (K) and the maximum coverage (N). Historically two types of adsorption sites have been recognized on apatite, calcium and phosphate surface sites. Negatively charged molecules were considered to bind to calcium sites and positively charged ones to phosphate sites.

For example, bisphosphonate molecules (BPs) present a high affinity for apatite, and were found to prevent mineral dissolution and bone resorption. BPs are thus well established as successful antiresorptive agents for the prevention and treatment of bone diseases such as osteoporosis [199, 200]. However although it is commonly admitted that BPs attach to bone mineral crystals, a very few studies have been carried out on these aspects. The adsorption isotherm of a bisphosphonate, risedronate, onto apatitic calcium phosphates is reported in Fig. **10**.

In the case of risedronate adsorbed onto hydroxyapatite (HA), the isotherm is Langmuirian in shape. Interestingly the follow-up of the variation of the mineral ions content of the solution indicates that the adsorption of risedronate is associated with a release of phosphate ions whereas the calcium concentration remains almost unchanged [196]. Similar adsorption characteristics were reported for the interaction of others bisphosphonates on calcium phosphates [201]. Considering that the dissolution of HA is congruent (*i.e.* the Ca/P ratio is the same for the solid and the solution), the analysis of the suspensions obtained after adsorption revealed that the binding of one risedronate ion onto the HA surface was accompanied by the release of one phosphate ion in the solutionThe adsorption reaction can thus be described as an ion exchange process with a one-to-one substitution in the case of the risedronate (Fig. **11**):

$$[Ri^{3-}]_{sol} + [PO_4^{3-}]_{surf} \Leftrightarrow [Ri^{3-}]_{surf} + [PO_4^{3-}]_{sol} \quad \textbf{(1)}$$

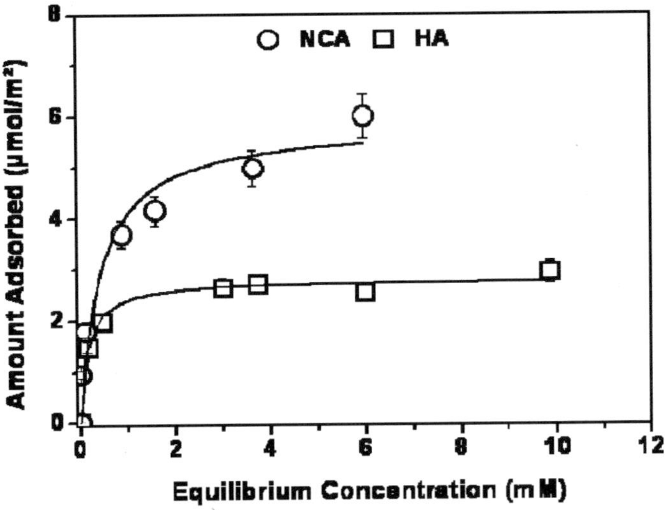

Figure 10: Adsorption isotherms of risedronate onto nanocrystalline carbonated apatite (NCA) and well crystallised hydroxyapatite (HA) at pH 7. 4 and T = 37°C (from Ref. [218] with permission).

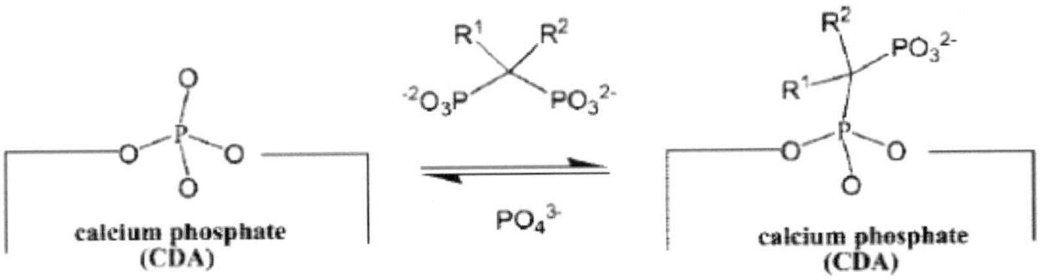

Figure 11: Schematic representation of chemical association of BPs onto the surface of calcium deficient apatite (CDA) particles, *via* BP/phosphate exchange process (from Ref. [202] with permission).

It is interesting to notice that the amount of BPs molecules adsorbed at saturation can vary considerably depending on different parameters such as the temperature, pH, or initial phosphate and calcium contents in the solution [202]. The maximum amount adsorbed is expected to correspond to the saturation of the surface adsorption sites, and such variations suppose that the maximum number of sites which can be occupied (N) varies according to these various factors. Considering the proposed ion exchange reactions, the chemical equilibriums involved cannot explain a variation of the adsorption limit: when the concentration of absorbate increases to large values, all sites should be occupied. Sterical considerations, strength of attachment and surface mobility of the adsorbed bioactive molecules have then to be considered and could be involved in the reorganization of such agents on the apatitic surface [196].

Antibiotics interaction on calcium phosphate surface displays similar behavior to BPs. The nature of the charged molecules plays a significant role on the adsorption process. It appears from several studies that antibiotics containing carboxylic groups in their chemical structure were better incorporated into calcium phosphates biomaterials than others [203]. For example it has been published that antibiotics containing carboxylic groups like cephalothin, have better interaction with calcium at the surface on the apatite coating applied onto titanium implants, resulting in a better binding and higher incorporation into the calcium phosphate coating [203]. Other studies have shown that carboxyl-containing molecules are good binders of calcium [198, 204].

Concerning bone morphogenetic proteins (BMPs) growth factors, only a very few studies have been made on the adsorption of these molecules on apatitic calcium phosphates. The study of interaction between such proteins and apatitic surfaces is of great biological and medical interest as BMPs have been attempted to be applied for the reconstruction of bone defects resulting from trauma, surgical resection of tumors, and congenital anomalies in orthopedic and maxillofacial surgery. Few works show a strong affinity of growth factors for apatite surfaces, enhanced in the simultaneous presence of calcium ions in the solution as in the case of the adsorption of other proteins [205, 206]. It was published for the adsorption of BMP-2 growth factor onto HA that the isotherm is Langmuirian in shape [177], and that adsorbate-adsorbent interactions can take place during protein adsorption, depending on the protein structure and the nature of the surface of the adsorbent [177]. A computed theoretical modeling based on the interactions of the (001) HA surface with functional groups of BMP-2 [207] has recently put evidenced three types of interactions involved in the adsorption process with this type of protein: an electrostatic interaction of charged COO^- groups from the protein with Ca^{2+} ions on the apatite surface, and water-bridged H-bonds between OH and NH_2 residues of BMP-2 and with PO_4^{3-} ions on the surface [144].

For all this mechanism of drugs adsorption onto apatitic surface, several processes can disturb the adsorption equilibrium and have to be taken into account: the possible precipitation of calcium (or phosphate) salts, the dissolution equilibrium of the apatite and the surface hydrolysis reactions of apatite. For example, with zoledronate, a BPs molecule, a formation of a crystalline zoledronate complex onto calcium phosphate surface was observed [201]. In the case of risedronate adsorbed onto HA (accompanied by the release of phosphate ions in the solution), it should be noted that although the solubility equilibrium of HA would lead to a decrease of the free calcium ions in the solution, this is not observed (Fig. **11**) and the relative stability (even the slight increase) of the calcium content with the increase of risedronate in solution can be interpreted as an indication for the probable formation of ion pairs or complexes between risedronate molecules and calcium ions in solution. Concerning the anti-neoplasic drug cisplatin interactions with bone substitutes, it has been reported that phosphate species are able to displace chloride ions from cisplatin molecules to form phosphate-complexes, which may be more reactive toward the apatite surface [208].

At last the adsorption of bioactive molecules onto apatitic surfaces appears generally irreversible with respect to dilution, as observed for proteins [206]. Washing an apatitic sample with adsorbed drugs accompanied by an ion exchange process does not create any reverse exchange possibility. No desorption should be expected by simple dilution of the adsorption solution or even upon washing of the samples, except for faint release effects due to the dissolution of the apatite [196]. This behavior could be useful in view of the controlled release of such agents from apatitic biomaterials *in vivo*.

5.2. Adsorption on Nanocrystalline Apatites

To better understand the adsorption behaviour of bioactive molecules *in vivo*, it is important to consider nanocrystalline apatites as a better model of bone material than HA, and bone tissue as a dynamic system. The adsorption behavior can be affected not only by the chemical properties of drugs but also by the physico-chemical characteristics of the apatitic support. Very few studies have been performed on the adsorption of organic molecules on nanocrystalline apatites as model of bone material. For the adsorption of the anti-neoplasic drug cisplatin, the greater activity of the poorly crystalline samples in the cisplatin adsorption process was observed, compared to well-crystallized apatitic HA [208]. The adsorption process does not differ strongly from well crystallized apatites, however an interesting aspect is the variation of the adsorption parameters with the characteristics of the nanocrystals. Adsorption on nanocrystalline apatite of bioactive agents indicates an increase of the affinity constant and a decrease of the amount adsorbed as maturation time increases, as we can see in Fig. **10** for risedronate adsorption [121].

The increase of the affinity constant may be correlated with an increase of interfacial energy of nanocrystals during maturation in aqueous media [121]. The hydrated layer probably reduces the surface energy of the nanocrystals in aqueous media [196]. Another explanation could be that ions in the hydrated layer become less exchangeable as the maturation time increases, which is effectively observed [135, 196]. It showed that the maturation of these compounds was directly related to the decrease in the concentration of non-apatitic phosphate groups. Moreover, the morphology and size of the crystals also account for the higher reactivity of these materials.

The decrease of the number of adsorption sites with maturation time can be interpreted as a decrease of potentially exchangeable ions in the hydrated layer or a change in the equilibrium constants [121]. It can be seen that the ability of a poorly crystalline apatite to adsorb is related to its number of non-apatiticsites and its low crystallinity. As the adsorption process can be explained by ion exchange process at the surface of the apatitic support, the amount of bioactive molecules adsorbed could be related to the amount of labile mineral ions available in the hydrated layer and then to the extension of the hydrated layer. As the maturation proceeds, the amount of such exchangeable species diminishes and the adsorption capacity of the solid decreases [139].

5.3. Application to Drug Delivery

Apatites and other calcium phosphates are widely used as biomaterials and may find several applications in implant drug device. In fact, as drug carriers, calcium phosphate nanoparticles have some advantageous properties. They are dissolved at low pH (around 4), *e.g.* in lysosomes after the cellular intake or in the environment of solid tumors, thereby releasing incorporated drugs or biomolecules. Their size can be controlled by stabilizing agents, such as polymers, nucleic acids, or phospholipid derivatives. The nanoparticles can also be made to fluoresce by the incorporation of lanthanide ions such as europium (see for example ref. [149]), and they can also act as carriers for different drugs [52]. For example one of the most efficient ways to improve the bone forming ability of biomaterials is their association with bone morphogenetic proteins [144]. Thus the choice of the process used to prepare implantable bone biomaterials has an influence on the release of the associated bioactive molecules. Various techniques associating drugs with a calcium phosphate biomaterial have been reported: adsorption and impregnation allow the therapeutic agent to be incorporated at the surface of the biomaterial, whereas centrifugation and vacuum based-techniques enable it to enter into pores of biomaterials [209, 210].

Generally bioactive molecules, such as growth factors, are incorporated in biomaterials by simple impregnation followed by drying, and the type of bonding with the substrate and the release rate are often undetermined [211]. It is suspected that such associations do not allow the chemical bonding of the growth factor to the biomaterial and thus the release rate is often difficult to control. For example, precipitation and clustering of the growth factor molecules may occur and the release is only determined by local dissolution and diffusion rules [144]. The uncontrolled release of growth factors has, in some instances, been related to an accelerated resorption of bone tissue and of the implant [144]. Since growth factors agents can stimulate the degradation as well as the formation of bone (depending on their local concentrations), they could impair the osteoconductivity of the coated-implant surface [212]. Similarly bisphosphonates molecules, by affecting bone remodeling, could also block the bone repair process: the drug at too high concentration could have detrimental effects on the fixation of the implant over longer periods of time. Zoledronate grafted to HA coating on titanium implants shows a dose-dependent effect on the inhibition of resorption activity according to the amount of zoledronate loaded [213]. Local and slow administration of anti neoplasic drug as methotraxate (MTX) is also useful to avoid systemic side effects, and because its time effect (the sensitivity of cells to this drug increases with time) is greater than dose effect [215].

Adsorption on the contrary leads to stable association and control of the amount of bioactive molecules contained in the solid implant, and thus of the dose released. Generally, the release is rather low because most of the bioactive molecules adsorbed are irreversibly bound and they are not spontaneously released in a cell culture media [196]. They can only be displaced by mineral ions and/or soluble proteins with a stronger affinity for apatite surfaces but in a predicable manner, or by cell activity [196]. This characteristic has been observed for various growth factors like bone morphogenetic protein (BMP-2) or vascular endothelial growth factor (VEGF) [144, 196, 215], anti osteoporisis agents [213, 216-218] and anticancer drugs as methotraxate and cisplatin [208, 214]. It has been reported that slow release of MTX from calcium phosphate is not only due to the porosity like in most of the cases, but mainly due to the adsorption of MTX [214]. Fig. **12** illustrates antibiotics release incorporated in apatitic calcium phosphate coatings on titanium implant using biomimetic method [203]. A local release system of antibiotics could be used to prevent post-surgical infections favoring early osteo integration of prosthesis. Since some of the antibiotics containing carboxylic groups in their chemical structure, like cephalothin, were better adsorbed than others into calcium phosphate biomaterials, these molecules are slowly released from the carrier [204]. Moreover the high binding capability of apatitic support for a wide range of therapeutic agents allows its surface functionalisation with linking agents, such as BPs molecules, to anchor biologically active molecules which can be released, breaking the linkage as a consequence of external stimuli or internal chemical factors, such as pH and ionic force variation due to physiological or pathological biological process [52, 179, 219]. For example, some works have investigated the adsorption and "smart release" of antitumoral platinum complexes containing BPs onto apatitic nanocrystals as bone specific drug delivery devices to be used for the treatment of bone tumors upon local implantation [219].

It is important to notice that, as the adsorption properties of the bioactive molecules are strongly dependent on the nature of the calcium phosphate support and maturation stage, it is probable, although this has not been systematically investigated, that the release rate is related to the physico-chemical properties of the apatite samples. For example the cisplatin release rate can be controlled by use of a more crystalline apatite for a faster release rate or a less crystalline material for a slower rate [208]. Similarly some authors show that the coating of sintered biphasic calcium phosphate granules (HA and β-TCP) with nanocrystalline apatite improve BMP-2 growth factors adsorption (the rate of adsorption was increased), [144]. Only a low quantity of protein was spontaneously released in cell culture medium during the first weeks (Fig. **13**).

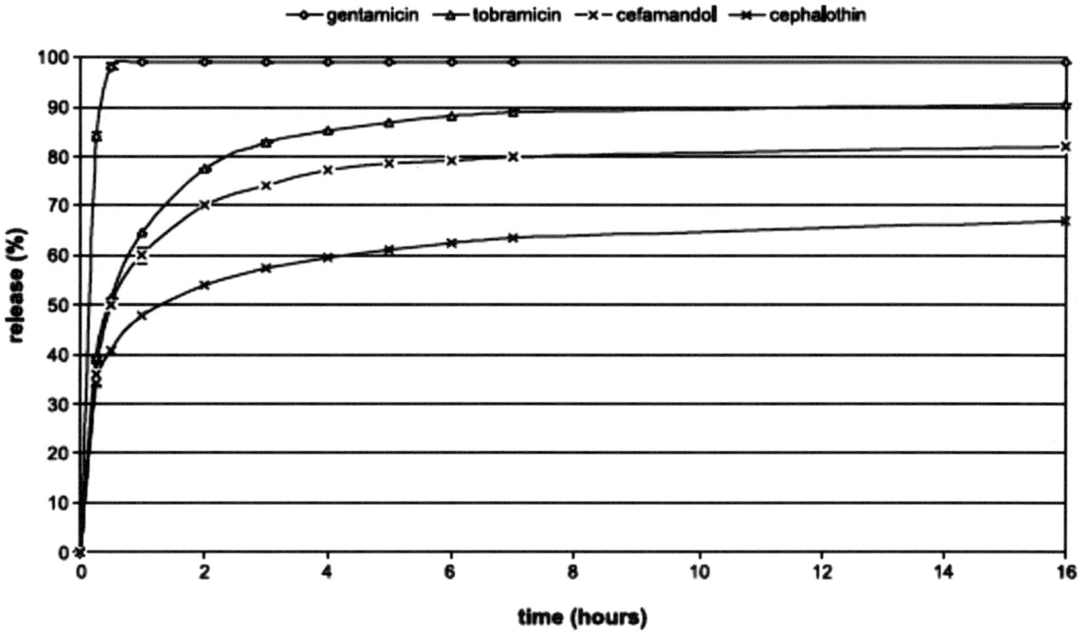

Figure 12: Antibiotic release from carbonated apatitic calcium phosphate coatings on titanium implantsinto phosphate buffered saline at pH 7. 3 (from Ref. [203] with permission).

The adsorption and release of bioactive molecules are strongly affected not only by the chemical properties of the drug molecule, but also by the chemical and structural characteristics of the HA substrates. The adsorption and release of cisplatin, alendronate and di (ethylendiamineplatinum)medronate have been investigated using two biomimetic synthetic HA nanocrystal materials with either plate-shaped or needle-shaped morphologies and with different physico-chemical surface properties by Palazzo *et al.* [220]. These bioactive molecules were chosen in order to compare the behaviour of metal based drugs to that of a classical organic drug (alendronate), evaluating the effect of the drug molecule overall charge in influencing the drug affinity for apatite nanocrystals with variable structural and chemical different properties. The HA surface area and surface charge (as well as other parameters such as Ca/P ratio *etc.*), as well as the charge on the adsorbed molecules and their mode of interaction with the HA surface, influence the adsorption and release kinetics of the three drugs investigated. The results demonstrated that HA nanocrystals and anti-tumour drugs can be selected in such a way that the bioactivity of the drug-HA conjugate could be tailored for specific therapeutic applications.

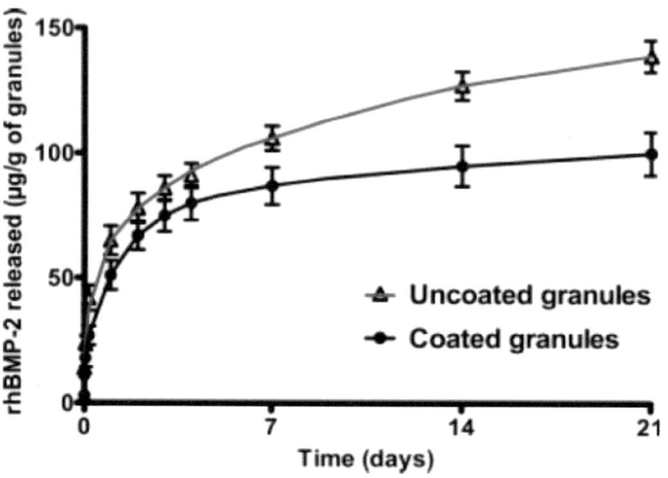

Figure 13: Cumulative Release of BMP-2 in DMEM complete culture medium from adsorbed BMP-2 on nanocrystalline-apatite coated and uncoated Biphasic Calcium Phosphate (from reference (144) with permission).

The adsorption of two different platinum complexes with cytotoxic activity, bis-{ethylenediamineplatinum (II)}-2-amino-1-hydroxyethane-1, 1-diyl-bisphosphonate (A) and bis-{ethylenediamineplatinum (II)}medronate (B), on the synthesized biomimetic nanocrystals has been investigated by Iafisco *et al.* [219]. Both complexes contain a geminal bisphosphonate but, in addition, complex A contains a charged ammonium group (Fig. **14**).

This structural difference dramatically affects the affinity of A and B towards HA nanocrystals, complex A having a greater affinity for calcium phosphate nanocrystals. The release profiles of the platinum complexes from the HA nanoparticles follow an inverted trend (complex B > complex A) when compared with the adsorption process. Most probably the less effective desorption in the case of complex A could be due to the aminic group present on the bisphosphonate which remains anchored to the HA matrix, coordinating and holding some of the ethylenediamineplatinum (Pt (en)) residues. Unmodified and HA-adsorbed Pt complexes were tested for their cytotoxicity towards human cervix carcinoma cells (HeLa). The HA-loaded Pt complexes were more cytotoxic than the unmodified compounds A and B and their cytotoxicity was comparable to that of dichloridoethylenediamineplatinum [$PtCl_2$ (en)] thus indicating a common active species. The above results demonstrate that HA nanocrystals and antitumor drugs can be conjugated in such a way to yield a smart bone filler delivery system, acting both as bone substitutes and as platinum drug releasing agents with the final goal of locally inhibiting the tumour re-growth and reducing the systemic toxicity. The one here described not only can ensure a prolonged release of active species but also improve the performance of the unmodified drug. Moreover, these results suggest the possibility of using the chemical-physical differences of HA nanocrystals, above all degree of crystallinity, crystal size and surface area, in order to strongly tailor the Pt complex release kinetics. Considering the biomimetic apatite nanocrystals functionalization effects, an attracting goal could be to obtain a drug delivery process characterized by a stimuli responsive kinetic. This aim induces to functionalize the surface of apatite nanocrystals with different linking agents, such as bisphosphonates, to anchor biologically active molecules which can be released breaking the linkage as a consequence of external stimuli or internal chemical factors, such as pH and ionic force variation due to physiological or pathological biological process.

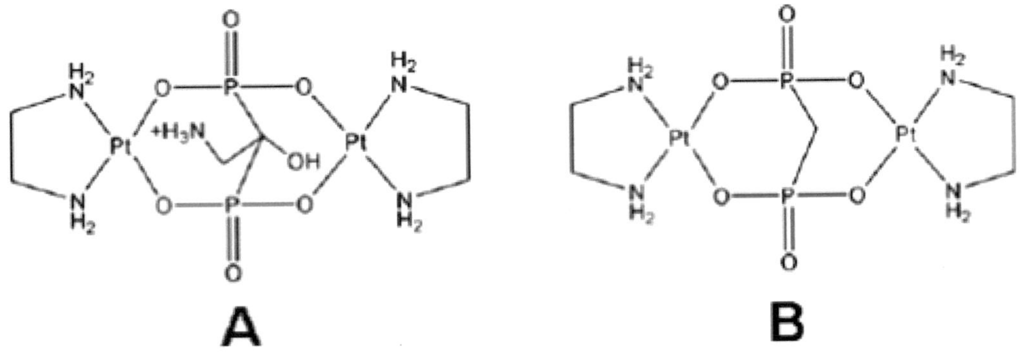

Figure 14: Sketches of bis-{ethylenediamineplatinum(II)}-2-amino-1-hydroxyethane-1,1-diyl-bisphosphonate (A) and bis-{ethylenediamineplatinum (II)}medronate (B).

6. DESCRIPTION OF SOME *IN VITRO* AND *IN VIVO* EXPERIMENTS OF FUNCTIONALIZED CALCIUM PHOSPHATES - TOWARD DRUG DELIVERY DEVICES

6.1. Example of Bisphosphonates Adsorption and Release from Calcium Phosphates

Today bisphosphonates (BPs) are well established as successful antiresorptive agents for the prevention and treatment of post-menopausal osteoporosis [199, 200]. BPs were first studied over 30 years ago as synthetic analogous of inorganic pyrophosphate, characterized by a non-hydrolysable P-C-P structure with two phosphonic acid groups bonded to the same carbon (Fig. **15**). Their mechanisms of action are at the same time physicochemical effects and biological activity: like pyrophosphate, BPs present a high affinity to adsorb onto apatite crystals due to the high attraction of phosphonate groups for calcium, and thereby prevent mineral dissolution and bone resorption, inhibiting osteoclasts activity [200-221]. The P-C-

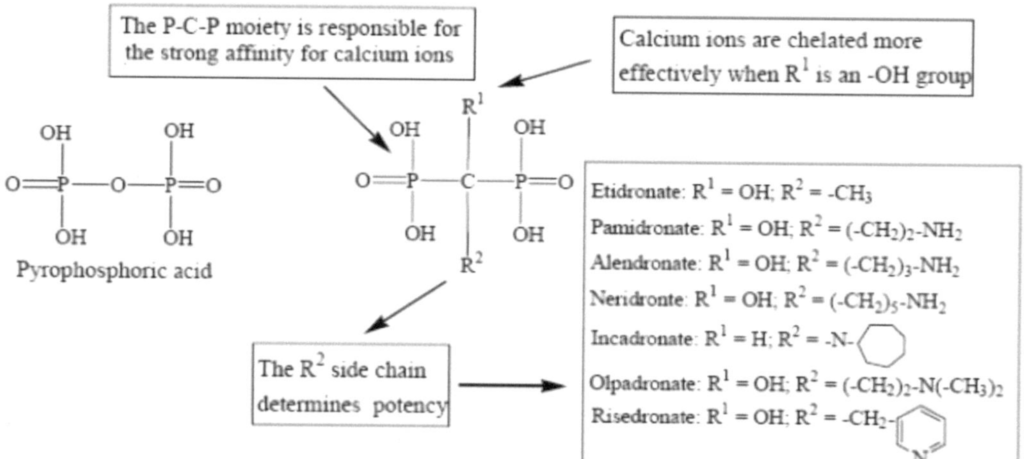

Figure 15: Structure of the endogenous pyrophosphonate, and its synthetic analogue, bisphosphonate (BP), which exhibit a strong bone affinity. The geminalcarbon in BPs typically contains two separate substituents, R^1 and R^2, which may significantly affect both the mineral affinity and the pharmacological activity. Examples of BP class of compounds currently used in a clinical setting. The compounds have been categorized into two classes, based on the presence of an amino group in the R^2 side-chain. The amino-BPs typically exhibit a higher potency in antiresorptive effects, the primary clinical utility of BPs. Most of the BPs contain a geminal -OH group that enhances the mineral affinity of the compound.

P backbone is responsible for the strong affinity for bone mineral and inhibitory effects on bone dissolution [196]. The ability of the BPs to bind to bone mineral is increased when the R_1functional groupis able to co-ordinate to calcium, as hydroxyl group (OH), [199]. Varying the R_2functional group can influence antiresorptive potency, those containing a nitrogen atom within a heterocyclic ring (such as risedronate, zoledronate) have the most potent antiresorptive effect [222-225].

Preclinical studies with different BPs show that these antiresorptive agents not only inhibit bone loss, but can also increase bone mass and resistance to fracture [226, 227]. The classic bisphosphonates treatment is the systemic way by oral administration or intravenous injection. However, systemic use of BPs can result in undesirable side-effects like fever [228], gastrointestinal disturbances as ulcers [229], or osteonecrosis of the jaw especially observed by intravenous preparations [230, 231]. Moreover low bio availabilility is commonly observed for oral administration [232]. On this basis, the development of strategies for local administration of BPs within osteoporotic sites becomes even more interesting.

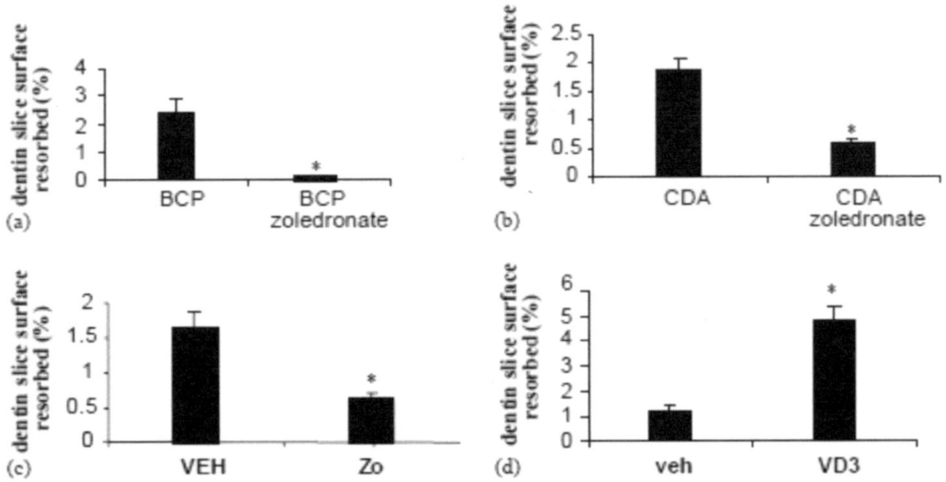

Figure 16: Cell preparation was then cultured for 4 days either in the presence of BCP or zoledronate-loaded BCP (a), CDA or zoledronate-loaded CDA (b), 10^{-6} mol. L^{-1} zoledronate or its vehicle (VEH) (c) or 10^{-8} mol. L^{-1} vitamin D3 or it vehicle (d). Resorption activity of osteoclasts is expresses as the percentage of dentin slice surface resorbed (from Ref. [201] with permission).

Few studies have tried to evaluate the effect of local bisphosphonates (BPs) delivery from implants. Josse *et al.* studied associations of zoledronate onto different types of calcium phosphate matrices as calcium-deficient apatites (CDA) and β tricalcium phosphate (BCP) [202]. When zoledronate was grafted onto calcium phosphate powders, the adsorption process was associated with the release of phosphate ions in the solution from the apatitic structure. At high BPs concentrations, the formation of a crystalline zoledronate complex on the surface of the calcium phosphate support was evidenced by SEM observation and solid-state NMR. The evaluation of the biological activity of zoledronate-loaded calcium phosphate materials, resulting in the inhibition of osteoclastic activity, was shown using a specific *in vitro* bone resorption model (Fig. **16**). ; a total rabbit bone cell preparation was established on dentin slices.

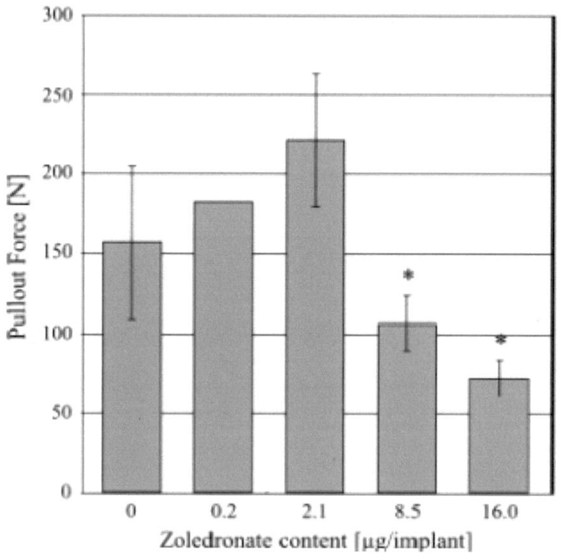

Figure 17: Pullout force as function of zoledronate concentration (mean ± SEM). At low zoledronate concentrations, the pullout force increases with increasing zoledronate content of the coating. The pullout force reaches a maximum with a zoledronate content of 2.1μg/implant. By further increasing the zoledronate content of the coating, the pullout force decreases and reaches levels lower than when no zoledronate is present (from Ref. [213] with permission).

Peter *et al.* [213] chemically grafted zoledronate on a HA coating of titanium implants. The implants were immersed in zoledronate solution of variable concentration to favour BPs adsorption. *In vivo* experiments established that when the implants were inserted in rats, the mechanical fixation of the orthopaedic implants measured using a tensile testing machinewas increased by the local presence of zoledronate (Fig. **17**). Moreover the increase in peri-implant density appears to be zoledronate concentration-dependent: at higher zoledronate concentrations the mechanical stability decreases and a zoledronate concentration leading to an optimal implant fixation can thus be evidenced. Several studies have reported that BPs at high concentration could have negative effects on the fixation of implants over longer periods of time.

All of these studies demonstrate the interest of new biomaterials preparation for hard tissue repair with a local action of these antiresorptive agents. Local and slow BPs release from implants seems to be able to influence bone cells and promotes bone repair. Moreover calcium phosphate-based biomaterials appear to be good candidates for releasing bioactive BPs molecules.

6.2. Example of Strontium Enrichment on Calcium Phosphates

The surface reactivity of calcium phosphate apatites and the exceptional capacity of the apatite structure to accommodate numerous types of ions (or vacancies) have been at the origin of several medically-oriented applications, in particular in the bone tissue engineering field. Of special interest is the possibility to enrich such systems with biologically-active ions which exhibit a specific role once *in vivo*. Among such ions, Sr^{2+} is of particular interest. This minor constituent of bone has indeed been reported to play an important role in biological processes such as homeostasis and for treating some pathologies Marie *et al.* [233] showed that Sr^{2+} ions had clear pharmacological effects on bone, including antiresorptive and anabolic activity, thus leading to a potential interest for the treatment of osteoporosis and other osteopenic pathologies.

Strontium-containing systems, such as Sr anelate for example, are indeed presently found in the medical armamentarium for treating such pathologies [234]. In a similar view, associations between hydroxyapatite and strontium ions have also been reported. In recent studies, some authors [235, 236] showed that a bone cement formulation based on Sr-substituted hydroxyapatite (10at.% calcium ions are substituted by Sr) powder coupled with bisphenol-α-glycidyldimethacrylate (Bis-GMA)-based resin led to appreciably good *in vivo* outcome in the hip bone surgery in rabbits. The bone-bonding ability was also compared to that of a polymethylmethacrylate (PMMA) bone cement, in the case of unilateral hip replacement in rabbits. The authors reported a widespread osteo integration of the Sr-HA cement with cancellous bone and intimate contact was found from microscopy analyses between the Sr-HA cement and cortical bone without the formation of fibrous tissue. The overall affinity of bone for theSr-HA cement (evaluated from the ratio of the length of bone directly in contact with the implant without an intervening fibrous membrane over the total length of the bone-implant interface [237[) was found to be noticeably greater (ca. 85%) than the affinity for the control PMMA cement (ca. 3%) [238]. Moreover, a fibrous layer was observed between the PMMA cement and bone, and PMMA was found to lead to some inflammatory response in the vicinity of the implant. However, it should be noted that strontium was selected in these works mainly in relation with its effects on apatite dissolution rate (due to lattice distortion effects [239]), while the potential effect on cells was not directly commented.

In a similar approach, Wong *et al.* [240] prepared Sr-loaded hydroxyapatite/polyetheretherketone composites (Sr-HA/PEEK) by compression molding, in view of load-bearing orthopaedic applications. These authors examined the ability of these systems to form nanocrystalline apatite when immersed in SBF, and drew quantitative evaluations of MG63 osteoblast cells-mediated mineralization *in vitro*. The Sr-HA/PEEK composites were found to offer a better bone-bonding ability than the strontium-free control or the PEEK itself, as indicated from alizarin red staining (a common histo chemical reactant used to stain for calcium-based deposits and evaluate osteoblasts culture mineralization [241]), therefore providing an enhanced bioactivity, and this effect was linked to the presence of strontium ions (Fig. **18**) [240]. However, it should be noted that no difference was found between such Sr-loaded and Sr-free composites in terms of cell proliferation and alkaline phosphatase activity: in all cases, composites allowed good MG63 cell proliferation (reaching confluence around day 7) and did not induce cytotoxicity. In another study, Wong *et al.* [242] investigated the mineralization process leading to osteo integrationin the case of another formulation of Sr-containing HA mineral-organic cement (implying silica and an organic resin based on Bis-GMA) after injection into cancellous bone *in vivo* (rabbit ilium). According to these authors, the osteo integration of this cement was achieved after 3 months, and the dissolution of Sr-HA during the bone remodelling process was thought to stimulate bone formation. The newly-formed bone was found to bind tightly to the Sr-HA cement, with collagen fibers inserted after 6 months.

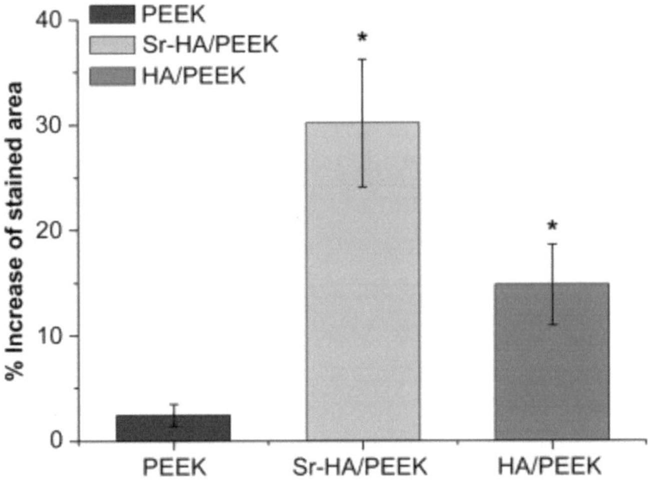

Figure 18: Percentage increase at day 14 of stained area by alizarin red (follow-up of MG63 culture mineralization) on PEEK, 25 vol% Sr-HA/PEEK and 25 vol% HA/PEEK composites (reproduced with permission from Ref. [240]).

The study of other Sr-enriched apatite systems was also reported in the literature. For example, the preparation of an injectable purely-mineral bone cement based on a mixture of brushite (DCPD, or $CaHPO_4$. $2H_2O$), calcium oxide and strontium carbonate was reported recently [243]. According to the authors, the cement obtained after mixing with water exhibited a good injectability, while showing compressive strength close to 20 MPa adapted to bone repair applications in non-load bearing sites or for reinforcing osteoporotic vertebrae. The final state of the cement after setting was found to be a strontium-calcium-deficient carbonate apatite, and the authors reported a radio-opacity for the resulting cement three times greater than that of cortical bone (due to Sr^{2+} ions), enabling the use of such systems for enhanced radio-opacity after implantation *in vivo* (e.g. in spinal surgery).

Other investigations of Sr-enriched calcium phosphate compounds have also been reported. For instance, the effect of strontium on the setting properties and *in vitro* bio activity of a gelatine (15 wt.%) - calcium phosphate (mainly α-tri calcium phosphate) bone cement have been investigated [244]. In this work, strontium was added to the mixture as strontium chloride, and the presence of this ion was found to affect the setting time of the cement (increasing with the Sr content). Also, the rate of transformation of the α-tricalcium phosphate into calcium-deficient apatite also increased upon strontium enrichment. Finally, the authors reported from SEM observations that MG63 osteoblast-like cells grown on the cement exhibited a normal morphology and an excellent rate of proliferation. Furthermore, the presence of strontium was found to stimulate alkaline phosphatase (ALP) activity, collagen type I, osteocalcin, and osteoprotegerin expression, playing an important role in the bone healing process. It can be noted here that quite similar satisfying results were obtained in the case of titanium implants coated with plasma-sprayed Sr-enriched HA [245], where the attachment of osteo progenitor cells (OPC1) cells was promoted by the presence of Sr, and for which an increasing amount of local contact points with the surface was evidenced as compared to Sr-free control. Additionally, Sr was then found to stimulate OPC1 cell differentiation and ALP expression.

The development of other calcium-phosphate bone cements enriched with Sr was also investigated, for instance starting from tetracalcium phosphate and monetite (DCPA) and leading (as final state after setting of the cement) to a single-phased Sr-enriched HA [246]. In this study, the authors examined in particular the *in vitro* biocompatibility and the *in vivo* degradability during intramuscular and femur implantation in the rabbit. *In vitro* data showed that the Sr-doped bone cement exhibited a satisfying biocompatibility, but the Sr/ (Sr+Ca) molar ratio was found to play a important effect (a molar ratio of 5% showed higher biocompatibility than 0 or 10%). *In vivo* results then showed that Sr-enriched samples led to noticeably faster degradation than the Sr-free control. In addition, the authors concluded from observations by optical transmission and energy dispersion X-ray analyses that Sr actively took part at the bone-implant interface.

Generally speaking, the *in vivo* data reported for Sr-doped apatite-based systems in terms of bone regeneration are very positive, and the Sr-loaded implants are commonly found to form a high-quality interface with the surrounding bone tissue. This is for example illustrated in Fig. **19**.

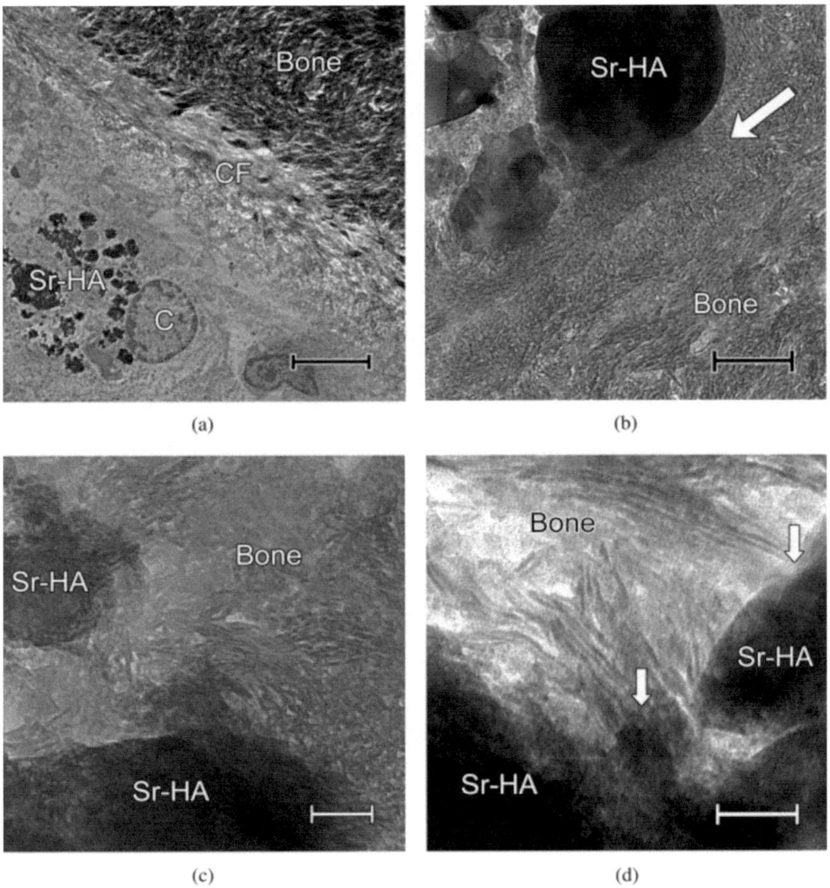

Figure 19: Transmission electron micrographs of bone-Sr-HA-based cement interface. (a) After 1 month, bone remodeling was observed (rabbit ilium). Collagen fibers (CF) were observed on the bone surface with cells (C) next to the Sr-HA cement (bar = 4 m). (b) After 3 months, direct contact of bone and Sr-HA cement was achieved (bar = 200 nm). A new bone region (white arrow) was grown onto the Sr-HA cement. (c) After 6 months, the Sr-HA cement was completely surrounded by mature bone (bar = 50 nm). (d) Collagen fibers inserted perpendicularly into the Sr-HA cement at 6 months (bar = 50 nm) (reproduced with permission from Ref. [242]).

Such data therefore illustrate, in terms of Sr^{2+} ion enrichment in this example, the wide possibilities offered by calcium phosphate-based systems in view of conferring attractive additional properties to advanced biomaterials. In particular, the possibility to tailor, by way of the modification of synthesis parameters, the physico-chemical characteristics of Sr-loaded calcium phosphate (and typically their solubility) may be seen as an interesting way to control the release rate of such ions after implantation (see for example [247]). In this case, the physico-chemical study of the incorporation of such ions into or onto the calcium phosphate matrix (often apatitic in the literature) remains a pre-requisite, especially when the incorporation of biologically-active ions is undergone *via* ion surface exchanges [116], which may generally be satisfyingly described by a Langmuir-like mathematical modelling.

ACKNOWLEDGEMENTS

CD and SS thank the Agence Nationale de la Recherche (ANR) for support in the scope of project ANR « Nanobiocer » (ref. 2007 BLAN 373), and the European Commission for support within the « Autobone » project (NMP3-CT-2003-505711-1). MI would like to thank Regione Piemonte (Progetto Ricerca Sanitaria Finalizzata 2009, Grant no: 30258/DB2001), the Inter University Consortium for Research on Chemistry of Metals in Biological Systems (C. I. R. C. M. S. B) and Ministerodell' Istruzione, dell'Università e della Ricerca (MIUR) (Azioni Integrate Italia-Spagna, Grant no: IT10M4FC89) for providing financial support. JGM acknowledges Spanish MINECO for support (Factoría de Cristalización, Spanish-Italian Integrated Action IT2009-00028, and CRYSFUNBIO ref. MAT2011-28543).

CONFLICT OF INTEREST

The author(s) confirm that this chapter content has no conflict of interest.

REFERENCES

[1] Dorozhkin SV. Bioceramics of calcium orthophosphates. Biomaterials 2010 Mar;31(7):1465-85.
[2] Dorozhkin SV. Nanosized and nanocrystalline calcium orthophosphates. Acta Biomater 2010 Mar;6(3):715-34.
[3] Dorozhkin SV. Calcium orthophosphates. J Mater Sci 2007;42(4):1061-95.
[4] Loveridge N. Bone: More than a stick. J Anim Sci 1999;77 Suppl 2:190-6.
[5] Reddi AH. Symbiosis of Biotechnology and Biomaterials - Applications in Tissue Engineering of Bone and Cartilage. J Cell Biochem 1994;56(2):192-5.
[6] Cowles EA, DeRome ME, Pastizzo G, Brailey LL, Gronowicz GA. Mineralization and the expression of matrix proteins during *in vivo* bone development. Calcified Tissue Int 1998;62(1):74-82.
[7] Tzaphlidou M. Bone Architecture: Collagen Structure and Calcium/Phosphorus Maps. J Biol Phys 2008;34(1-2):39-49.
[8] Urist MR. Bone: Formation by autoinduction (Reprinted from Science, vol 150, pg 893-899, 1965). Clin Orthop Relat R 2002;(395):5-10.
[9] Wozney JM. The Bone Morphogenetic Protein Family and Osteogenesis. Mol Reprod Dev 1992;32(2):160-7.

[10] Manolagas SC. Birth and death of bone cells: Basic regulatory mechanisms and implications for the pathogenesis and treatment of osteoporosis. J Aging Phys Activ 2000;8(3):248-.

[11] Legeros RZ, Craig RG. Strategies to Affect Bone Remodeling - Osteointegration. J Bone Miner Res 1993;8 Suppl 2:S583-S96.

[12] Boskey AL. Mineralization of bones and teeth. Elements 2007;3(6):385-91.

[13] Palmer LC, Newcomb CJ, Kaltz SR, Spoerke ED, Stupp SI. Biomimetic systems for hydroxyapatite mineralization inspired by bone and enamel. Chem Rev 2008;108(11):4754-83.

[14] Weiner S, Wagner HD. The material bone: Structure mechanical function relations. Annu Rev Mater Sci 1998;28:271-98.

[15] Fernandez-Moran H, Engstrom A. Electron microscopy and x-ray diffraction of bone. Biochim Biophys Acta 1957;23(2):260-4.

[16] Traub W, Arad T, Weiner S. Three-dimensional ordered distribution of crystals in turkey tendon collagen fibers. Proc Natl Acad Sci U S A. 1989;86(24):9822-6.

[17] Feng JQ, Ward LM, Liu S, Lu Y, Xie Y, Yuan B, *et al.* Loss of DMP1 causes rickets and osteomalacia and identifies a role for osteocytes in mineral metabolism. Nat Genet 2006;38(11):1310-5.

[18] Termine JD, Kleinman HK, Whitson SW, Conn KM, McGarvey ML, Martin GR. Osteonectin, a bone-specific protein linking mineral to collagen. Cell 1981;26(1 Pt 1):99-105.

[19] Cui FZ, Li Y, Ge J. Self-assembly of mineralized collagen composites. Mat Sci Eng R 2007;57(1-6):1-27.

[20] Olszta MJ, Cheng XG, Jee SS, Kumar R, Kim YY, Kaufman MJ, *et al.* Bone structure and formation: A new perspective. Mat Sci Eng R 2007;58(3-5):77-116.

[21] Burger C, Zhou HW, Wang H, Sics I, Hsiao BS, Chu B, *et al.* Lateral packing of mineral crystals in bone collagen fibrils. Biophys J 2008;95(4):1985-92.

[22] Wahl DA, Czernuszka JT. Collagen-hydroxyapatite composites for hard tissue repair. Eur Cells Mater 2006;11:43-56.

[23] Fratzl P, Gupta HS, Paschalis EP, Roschger P. Structure and mechanical quality of the collagen-mineral nano-composite in bone. J Mater Chem 2004;14(14):2115-23.

[24] Whittaker DK. Structural variations in the surface zone of human tooth enamel observed by scanning electron microscopy. Arch Oral Biol 1982;27(5):383-92.

[25] Tamerler C, Sarikaya M. Molecular biomimetics: Genetic synthesis, assembly, and formation of materials using peptides. Mrs Bull 2008;33(5):504-10.

[26] Arsenault AL, Robinson BW. The Dentino-Enamel Junction - a Structural and Microanalytical Study of Early Mineralization. Calcified Tissue Int 1989;45(2):111-21.

[27] White SN, Luo W, Paine ML, Fong H, Sarikaya M, Snead ML. Biological organization of hydroxyapatite crystallites into a fibrous continuum toughens and controls anisotropy in human enamel. J Dent Res 2001;80(1):321-6.

[28] Imbeni V, Kruzic JJ, Marshall GW, Marshall SJ, Ritchie RO. The dentin-enamel junction and the fracture of human teeth. Nat Mater 2005;4(3):229-32.

[29] Paine ML, White SN, Luo W, Fong H, Sarikaya M, Snead ML. Regulated gene expression dictates enamel structure and tooth function. Matrix Biol 2001;20(5-6):273-92.

[30] Wen HB, Moradian-Oldak J, Fincham AG. Dose-dependent modulation of octacalcium phosphate crystal habit by amelogenins. J Dent Res 2000;79(11):1902-6.

[31] Cui FZ, Ge J. New observations of the hierarchical structure of human enamel, from nanoscale to microscale. J Tissue Eng Regen M 2007;1(3):185-91.

[32] Wang RZ, Weiner S. Strain-structure relations in human teeth using Moire fringes. J Biomech 1998;31(2):135-41.

[33] Kodaka T, Nakajima F, Higashi S. Structure of the So-Called Prismless Enamel in Human Deciduous Teeth. Caries Rcs 1989;23(5):290-6.
[34] Daculsi G, Bouler JM, LeGeros RZ. Adaptive crystal formation in normal and pathological calcifications in synthetic calcium phosphate and related biomaterials. Int Rev Cytol 1997;172:129-91.
[35] Kazama JJ, Amizuka N, Fukagawa M. Ectopic calcification as abnormal biomineralization. Ther Apher Dial 2006;10:S34-S8.
[36] Brancaccio D, Cozzolino M. The mechanism of calcium deposition in soft tissues. Contrib Nephrol 2005;149:279-86.
[37] Achilles W, Jockel U, Schaper A, Burk M, Riedmiller H. *In vitro* Formation of Urinary Stones - Generation of Spherulites of Calcium-Phosphate in Gel and Overgrowth with Calcium-Oxalate Using a New Flow Model of Crystallization. Scanning Microscopy 1995;9(2):577-86.
[38] Hayashizaki J, Ban S, Nakagaki H, Okumura A, Yoshii S, Robinson C. Site specific mineral composition and microstructure of human supra-gingival dental calculus. Arch Oral Biol 2008;53(2):168-74.
[39] Marra SP, Daghlian CP, Fillinger MF, Kennedy FE. Elemental composition, morphology and mechanical properties of calcified deposits obtained from abdominal aortic aneurysms. Acta Biomater 2006;2(5):515-20.
[40] Weissen-Plenz G, Nitschke Y, Rutsch F. Mechanisms of Arterial Calcification: Spotlight on the Inhibitors. Adv Clin Chem 2008;46:263-93.
[41] Laird DF, Mucalo MR, Yokogawa Y. Growth of calcium hydroxyapatite (Ca-HAp) on cholesterol and cholestanol crystals from a simulated body fluid: A possible insight into the pathological calcifications associated with atherosclerosis. J Colloid Interf Sci 2006;295(2):348-63.
[42] Wesson JA, Ward MD. Pathological biomineralization of kidney stones. Elements 2007;3(6):415-21.
[43] Legeros RZ, Orly I, Legeros JP, Gomez C, Kazimiroff J, Tarpley T, *et al.* Scanning Electron-Microscopy and Electron-Probe Microanalyses of the Crystalline Components of Human and Animal Dental Calculi. Scanning Microscopy 1988;2(1):345-56.
[44] Kirsch T. Determinants of pathological mineralization. Curr Opin Rheumatol2006;18(2):174-80.
[45] Roveri N, Battistella E, Bianchi CL, Foltran I, Foresti E, Iafisco M, *et al.* Surface Enamel Remineralization: Biomimetic Apatite Nanocrystals and Fluoride Ions Different Effects J Nanomater. 2009:-.
[46] Dorozhkin SV. Calcium orthophosphate cements for biomedical application. J Mater Sci 2008;43(9):3028-57.
[47] Jandt KD. Evolutions, revolutions and trends in biomaterials science - A perspective. Adv Eng Mater 2007;9(12):1035-50.
[48] LeGeros RZ. Calcium Phosphate-Based Osteoinductive Materials. Chem Rev 2008;108(11):4742-53.
[49] Rivera EM, Araiza M, Brostow W, Castano VM, Diaz-Estrada JR, Hernandez R, *et al.* Synthesis of hydroxyapatite from eggshells. Mater Lett 1999;41(3):128-34.
[50] Balazsi C, Weber F, Kover Z, Horvath E, Nemeth C. Preparation of calcium-phosphate bioceramics from natural resources. J Eur Ceram Soc 2007;27(2-3):1601-6.
[51] Murugan R, Ramakrishna S. Crystallographic study of hydroxyapatite bioceramics derived from various sources. Cryst Growth Des 2005;5(1):111-2.

[52] Roveri N, Palazzo B, Iafisco M. The role of biomimetism in developing nanostructured inorganic matrices for drug delivery. Expert Opin Drug Del 2008;5(8):861-77.
[53] Lecomte A, Gautier H, Bouler JM, Gouyette A, Pegon Y, Daculsi G, *et al.* Biphasic calcium phosphate: A comparative study of interconnected porosity in two ceramics. J Biomed Mater Res B 2008;84B(1):1-6.
[54] Daculsi G, Uzel AP, Weiss P, Goyenvalle E, Aguado E. Developments in injectable multiphasic biomaterials. The performance of microporous biphasic calcium phosphate granules and hydrogels. J Mater Sci-Mater M 2010;21(3):855-61.
[55] Daculsi G, Laboux O, Malard O, Weiss P. Current state of the art of biphasic calcium phosphate bioceramics. J Mater Sci-Mater M 2003;14(3):195-200.
[56] Sánchez-Salcedo S, Arcos D, Vallet-Regí M. Upgrading calcium phosphate scaffolds for tissue engineering applications. Key Eng Mater 2008;377:19-42.
[57] Sayer M, Stratilatov AD, Reid J, Calderin L, Stott MJ, Yin X, *et al.* Structure and composition of silicon-stabilized tricalcium phosphate. Biomaterials 2003;24(3):369-82.
[58] da Silva RV, Bertran CA, Kawachi EY, Camilli JA. Repair of cranial bone defects with calcium phosphate ceramic implant or autogenous bone graft. J Craniofac Surg 2007;18(2):281-6.
[59] Astala R, Calderin L, Yin X, Stott MJ. Ab initio simulation of Si-doped hydroxyapatite. Chem Mater 2006;18(2):413-22.
[60] Khalil KA, Kim HY, Kim SW, Kim KW. Observation of toughness improvement of the hydroxyapatite bioceramics densified using high-frequency induction heat sintering. Int J Appl Ceram Tec 2007;4(1):30-7.
[61] Kobayashi S, Kawai W, Wakayama S. The effect of pressure during sintering on the strength and the fracture toughness of hydroxyapatite ceramics. J Mater Sci-Mater M 2006;17(11):1089-93.
[62] Narasaraju TSB, Phebe DE. Some physico-chemical aspects of hydroxylapatite. J Mater Sci 1996;31(1):1-21.
[63] Suchanek W, Yoshimura M. Processing and properties of hydroxyapatite-based biomaterials for use as hard tissue replacement implants. J Mater Res 1998;13(1):94-117.
[64] Ramesh S, Tan CY, Bhaduri SB, Teng WD, Sopyan I. Densification behaviour of nanocrystalline hydroxyapatite bioceramics. J Mater Process Technol 2008;206(1-3):221-30.
[65] Linhart W, Briem D, Amling M, Rueger JM, Windolf J. Mechanical failure of porous hydroxyapatite ceramics 7.5 years after implantation in the proximal tibial. Unfallchirurg. 2004;107(2):154-7.
[66] Dorozhkin SV. Calcium orthophosphate-based biocomposites and hybrid biomaterials. J Mater Sci 2009;44(9):2343-87.
[67] Best SM, Porter AE, Thian ES, Huang J. Bioceramics: Past, present and for the future. J Eur Ceram Soc 2008;28(7):1319-27.
[68] Mastrogiacomo M, Scaglione S, Martinetti R, Dolcini L, Beltrame F, Cancedda R, *et al.* Role of scaffold internal structure on *in vivo* bone formation in macroporous calcium phosphate bioceramics. Biomaterials 2006;27(17):3230-7.
[69] Karageorgiou V, Kaplan D. Porosity of 3D biomaterial scaffolds and osteogenesis. Biomaterials 2005;26(27):5474-91.
[70] Lu JX, Flautre B, Anselme K, Hardouin P, Gallur A, Descamps M, *et al.* Role of interconnections in porous bioceramics on bone recolonization *in vitro* and *in vivo*. J Mater Sci-Mater M 1999;10(2):111-20.

[71] Tancret F, Bouler JM, Chamousset J, Minois LM. Modelling the mechanical properties of microporous and macroporous biphasic calcium phosphate bioceramics. J Eur Ceram Soc 2006;26(16):3647-56.
[72] Walsh D, Tanaka J. Preparation of a bone-like apatite foam cement. J Mater Sci-Mater M 2001;12(4):339-43.
[73] Chevalier E, Chulia D, Pouget C, Viana M. Fabrication of porous substrates: A review of processes using pore forming agents in the biomaterial field. J Pharm Sci-Us. 2008;97(3):1135-54.
[74] Gonzalez-McQuire R, Green D, Walsh D, Hall S, Chane-Ching JY, Oreffo ROC, *et al.* Fabrication of hydroxyapatite sponges by dextran sulphate/amino acid templating. Biomaterials 2005;26(33):6652-6.
[75] Liu DM. Fabrication of hydroxyapatite ceramic with controlled porosity. J Mater Sci-Mater M 1997;8(4):227-32.
[76] Walsh D, Boanini E, Tanaka J, Mann S. Synthesis of tri-calcium phosphate sponges by interfacial deposition and thermal transformation of self-supporting calcium phosphate films. J Mater Chem 2005;15(10):1043-8.
[77] Charriere E, Lemaitre J, Zysset P. Hydroxyapatite cement scaffolds with controlled macroporosity: fabrication protocol and mechanical properties. Biomaterials 2003;24(5):809-17.
[78] Araiza MA, Gomez-Morales J, Clemente RR, Castano VM. Conversion of the echinoderm Mellita eduardobarrosoi calcite skeleton into porous hydroxyapatite by treatment with phosphated boiling solutions. J Mater Synth Proces 1999;7(4):211-9.
[79] Sopyan I, Mel M, Ramesh S, Khalid KA. Porous hydroxyapatite for artificial bone applications. Sci Technol Adv Mat 2007;8(1-2):116-23.
[80] Brown WE, Chow LC, inventors; Dental restorative cement pastes.1985.
[81] LeGeros, RZ, Chohayeb, A, Shulman, A. Apatitic Calcium Phosphates: Possible. Dental Restorative Materials. J Dent Res 1982;61:343-7.
[82] Gbureck U, Barralet JE, Spatz K, Grover LM, Thull R. Ionic modification of calcium phosphate cement viscosity. Part I: hypodermic injection and strength improvement of apatite cement. Biomaterials 2004;25(11):2187-95.
[83] Tagaya M, Goto H, Iinuma M, Wakamatsu N, Tamura Y, Doi Y. Development of self-setting Te-CP/alpha-TCP cement for pulpotomy. Dent Mater J 2005;24(4):555-61.
[84] Driessens FCM, Planell JA, Boltong MG, Khairoun I, Ginebra MP. Osteotransductive bone cements. P I Mech Eng H 1998;212(H6):427-35.
[85] Driessens FCM, Boltong MG, Bermudez O, Planell JA. Formulation and Setting Times of Some Calcium Orthophosphate Cements - a Pilot-Study. J Mater Sci-Mater M 1993;4(5):503-8.
[86] Xia ZD, Grover LM, Huang YZ, Adamopoulos LE, Gbureck U, Triffitt JT, *et al. In vitro* biodegradation of three brushite calcium phosphate cements by a macrophage cell-line. Biomaterials 2006;27(26):4557-65.
[87] Kudrman J, Fousek J, Brezina V, Mikova R, Vesely J. Titanium alloys for implants in medicine. Kovove Mater 2007;45(4):199-208.
[88] Niinomi M. Mechanical biocompatibilities of titanium alloys for biomedical applications. J Mech Behav Biomed Mater 2008;1(1):30-42.
[89] Raikar GN, Gregory JC, Ong JL, Lucas LC, Lemons JE, Kawahara D, *et al.* Surface Characterization of Titanium Implants. J Vac Sci Technol A 1995;13(5):2633-7.

[90] Manara S, Paolucci F, Palazzo B, Marcaccio M, Foresti E, Tosi G, *et al.* Electrochemically-assisted deposition of biomimetic hydroxyapatite-collagen coatings on titanium plate. Inorg Chim Acta 2008;361(6):1634-45.

[91] Wennerberg A, Albrektsson T. Structural influence from calcium phosphate coatings and its possible effect on enhanced bone integration. Acta Odontol Scand 2009;67(6):333-40.

[92] Schouten C, Meijer GJ, van den Beucken JJ, Leeuwenburgh SC, de Jonge LT, Wolke JG, *et al.* *In vivo* bone response and mechanical evaluation of electrosprayed CaP nanoparticle coatings using the iliac crest of goats as an implantation model. Acta Biomater 2010;6(6):2227-36.

[93] de Jonge LT, Leeuwenburgh SC, van den Beucken JJ, te Riet J, Daamen WF, Wolke JG, *et al.* The osteogenic effect of electrosprayed nanoscale collagen/calcium phosphate coatings on titanium. Biomaterials 2010;31(9):2461-9.

[94] Kokubo T, Takadama H. How useful is SBF in predicting *in vivo* bone bioactivity? Biomaterials 2006;27(15):2907-15.

[95] Degroot K, Geesink R, Klein CPAT, Serekian P. Plasma Sprayed Coatings of Hydroxylapatite. J Biomed Mater Res1987;21(12):1375-81.

[96] Paital SR, Dahotre NB. Calcium phosphate coatings for bio-implant applications: Materials, performance factors, and methodologies. Mat Sci Eng R 2009;66(1-3):1-70.

[97] Gómez Morales J, Rodríguez Clemente R, Armas B, Combescure C, Berjoan R, Cubo J, *et al.* Controlled Nucleation and Growth of Thin Hydroxyapatite Layers on Titanium Implants by Using Induction Heating Technique. Langmuir 2004;20(13):5174-8.

[98] MacDonald DE, Betts F, Stranick M, Doty S, Boskey AL. Physicochemical study of plasma-sprayed hydroxyapatite-coated implants in humans. J Biomed Mater Res 2001;54(4):480-90.

[99] Lusquinos F, De Carlos A, Pou J, Arias JL, Boutinguiza M, Leon B, *et al.* Calcium phosphate coatings obtained by Nd : YAG laser cladding: Physicochemical and biologic properties. J Biomed Mater Res A 2003;64A(4):630-7.

[100] Yang YZ, Kim KH, Ong JL. Review on calcium phosphate coatings produced using a sputtering process - an alternative to plasma spraying. Biomaterials 2005;26(3):327-37.

[101] Wei M, Ruys AJ, Swain MV, Kim SH, Milthorpe BK, Sorrell CC. Interfacial bond strength of electrophoretically deposited hydroxyapatite coatings on metals. J Mater Sci-Mater M 1999;10(7):401-9.

[102] Wang DG, Chen CZ, Ting H, Lei TQ. Hydroxyapatite coating on Ti6Al4V alloy by a sol-gel method. J Mater Sci-Mater M 2008;19(6):2281-6.

[103] Hashimoto Y, Kawashima M, Hatanaka R, Kusunoki M, Nishikawa H, Hontsu S, *et al.* Cytocompatibility of calcium phosphate coatings deposited by an ArF pulsed laser. J Mater Sci-Mater M 2008;19(1):327-33.

[104] Muller L, Conforto E, Caillard D, Muller FA. Biomimetic apatite coatings - Carbonate substitution and preferred growth orientation. Biomol Eng 2007;24(5):462-6.

[105] Choi JM, Kim HE, Lee IS. Ion-beam-assisted deposition (IBAD) of hydroxyapatite coating layer on Ti-based metal substrate. Biomaterials 2000;21(5):469-73.

[106] Leeuwenburgh SC, Wolke JG, Siebers MC, Schoonman J, Jansen JA. *In vitro* and *in vivo* reactivity of porous, electrosprayed calcium phosphate coatings. Biomaterials 2006;27(18):3368-78.

[107] de Jonge LT, Leeuwenburgh SC, Wolke JG, Jansen JA. Organic-inorganic surface modifications for titanium implant surfaces. Pharm Res 2008;25(10):2357-69.

[108] Shepperd JAN, Apthorp H. A contemporary snapshot of the use of hydroxyapatite coating in orthopaedic surgery. J Bone Joint Surg Br 2005;87B(8):1046-9.

[109] Rey C, Combes C, Drouet C, Glimcher MJ. Bone mineral: update on chemical composition and structure. Osteoporosis Int 2009;20(6):1013-21.
[110] Grynpas MD, Bonar LC, Glimcher MJ. X-ray diffraction radial distribution function studies on bone mineral and synthetic calcium phosphates. J Mater Sci 1984;19(3):723-36.
[111] Eichert D. Ph.D. Thesis, INPT France; 2001.
[112] Legros R, Balmain N, Bonel G. Age-related changes in mineral of rat and bovine cortical bone. Calcif Tissue Int 1987;41(3):137-44.
[113] Paschalis EP, DiCarlo E, Betts F, Sherman P, Mendelsohn R, Boskey AL. FTIR microspectroscopic analysis of human osteonal bone. Calcified Tissue Int 1996;59(6):480-7.
[114] Rey C, Collins B, Goehl T, Dickson IR, Glimcher MJ. The Carbonate Environment in Bone-Mineral - a Resolution-Enhanced Fourier-Transform Infrared-Spectroscopy Study. Calcified Tissue Int 1989;45(3):157-64.
[115] Rey C, Shimizu M, Collins B, Glimcher MJ. Resolution-enhanced Fourier transform infrared spectroscopy study of the environment of phosphate ions in the early deposits of a solid phase of calcium-phosphate in bone and enamel, and their evolution with age. I: Investigations in the upsilon 4 PO4 domain. Calcif Tissue Int 1990;46(6):384-94.
[116] Combes C, Rey C, Mounic S. Identification and evaluation of HPO4 ions in biomimetic poorly crystalline apatite and bone mineral. Key Eng Mater 2001;192-195:143-6.
[117] Drouet C, Carayon MT, Combes C, Rey C. Surface enrichment of biomimetic apatites with biologically-active ions Mg2+ and Sr2+: A preamble to the activation of bone repair materials. Mat Sci Eng C-Bio S 2008;28(8):1544-50.
[118] Eichert C, Drouet C, Sfihi H, Rey C, Combes C. Nanocrystalline apatite-based biomaterials: synthesis, processing and characterization. In: Kendall JB, editor. Biomaterials Research Advances Nova Science Publishers; 2007. p. 93-143.
[119] Hina A. Ph.D. Thesis, INPT France; 1996.
[120] Sfihi H, Rey C. 1-D and 2-D double heteronuclear magnetic resonance study of the local structure of type B carbonate fluoroapatite. In: Fraissard F, Lapina B, editors. Magnetic Resonance in Colloid and Interface Science: Kluwer Academic Publishers.; 2002. p. 409-18.
[121] Rey C, Combes C, Drouet C, Sfihi H, Barroug A. Physico-chemical properties of nanocrystalline apatites: Implications for biominerals and biomaterials. Mat Sci Eng C-Bio S 2007;27(2):198-205.
[122] Cho G, Wu Y, Ackerman JL. Detection of hydroxyl ions in bone mineral by solid-state NMR spectroscopy. Science 2003;300(5622):1123-7.
[123] Cho G, Yesinowski JP. H-1 and F-19 multiple-quantum NMR dynamics in quasi-one-dimensional spin clusters in apatites. J Phys Chem-Us 1996;100(39):15716-25.
[124] Miquel JL, Facchini L, Legrand AP, Rey C, Lemaitre J. Solid-State Nmr to Study Calcium-Phosphate Ceramics. Colloids Surf 1990;45:427-33.
[125] Aue WP, Roufosse AH, Glimcher MJ, Griffin RG. Solid-state phosphorus-31 nuclear magnetic resonance studies of synthetic solid phases of calcium phosphate: potential models of bone mineral. Biochemistry 1984;23(25):6110-4.
[126] Pourpoint F, Gervais C, Bonhomme-Coury L, Azaïs T, Coelho C, Mauri F, *et al.* Calcium phosphates and hydroxyapatite: solid-state NMR experiments and first-principles calculations. Appl Magn Reson 2007;32(4):435-57.
[127] Roufosse AH, Aue WP, Roberts JE, Glimcher MJ, Griffin RG. Investigation of the mineral phases of bone by solid-state phosphorus-31 magic angle sample spinning nuclear magnetic resonance. Biochemistry 1984;23(25):6115-20.

[128] Roberts JE, Bonar LC, Griffin RG, Glimcher MJ. Characterization of very young mineral phases of bone by solid state 31phosphorus magic angle sample spinning nuclear magnetic resonance and X-ray diffraction. Calcif Tissue Int 1992;50(1):42-8.

[129] Wu Y, Glimcher MJ, Rey C, Ackerman JL. A unique protonated phosphate group in bone mineral not present in synthetic calcium phosphates. Identification by phosphorus-31 solid state NMR spectroscopy. J Mol Biol 1994;244(4):423-35.

[130] Jarlbring M, Sandstrom DE, Antzutkin ON, Forsling W. Characterization of active phosphorus surface sites at synthetic carbonate-free fluorapatite using single-pulse H-1, P-31, and T-31 CP MAS NMR. Langmuir 2006;22(10):4787-92.

[131] Kaflak A, Kolodziejski W. Kinetics H-1 -> P-31 NMR cross-polarization in bone apatite and its mineral standards. Magn Reson Chem 2008;46(4):335-41.

[132] Sakhno Y, Bertinetti L, Iafisco M, Tampieri A, Roveri N, Martra G. Surface Hydration and Cationic Sites of Nanohydroxyapatites with Amorphous or Crystalline Surfaces: A Comparative Study. J Phys Chem C 2010;114(39):16640-8.

[133] Drouet C, Bosc F, Banu M, Largeot C, Combes C, Dechambre G, *et al.* Nanocrystalline apatites: From powders to biomaterials. Powder Technol 2009;190(1-2):118-22.

[134] Cazalbou S, Eichert D, Ranz X, Drouet C, Combes C, Harmand MF, *et al.* Ion exchanges in apatites for biomedical application. J Mater Sci-Mater M 2005;16(5):405-9.

[135] Cazalbou S, Combes C, Eichert D, Rey C, Glimcher MJ. Poorly crystalline apatites: evolution and maturation *in vitro* and *in vivo*. J Bone Miner Metab 2004;22(4):310-7.

[136] Cazalbou S, Combes C, Eichert D, Rey C. Adaptative physico-chemistry of bio-related calcium phosphates. J Mater Chem 2004;14(14):2148-53.

[137] Cazalbou S. Ph.D. Thesis, INPT France; 2000.

[138] Midy V, Rey C, Bres E, Dard M. Basic fibroblast growth factor adsorption and release properties of calcium phosphate. J Biomed Mater Res1998;41(3):405-11.

[139] Quizat S, Barroug A, Legrouri A, Rey C. Adsorption of bovine serum albumin on poorly crystalline apatite: Influence of maturation. Mater Res Bull 1999;34(14-15):2279-89.

[140] Rey C, Kim HM, Glimcher MJ. Maturation of poorly crystalline synthetic and biological apatites. In: Brown PW, Constantz B, editors. Hydroxyapatite and Related Materials: Boca Raton, CRC Press; 1994. p. 181-7.

[141] Bertinetti L, Drouet C, Combes C, Rey C, Tampieri A, Coluccia S, *et al.* Surface Characteristics of Nanocrystalline Apatites: Effect of Mg Surface Enrichment on Morphology, Surface Hydration Species, and Cationic Environments. Langmuir 2009;25(10):5647-54.

[142] Bracci B, Torricelli P, Panzavolta S, Boanini E, Giardino R, Bigi A. Effect of Mg2+, Sr2+, and Mn2+ on the chemico-physical and *in vitro* biological properties of calcium phosphate biomimetic coatings. J Inorg Biochem 2009;103(12):1666-74.

[143] Habibovic P, Sees TM, van den Doel MA, van Blitterswijk CA, de Groot K. Osteoinduction by biomaterials--physicochemical and structural influences. J Biomed Mater Res A 2006;77(4):747-62.

[144] Autefage H, Briand-Mesange F, Cazalbou S, Drouet C, Fourmy D, Goncalves S, *et al.* Adsorption and Release of BMP-2 on Nanocrystalline Apatite-Coated and Uncoated Hydroxyapatite/beta-Tricalcium Phosphate Porous Ceramics. J Biomed Mater Res B 2009;91B(2):706-15.

[145] Grossin D, Rollin-Martinet S, Estournes C, Rossignol F, Champion E, Combes C, *et al.* Biomimetic apatite sintered at very low temperature by spark plasma sintering: Physico-chemistry and microstructure aspects. Acta Biomater 2010;6(2):577-85.

[146] Banu M. Ph.D. Thesis, INPT France; 2005.
[147] Drouet C, Largeot C, Raimbeaux G, Estournes C, Dechambre G, Combes C, *et al.* Bioceramics: spark plasma sintering (SPS) of calcium phosphates. Adv Sci Technol 2006;49:45-50.
[148] Uddin MH, Matsumoto T, Ishihara S, Nakahira A, Okazaki M, Sohmura T. Apatite containing aspartic acid for selective protein loading. J Dent Res 2010;89(5):488-92.
[149] Mondejar SP, Kovtun A, Epple M. Lanthanide-doped calcium phosphate nanoparticles with high internal crystallinity and with a shell of DNA as fluorescent probes in cell experiments. J Mater Chem 2007;17(39):4153-9.
[150] Al-Kattan A, Dufour P, Dexpert-Ghys J, Drouet C. Preparation and Physicochemical Characteristics of Luminescent Apatite-Based Colloids. J Phys Chem C 2010;114(7):2918-24.
[151] Bouladjine A, Al-Kattan A, Dufour P, Drouet C. New Advances in Nanocrystalline Apatite Colloids Intended for Cellular Drug Delivery. Langmuir 2009;25(20):12256-65.
[152] Chane-Ching JY, Lebugle A, Rousselot I, Pourpoint A, Pelle F. Colloidal synthesis and characterization of monocrystalline apatite nanophosphors. J Mater Chem 2007;17(28):2904-13.
[153] Saha SK, Banerjee A, Banerjee S, Bose S. Synthesis of nanocrystalline hydroxyapatite using surfactant template systems: Role of templates in controlling morphology. Mat Sci Eng C-Mater 2009;29(7):2294-301.
[154] Chander S, Fuerstenau DW. Solubility and interfacial properties of hydroxyapatite: a review. In: Misra DN, editor. Adsorption on, and surface chemistry of hydroxyapatite: Plenum Press: New York; 1984.p. 29-49.
[155] Tung MS, Skrtic D. Interfacial properties of hydroxyapatite, fluorapatite and octacalcium phosphate. In: Chow LC, Eanes ED, editors. Octacalcium phosphate, Monographs in oral science: Karger; 2001. p. 112-29.
[156] Saleeb FZ, De Bruyn PL. Surface properties of alkaline earth apatites. J Electroanal Chem 1972;37(1):99-118.
[157] Mangood A, Malkaj P, Dalas E. Hydroxyapatite crystallization in the presence of acetaminophen. J Cryst Growth 2006;290(2):565-70.
[158] Burke EM, Nancollas GH. Relation of lattice ion solution composition to octacalcium phosphate dissolution kinetics. Colloid Surface A 1999;150(1-3):151-60.
[159] Nancollas GH, Wu WJ. Biomineralization mechanisms: a kinetics and interfacial energy approach. J Cryst Growth 2000;211(1-4):137-42.
[160] Wu WJ, Nancollas GH. Determination of interfacial tension from crystallization and dissolution data: a comparison with other methods. Adv Colloid Interfac 1999;79(2-3):229-79.
[161] Pietronave S, Iafisco M, Locarno D, Rimondini L, Prat M. Functionalized nanomaterials for diagnosis and therapy of cancer. J Appl Biomater Biom 2009;7(2):77-89.
[162] Kasemo B, Gold J. Implant surfaces and interface processes. Adv Dent Res. 1999;13:8-20.
[163] El-Ghannam A, Ducheyne P, Shapiro IM. Effect of serum proteins on osteoblast adhesion to surface-modified bioactive glass and hydroxyapatite. J Orthopaed Res 1999;17(3):340-5.
[164] Kandori K, Masunari A, Ishikawa T. Study on adsorption mechanism of proteins onto synthetic calcium hydroxyapatites through ionic concentration measurements. Calcified Tissue Int 2005;76(3):194-206.
[165] Makrodimitris K, Masica DL, Kim ET, Gray JJ. Structure prediction of protein-solid surface interactions reveals a molecular recognition motif of statherin for hydroxyapatite. J Am Chem Soc 2007;129(44):13713-22.

[166] Baeza A, Izquierdo-Barba I, Vallet-Regi M. Biotinylation of silicon-doped hydroxyapatite: A new approach to protein fixation for bone tissue regeneration. Acta Biomater 2010;6(3):743-9.

[167] Iafisco M, Marchetti M, Morales JG, Hernandez-Hernandez MA, Ruiz JMG, Roveri N. Silica Gel Template for Calcium Phosphates Crystallization. Cryst Growth Des 2009;9(11):4912-21.

[168] Iafisco M, Morales JG, Hernandez-Hernandez MA, Garcia-Ruiz JM, Roveri N. Biomimetic Carbonate-Hydroxyapatite Nanocrystals Prepared by Vapor Diffusion. Adv Eng Mater 2010;12(7):B218-B23.

[169] Koutsopoulos S, Dalas E. The crystallization of hydroxyapatite in the presence of lysine. J Colloid Interf Sci 2000;231(2):207-12.

[170] Koutsopoulos S, Dalas E. Hydroxyapatite crystallization in the presence of serine, tyrosine and hydroxyproline amino acids with polar side groups. J Cryst Growth 2000;216(1-4):443-9.

[171] Koutsopoulos S, Dalas E. Hydroxyapatite crystallization in the presence of amino acids with uncharged polar side groups: Glycine, cysteine, cystine, and glutamine. Langmuir 2001;17(4):1074-9.

[172] Gonzalez-McQuire R, Chane-Ching JY, Vignaud E, Lebugle A, Mann S. Synthesis and characterization of amino acid-functionalized hydroxyapatite nanorods. J Mater Chem 2004;14(14):2277-81.

[173] Boanini E, Torricelli P, Gazzano M, Giardino R, Bigi A. Nanocomposites of hydroxyapatite with aspartic acid and glutamic acid and their interaction with osteoblast-like cells. Biomaterials 2006;27(25):4428-33.

[174] Rosseeva EV, Golovanova OA, Frank-Kamenetskaya OV. The influence of amino acids on the formation of nanocrystalline hydroxyapatite. Glass Phys Chem 2007;33(3):283-6.

[175] Jack KS, Vizcarra TG, Trau M. Characterization and surface properties of amino-acid-modified, carbonate-containing hydroxyapatite particles. Langmuir 2007;23(24):12233-42.

[176] Palazzo B, Walsh D, Iafisco M, Foresti E, Bertinetti L, Martra G, *et al.* Amino acid synergetic effect on structure, morphology and surface properties of biomimetic apatite nanocrystals. Acta Biomater 2009;5(4):1241-52.

[177] Boix T, Gomez-Morales J, Torrent-Burgues J, Monfort A, Pulgdomenech P, Rodriguez-Clemente R. Adsorption of recombinant human bone morphogenetic protein rhBMP-2m onto hydroxyapatite. J Inorg Biochem 2005;99(5):1043-50.

[178] Iafisco M, Sabatino P, Lesci IG, Prat M, Rimondini L, Roveri N. Conformational modifications of serum albumins adsorbed on different kinds of biomimetic hydroxyapatite nanocrystals. Colloids Surf B. 2010;81(1):274-84.

[179] Iafisco M, Palazzo B, Falini G, Foggia MD, Bonora S, Nicolis S, *et al.* Adsorption and conformational change of myoglobin on biomimetic hydroxyapatite nanocrystals functionalized with alendronate. Langmuir 2008;24(9):4924-30.

[180] El Shafei GMS, Moussa NA. Adsorption of some essential amino acids on hydroxyapatite. J Colloid Interf Sci 2001;238(1):160-6.

[181] Wu LM, Forsling W, Schindler PW. Surface Complexation of Calcium Minerals in Aqueous-Solution.1. Surface Protonation at Fluorapatite Water Interfaces. J Colloid Interf Sci 1991;147(1):178-85.

[182] Yin G, Liu Z, Zhan J, Ding FX, Yuan NJ. Impacts of the surface charge property on protein adsorption on hydroxyapatite. Chem Eng J 2002;87(2):181-6.

[183] Kawasaki K, Kambara M, Matsumura H, Norde W. A comparison of the adsorption of saliva proteins and some typical proteins onto the surface of hydroxyapatite. Colloids Surf B 2003;32(4):321-34.

[184] Smith DA, Connell SD, Robinson C, Kirkham J. Chemical force microscopy: applications in surface characterisation of natural hydroxyapatite. Anal Chim Acta 2003;479(1):39-57.
[185] Stayton PS, Drobny GP, Shaw WJ, Long JR, Gilbert M. Molecular recognition at the protein-hydroxyapatite interface. Crit Rev Oral Biol M 2003;14(5):370-6.
[186] Schwartz SS, Hay DI, Schluckebier SK. Inhibition of calcium phosphate precipitation by human salivary statherin: structure-activity relationships. Calcif Tissue Int 1992;50(6):511-7.
[187] Hay DI, Smith DJ, Schluckebier SK, Moreno EC. Relationship between concentration of human salivary statherin and inhibition of calcium phosphate precipitation in stimulated human parotid saliva. J Dent Res 1984;63(6):857-63.
[188] Douglas WH, Reeh ES, Ramasubbu N, Raj PA, Bhandary KK, Levine MJ. Statherin - a Major Boundary Lubricant of Human Saliva. Biochem Bioph Res Co 1991;180(1):91-7.
[189] Moreno EC, Kresak M, Hay DI. Adsorption of two human parotid salivary macromolecules on hydroxy-, fluorhydroxy- and fluorapatites. Arch Oral Biol 1978;23(7):525-33.
[190] Raj PA, Johnsson M, Levine MJ, Nancollas GH. Salivary Statherin - Dependence on Sequence, Charge, Hydrogen-Bonding Potency, and Helical Conformation for Adsorption to Hydroxyapatite and Inhibition of Mineralization. J Biol Chem 1992;267(9):5968-76.
[191] Gibson JM, Raghunathan V, Popham JM, Stayton PS, Drobny GP. A REDOR NMR study of a phosphorylated statherin fragment bound to hydroxyapatite crystals. J Am Chem Soc 2005;127(26):9350-1.
[192] Raghunathan V, Gibson JM, Goobes G, Popham JM, Louie EA, Stayton PS, *et al.* Homonuclear and heteronuclear NMR studies of a statherin fragment bound to hydroxyapatite crystals. J Phys Chem B 2006;110(18):9324-32.
[193] Gibson JM, Popham JM, Raghunathan V, Stayton PS, Drobny GP. A solid-state NMR study of the dynamics and interactions of phenylalanine rings in a statherin fragment bound to hydroxyapatite crystals. J Am Chem Soc 2006;128(16):5364-70.
[194] Goobes G, Goobes R, Shaw WJ, Gibson JM, Long JR, Raghunathan V, *et al.* The structure, dynamics, and energetics of protein adsorption-lessons learned from adsorption of statherin to hydroxyapatite. Magn Reson Chem 2008;45(S1):S32-S47.
[195] Fisher LW, Torchia DA, Fohr B, Young MF, Fedarko NS. Flexible structures of SIBLING proteins, bone sialoprotein, and osteopontin. Biochem Bioph Res Co 2001;280(2):460-5.
[196] Errassif F, Menbaoui A, Autefage H, Benaziz L, Ouizat S, Santran V, *et al.* Adsorption on Apatitic Calcium Phosphates: Applications to Drug Delivery In: Narayan R, Singh M, McKittrick J, editors. Advances in Bioceramics and Biotechnologies: Wiley-VCH Verlag GmbH & Co. KGaA; 2010. p. 159-174.
[197] Aoba T, Moreno EC. Adsorption of Phosphoserine onto Hydroxyapatite and Its Inhibitory Activity on Crystal-Growth. J Colloid Interf Sci 1985;106(1):110-21.
[198] Lopez-Macipe A, Gomez-Morales J, Rodriguez-Clemente R. The role of pH in the adsorption of citrate ions on hydroxyapatite. J Colloid Interf Sci 1998;200(1):114-20.
[199] Rodan GA, Fleisch HA. Bisphosphonates: Mechanisms of action. J Clin Invest 1996;97(12):2692-6.
[200] Russell RGG, Rogers MJ. Bisphosphonates: From the laboratory to the clinic and back again. Bone 1999;25(1):97-106.
[201] Josse S, Faucheux C, Soueldan A, Grimandi G, Massiot D, Alonso B, *et al.* Novel biomaterials for bisphosphonate delivery. Biomaterials 2005;26(14):2073-80.
[202] Roussiere H, Fayon F, Alonso B, Rouillon T, Schnitzler V, Verron E, *et al.* Reaction of zoledronate with beta-tricalcium phosphate for the design of potential drug device combined systems. Chem Mater 2008;20(1):182-91.

[203] Stigter M, Bezemer J, de Groot K, Layrolle P. Incorporation of different antibiotics into carbonated hydroxyapatite coatings on titanium implants, release and antibiotic efficacy. J Control Release 2004;99(1):127-37.
[204] Misra DN. Interaction of some alkali metal citrates with hydroxyapatite - Ion-exchange adsorption and role of charge balance. Colloid Surface A 1998;141(2):173-9.
[205] Barroug A, Fastrez J, Lemaitre J, Rouxhet P. Adsorption of succinylated lysozyme on hydroxyapatite. J Colloid Interf Sci 1997;189(1):37-42.
[206] Barroug A, Lernoux E, Lemaitre J, Rouxhet PG. Adsorption of catalase on hydroxyapatite. J Colloid Interf Sci 1998;208(1):147-52.
[207] Dong XL, Wang Q, Wu T, Pan HH. Understanding adsorption-desorption dynamics of BMP-2 on hydroxyapatite (001) surface. Biophys J 2007;93(3):750-9.
[208] Barroug A, Kuhn LT, Gerstenfeld LC, Glimcher MJ. Interactions of cisplatin with calcium phosphate nanoparticles: *in vitro* controlled adsorption and release. J Orthopaed Res 2004;22(4):703-8.
[209] Gautier H, Daculsi G, Merle C. Association of vancomycin and calcium phosphate by dynamic compaction: *in vitro* characterization and microbiological activity. Biomaterials 2001;22(18):2481-7.
[210] Gautier H, Merle C, Auget JL, Daculsi G. Isostatic compression, a new process for incorporating vancomycin into biphasic calcium phosphate: comparison with a classical method. Biomaterials 2000;21(3):243-9.
[211] Alam MI, Asahina I, Ohmamiuda K, Takahashi K, Yokota S, Enomoto S. Evaluation of ceramics composed of different hydroxyapatite to tricalcium phosphate ratios as carriers for rhBMP-2. Biomaterials 2001;22(12):1643-51.
[212] Liu YL, Enggist L, Kuffer AF, Buser D, Hunziker EB. The influence of BMP-2 and its mode of delivery on the osteoconductivity of implant surfaces during the early phase of osseointegration. Biomaterials 2007;28(16):2677-86.
[213] Peter B, Pioletti DP, Laib S, Bujoli B, Pilet P, Janvier P, *et al.* Calcium phosphate drug delivery system: influence of local zoledronate release on bone implant osteointegration. Bone 2005;36(1):52-60.
[214] Lebugle A, Rodrigues A, Bonnevialle P, Voigt JJ, Canal P, Rodriguez F. Study of implantable calcium phosphate systems for the slow release of methotrexate. Biomaterials 2002;23(16):3517-22.
[215] Midy V, Hollande E, Rey C, Dard M, Plouet J. Adsorption of vascular endothelial growth factor to two different apatitic materials and its release. J Mater Sci-Mater M 2001;12(4):293-8.
[216] Yoshinari M, Oda Y, Ueki H, Yokose S. Immobilization of bisphosphonates on surface modified titanium. Biomaterials 2001;22(7):709-15.
[217] McLeod K, Kumar S, Smart RSC, Dutta N, Voelcker NH, Anderson GI, *et al.* XPS and bioactivity study of the bisphosphonate pamidronate adsorbed onto plasma sprayed hydroxyapatite coatings. Appl Surf Sci 2006;253(5):2644-51.
[218] Errassifi F, Sarda S, Barroug A, Lebugle A, Legrouri A, Sfihi HA, *et al.* editors. Adsorption of a bisphosphonate onto apatitic calcium phosphates: Role of surface characteristics. Third Conference on the Valorization of Phosphates and Phosphorous compounds (COVAPHOS III) 2009; Marrakech.
[219] Iafisco M, Palazzo B, Marchetti M, Margiotta N, Ostuni R, Natile G, *et al.* Smart delivery of antitumoral platinum complexes from biomimetic hydroxyapatite nanocrystals. J Mater Chem 2009;19(44):8385-92.

[220] Palazzo B, Iafisco M, Laforgia M, Margiotta N, Natile G, Bianchi CL, *et al.* Biomimetic hydroxyapatite-drug nanocrystals as potential bone substitutes with antitumor drug delivery properties. Adv Funct Mater 2007;17(13):2180-8.

[221] Neves M, Gano L, Pereira N, Costa MC, Costa MR, Chandia M, *et al.* Synthesis, characterization and biodistribution of bisphosphonates Sm-153 complexes: correlation with molecular modeling interaction studies. Nucl Med Biol 2002;29(3):329-38.

[222] Coxon FP, Thompson K, Rogers MJ. Recent advances in understanding the mechanism of action of bisphosphonates. Curr Opin Pharmacol 2006;6(3):307-12.

[223] Rogers MJ, Watts DJ, Russell RGG. Overview of bisphosphonates. Cancer 1997;80(8):1652-60.

[224] Nancollas GH, Tang R, Phipps RJ, Henneman Z, Gulde S, Wu W, *et al.* Novel insights into actions of bisphosphonates on bone: Differences in interactions with hydroxyapatite. Bone 2006;38(5):617-27.

[225] Russell RGG. Determinants of structure-function relationships among bisphosphonates. Bone 2007;40(5):S21-S5.

[226] Lalla S, Hothorn LA, Haag N, Bader R, Bauss F. Lifelong administration of high doses of ibandronate increases bone mass and maintains bone quality of lumbar vertebrae in rats. Osteoporosis Int 1998;8(2):97-103.

[227] Balena R, Toolan BC, Shea M, Markatos A, Myers ER, Lee SC, *et al.* The Effects of 2-Year Treatment with the Aminobisphosphonate Alendronate on Bone Metabolism, Bone Histomorphometry, and Bone Strength in Ovariectomized Nonhuman-Primates. J Clin Invest 1993;92(6):2577-86.

[228] Monkkonen J, Simila J, Rogers MJ. Effects of tiludronate and ibandronate on the secretion of proinflammatory cytokines and nitric oxide from macrophages *in vitro*. Life Sci 1998;62(8):Pl95-Pl102.

[229] Elliott SN, McKnight W, Davies NM, MacNaughton WK, Wallace JL. Alendronate induces gastric injury and delays ulcer healing in rodents. Life Sci 1997;62(1):77-91.

[230] Kos M, Kuebler JF, Luczak K, Engelke W. Bisphosphonate-related osteonecrosis of the jaws: a review of 34 cases and evaluation of risk. J Craniomaxillofac Surg 2010;38(4):255-9.

[231] Kos M, Luczak K. Bisphosphonates promote jaw osteonecrosis through facilitating bacterial colonisation. Bioscience Hypotheses 2009;2(1):34-6.

[232] Hoffman A, Stepensky D, Ezra A, Van Gelder JM, Golomb G. Mode of administration-dependent pharmacokinetics of bisphosphonates and bioavailability determination. Int J Pharm 2001;220(1-2):1-11.

[233] Marie PJ, Ammann P, Boivin G, Rey C. Mechanisms of action and therapeutic potential of strontium in bone. Calcified Tissue Int 2001;69(3):121-9.

[234] Ringe JD. Strontium ranelate: an effective solution for diverse fracture risks. Osteoporos Int 2010;21 Suppl 2:S431-6.

[235] Chen QZ, Wong CT, Lu WW, Cheung KM, Leong JC, Luk KD. Strengthening mechanisms of bone bonding to crystalline hydroxyapatite *in vivo*. Biomaterials 2004;25(18):4243-54.

[236] Ni GX, Lu WW, Xu B, Chiu KY, Yang C, Li ZY, *et al.* Interfacial behaviour of strontium-containing hydroxyapatite cement with cancellous and cortical bone. Biomaterials 2006;27(29):5127-33.

[237] Hayashi K, Uenoyama K, Matsuguchi N, Nakagawa S, Sugioka Y. The affinity of bone to hydroxyapatite and alumina in experimentally induced osteoporosis. J Arthroplasty 1989;4(3):257-62.

[238] Ni GX, Lu WW, Chiu KY, Li ZY, Fong DY, Luk KD. Strontium-containing hydroxyapatite (Sr-HA) bioactive cement for primary hip replacement: an *in vivo* study. J Biomed Mater Res B Appl Biomater 2006;77(2):409-15.

[239] Chen D, Fu Y, Gu G, Qian Y, Gu Y. Preparation and solubility of the solid solution of strontium substituted hydroxyapatite. Chin J Biomed Eng. 2001;20(3):278-80.

[240] Wong KL, Wong CT, Liu WC, Pan HB, Fong MK, Lam WM, *et al.* Mechanical properties and *in vitro* response of strontium-containing hydroxyapatite/polyetheretherketone composites. Biomaterials 2009;30(23-24):3810-7.

[241] Gough JE, Jones JR, Hench LL. Nodule formation and mineralisation of human primary osteoblasts cultured on a porous bioactive glass scaffold. Biomaterials 2004;25(11):2039-46.

[242] Wong CT, Chen QZ, Lu WW, Leong JCY, Chan WK, Cheung KMC, *et al.* Ultrastructural study of mineralization of a strontium-containing hydroxyapatite (Sr-HA) cement *in vivo*. J Biomed Mater Res A 2004;70A(3):428-35.

[243] Romieu G, Garric X, Munier S, Vert M, Boudeville P. Calcium-strontium mixed phosphate as novel injectable and radio-opaque hydraulic cement. Acta Biomater 2010;6(8):3208-15.

[244] Panzavolta S, Torricelli P, Sturba L, Bracci B, Giardino R, Bigi A. Setting properties and *in vitro* bioactivity of strontium-enriched gelatin-calcium phosphate bone cements. J Biomed Mater Res A 2008;84(4):965-72.

[245] Xue WC, Hosick HL, Bandyopadhyay A, Bose S, Ding CX, Luk KDK, *et al.* Preparation and cell-materials interactions of plasma sprayed strontium-containing hydroxyapatite coating. Surf Coat Tech 2007;201(8):4685-93.

[246] Guo DG, Xu KW, Han Y. The influence of Sr doses on the *in vitro* biocompatibility and *in vivo* degradability of single-phase Sr-incorporated HAP cement. J Biomed Mater Res A 2008;86A(4):947-58.

[247] Landi E, Sprio S, Sandri M, Celotti G, Tampieri A. Development of Sr and CO3 co-substituted hydroxyapatites for biomedical applications. Acta Biomater 2008;4(3):656-63.

CHAPTER 3

Sonochemically Synthesized Materials for Biomedical Applications

Muthupandian Ashokkumar*

School of Chemistry, University of Melbourne, Parkville, VIC 3010, Australia

Abstract: The interaction between ultrasound and bubbles in liquids has led to the development of a number of applications. The scattering of sound waves by bubbles has been utilized in ultrasound contrast imaging. The absorption of sound energy by bubbles that exist in liquids can lead to acoustic cavitation, the growth and violent collapse of microbubbles. Extreme temperature conditions and highly reactive radicals are generated during acoustic cavitation, which have been utilized in a number of processes that include the synthesis of a variety of nanomaterials. Acoustic cavitation also generates strong shear forces and streaming effects in liquids that have been used for the deactivation of pathogens and modification of the properties of bioactive molecules. This chapter aims to provide an overview of the recent work that has been carried out in our laboratory that includes the ultrasonic and sonochemical synthesis of biofunctional metal nanoparticles, polymers, and microspheres.

Keywords: Sound waves, acoustic cavitation, sonochemical synthesis, ultrasonic synthesis, microspheres.

1. INTRODUCTION

The scientific community is continuously exploring various new methodologies for synthesizing functional materials. Conventionally, chemical precursors are mixed with appropriate catalysts during the synthetic process. While some synthetic reactions are driven under normal reaction conditions, several reactions require extreme temperature conditions [1-3]. It has also been shown that chemical and physical properties of the newly synthesized materials strongly depend upon the synthetic methodology [4-6]. For example, the addition of starch during the sonochemical synthesis of ZnO nanoparticles was shown to affect the morphology of the ZnO particles [4]. Tavakoli *et al.* [5] have reviewed the effect of synthetic procedures on the properties of metal nanoparticles. In this review,

***Address correspondence to Muthupandian Ashokkumar:** School of Chemistry, University of Melbourne, Parkville, VIC 3010, Australia; E-mail: masho@unimelb.edu.au

Lia Rimondini, Claudia Bianchi and Enrica Vernè (Eds)

the fundamental principles, advantages and disadvantages of several synthetic methods, such as, microemulsion, hydrothermal, sol-gel, sonochemical and microbial processes, have been discussed.

The demand for new materials for biomedical applications is increasing in recent years [7]. For example, advanced medical diagnostic imaging processes, such as MRI, sonography, *etc.*, require new biodegradable materials in order to achieve high resolution and excellent contrast [8-10]. Paradossi *et al.* [8] have synthesized poly(vinyl alcohol) based microstructures for biomedical applications. Grinstaff and Suslick [9] have synthesized air-filled microbubbles for sonographic applications. The ultrasonically generated superoxide radicals have been used to cross link serum albumins that resulted in the formation of stable air-filled microbubbles. Wang and Li [10] synthesized biocompatible polymeric porous materials for biomedical applications.

Among various new synthetic methodologies established, ultrasound-assisted and sonochemically synthesized materials show specific properties that are different to those synthesized by conventional methods [11-15]. Suslick and Price [11] have reviewed the applications of ultrasound to materials chemistry. In this review, a detailed discussion of acoustic cavitation process followed by several examples of ultrasonic synthesis of polymers, metals and biomaterials are provided. Ultrasound-assisted synthetic process involves the synthesis of materials by conventional methods in the presence of an ultrasonic field, where the mechanical (vibration and turbulence) effects enhance the mass transfer processes resulting in an increase in the rate of reaction and, in some cases, materials with different properties [16-18]. In a recent investigation on the effect of ultrasound on sonocrystallization of calcite, Sonawane *et al.* [17] showed that the rate of crystallization of calcite was substantially enhanced in the presence of ultrasound. In addition, the properties of the calcite nanoparticles were significantly affected by the sonication process. Similarly, the functional properties of starch granules were affected by sonication [18]. The viscosities of corn and rice starch pastes were significantly lowered by sonication at low frequency ultrasound. The physical effects of cavitation were found to be responsible for the viscosity reduction. While the physical effects are responsible for the changes observed in the previous examples, the sonochemical synthetic process involves the primary and secondary radicals, generated during acoustic cavitation process, in addition to the mechanical effects. Several studies are available [19-25] on the sonochemical synthesis of a variety of materials.

Having discussed that ultrasound and sonochemistry influence the properties of the materials synthesized, it is now necessary to understand the fundamental processes involved when ultrasound is applied into a reaction medium.

2. ACOUSTIC CAVITATION

Ultrasound, sound waves of frequencies beyond human hearing (typically > 18 kHz), is known to the general community as a scanning/imaging tool. For example, ultrasound has been used in diagnostic medicine for imaging purposes [26]. In order to enhance the quality of the images, stabilized gas bubbles (*ultrasound contrast agents*) are injected into the imaging region [26]. Since gas bubbles reflect sound waves, the contrast around the imaging body/section is significantly enhanced. However, under specific experimental conditions, gas bubbles can absorb the sound energy leading to a process known as *acoustic cavitation*. Acoustic cavitation has been used in a number of areas that include therapeutic medicine, contrast imaging, synthesis of nanomaterials, production of nanoemulsions, processing of food materials, waste-water treatment, *etc.* [26-31].

The term “cavitation” literally means the “creation” of a cavity. However, it is commonly associated with the formation, growth and collapse of gas/vapor filled bubbles (cavities) in a liquid. If the cavitation is generated by sound (acoustic) waves, the process is referred to as *acoustic cavitation*. Ultrasound, when passing through a liquid medium causes mechanical vibration in the liquid. If the liquid medium contains dissolved gas bubbles, which will be the case under normal conditions, they can be grown and collapsed by the action of the ultrasound. The combined effects of mechanical vibration, acoustic streaming and cavitation enhance the mass transfer processes. Sonoprocessing refers to the applications of ultrasound where the physical effects of ultrasound have been used: examples include emulsification, extraction and cleaning [29-31].

The collapse of the acoustic cavitation bubbles is also a near adiabatic process and generates temperatures in the range of thousands of degrees within the bubbles for a short period of time. Under this extreme temperature conditions, highly reactive radicals are generated. These radicals have been used to initiate a number of chemical reactions that include the synthesis of nanomaterials, polymers, degradation of organic pollutants, *etc.* [32-35]. In addition to generating the reactive radicals, the high temperature conditions also induce light emission from excited state molecules [36-38]. The light emission from the cavitation bubbles, sonoluminescence, is shown in Fig. **1**. Under specific experimental conditions, a hot-plasma is also generated within the cavitation bubbles. During the bubble collapse, other high energy events such as the generation of shock waves and microjets occur (Fig. **1**).

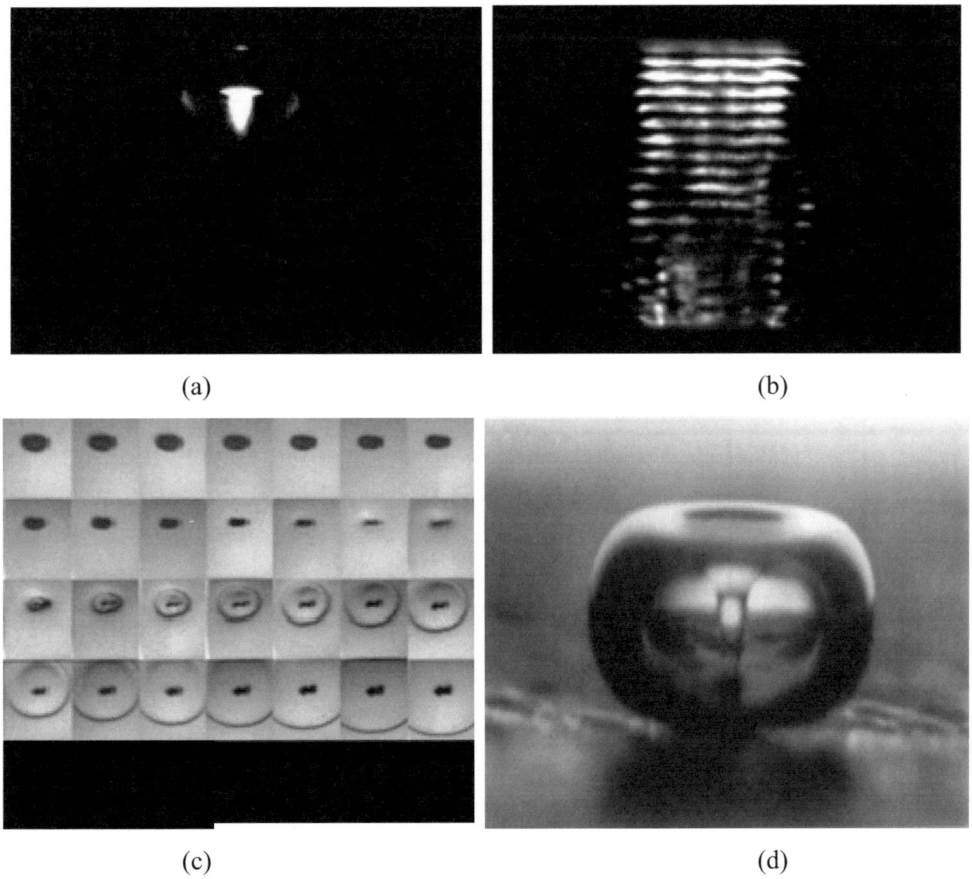

Figure 1: (a) Sonoluminescence from cavitation bubbles generated using a 20 kHz horn-type reactor where most of the acoustic energy is delivered at the tip of the horn. This leads to the "localized" cavitation activity at the tip of the horn. This localized activity is important for emulsification and extraction processes. (b) Sonoluminescence from cavitation bubbles generated

using a 440 kHz plate-type transducer/reactor where the cavitation activity is homogeneously distributed throughout the liquid due to standing wave formation.[5] The homogeneity is important for radical-induced chemical reactions. (c) High-energy shockwave generated (picture courtesy of Prof. Lauterborn University of Goettingen) due to the symmetric collapse of a bubble. (d) A microjet (speed ~100 m/s) generated (picture courtesy of Prof. Crum, University of Washington) due to the asymmetric collapse of a bubble.

3. SONOCHEMISTRY

Sonochemistry refers to the field of chemistry where chemical reactions are induced by the physical and chemical effects of acoustic cavitation [12]. As mentioned in the introduction section, very high temperatures (about 2000 - > 5000 K) are generated within the cavitation bubbles during the collapse. As discussed earlier, the physical effects (Fig. **1**) generated may influence the rate of chemical reactions due to enhanced mass transport processes. However, in most of the sonochemical reactions, the reactive chemical species directly involve in the chemical reactions, which is discussed below. The interior of the cavitation bubbles is filled with water vapor and gas molecules, such as, N_2 and O_2 (in air-saturated water) that evaporate/diffuse into the bubbles during bubble oscillations (growth phase). When subjected to the high temperature conditions generated during bubble collapse, the solvent and gas molecules undergo endothermic chemical reactions leading to the generation of a variety of highly reactive radicals. Reactions (1) – (4) show the formation of primary radicals ($^{\cdot}OH$ and $^{\cdot}H$) from water and the subsequent reactions of the primary radicals with solutes, such as, alcohols (RH) to generate secondary reducing radicals ($^{\cdot}R$) that can be used, for example, to reduce metal ions to form metal nanoparticles (*US* represents ultrasound). Reaction (4) shows the formation of gold (Au) by the reduction of Au(III) ions.

$$H_2O \xrightarrow{US} {}^{\cdot}OH + {}^{\cdot}H \quad \textbf{(1)}$$

$$RH + {}^{\cdot}OH\,({}^{\cdot}H) \xrightarrow{US} {}^{\cdot}R + H_2O\,(H_2) \quad \textbf{(2)}$$

$$RH \rightarrow \text{pyrolysis radicals and unstable products} \quad \textbf{(3)}$$

$$Au(III) + \text{Reducing species } ({}^{\cdot}H,\ {}^{\cdot}R,\ etc.) \rightarrow Au(0) \quad \textbf{(4)}$$

4. SONOCHEMICALLY SYNTHESIZED METAL PARTICLES IN BIOMEDICAL APPLICATIONS

Gold nanoparticles have been shown to possess antimicrobial activities. Several studies have been reported on the sonochemical synthesis of gold nanoparticles [25, 39-40]. The synthesis of metal nanoparticles using the sonochemical method has several advantages. We have developed a sonochemical synthetic methodology for controlling the size and size distribution of metal nanoparticles [39, 40]. Okitsu *et al.* [39] have shown that the size and size distribution of the gold nanoparticles are strongly dependent upon the rate of production of primary/secondary radicals. It is also possible to synthesize core/shell or alloy bimetallic nanoparticles that can be used in fuel cells and other devices [41, 42]. We have shown that Pt/Ru core/shell nanoparticles can be sonochemically synthesized [41]. Fig. **2** shows the transmission electron micrographs (TEM) of the sonochemically synthesized Pt/Ru core/shell and alloy particles. In a sequential reduction method, Pt nanoparticles are first produced followed by Ru(III) reduction. Under specific experimental conditions, Pt/Ru alloy particles can be produced.

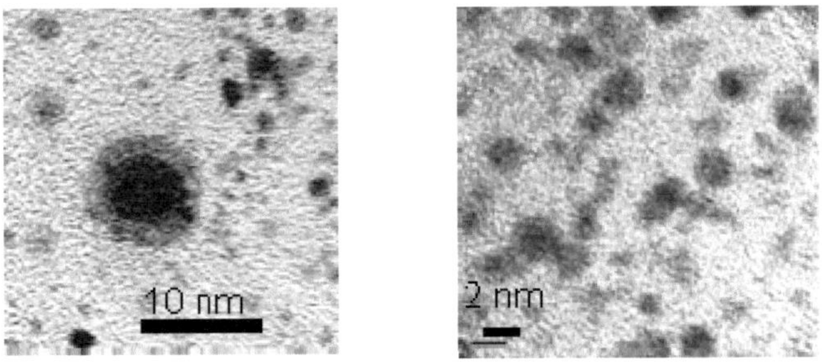

Figure 2: TEM images of the sonochemically synthesized bimetallic particles. Simultaneous and sequential reduction of platinum and ruthenium ions under different experimental conditions led to the formation of either core shell or alloy bimetallic nanoparticles [41].

We have recently shown that the sonochemically synthesized gold nanoparticles could be used in DNA biosensing [43]. This application stems from the unique optical and electronic properties of metal nanoparticles and their use in biomedical applications. We have observed the electronic interactions of sonochemically synthesized Au nanoparticles with Adenine Thymine-DNA (AT-DNA) and Guanine Cytosine-DNA (GC-DNA) using the photoluminescence properties.

In this study [43], about 15 nm average sized Au nanoparticles of narrow size distribution were synthesized by sonicating a solution of $HAuCl_4$ in water in the presence of polyvinylpyrrolidone as the stabilizing agent and 1-propanol as the reducing agent. Fig. **3** shows the TEM micrograph of the sonochemically synthesized Au nanoparticles.

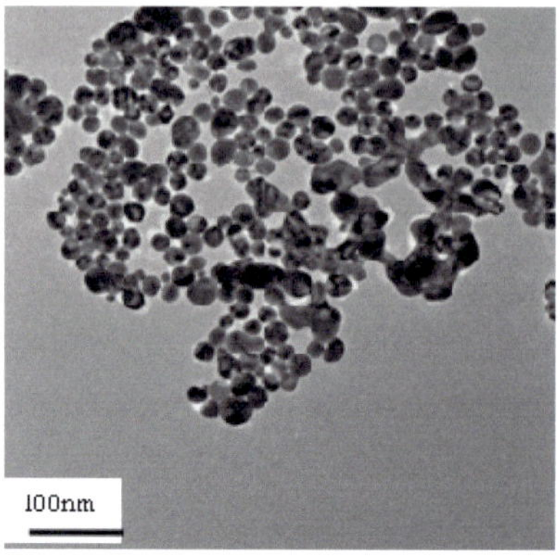

Figure 3: TEM image of sonochemically synthesized Au nanoparticles [43].

The Au nanoparticles were then mixed with AT-DNA and GC-DNA and their interactions were monitored by measuring their photoluminescence characteristics. The Au-DNA interactions were monitored by the changes in the gold emission band. As shown in Fig. **4**, the emission maximum of the photoluminescence spectrum of gold nanoparticles showed a red shift for both AT-DNA and GC-DNA samples.

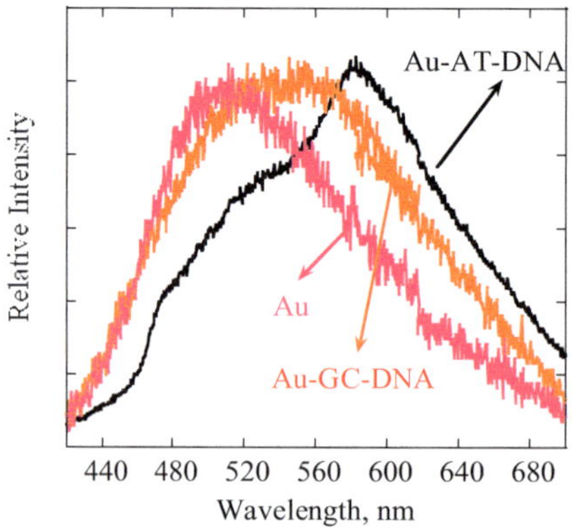

Figure 4: Normalized photoluminescence spectra (measured by confocal photoluminescence technique) of gold nanoparticles in the absence and presence of AT-DNA or GC-DNA [43].

The results also indicated that the metal nanoparticles-DNA interaction resulted in a significant change in the fluorescence characteristics for AT-DNA compared to GC-DNA. It is known that the metal-DNA interaction is Lewis acid-base interaction [43]. Au nanoparticles interact strongly through nitrogen atoms of the DNA. The absence of free nitrogen containing moiety in thymine base of the AT-DNA results in the interaction between the Au nanoparticles and the oxygen atoms of the carboxyl functional group. This difference in the interaction between Au nanoparticles and the N- and O- functional substituents of AT- and GC DNAs is responsible for the difference in the photoluminescence properties of these two systems shown in Fig. **4**. This study constitutes a preliminary step to using the sonochemically synthesized metal nanoparticles for biomedical imaging applications.

The interaction of sonochemically synthesized gold nanoparticles with serum albumins has also been investigated [44]. The sonochemically synthesized gold nanoparticles (Fig. **3**) were mixed with serum albumins and their interactions were evaluated using UV-visible and fluorescence spectroscopic techniques. It is known that the gold nanoparticles complex with serum albumins as per the following chemical equilibrium (Reaction 5):

$$\mathbf{SA + Au \xrightleftharpoons{K_{app}} [SA{-}Au]} \quad \mathbf{(5)}$$

The absorption spectra (Fig. **5**) clearly show that the absorption maxima of BSA and HSA shift to lower wavelengths due to the addition of sonochemically synthesized gold nanoparticles. Using Benesi-Hildebrand plots shown in Fig. **5** [44], association constants (K_{app}) for the systems, Au-BSA and Au-HSA were estimated to be 1.13 x 10^4 M^{-1} and 1.66 x 104 M^{-1}, respectively. These association constants indicate that HSA showed a greater affinity for gold nanoparticles. This observation was also further supported by the fluorescence quenching measurements. The fluorescence from the serum albumins was quenching by the gold nanoparticles due to the excited state energy transfer mechanism. From the Stern-Volmer quenching plots, the fluorescence quenching rate constants were found to be 6.58 x 10^{12} M^{-1} s^{-1} and 4.57 x 10^{12} M^{-1} s^{-1}, for BSA and HSA, respectively.

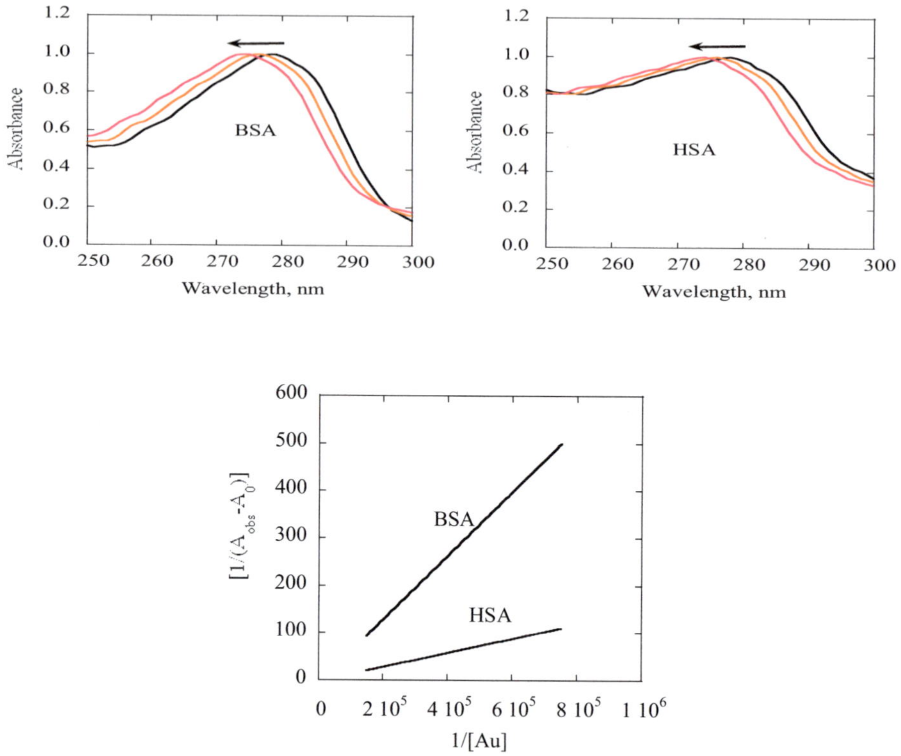

Figure 5: The changes to the absorption spectra of (a) BSA (top left) and (b) HSA (top right) with increasing addition of sonochemically synthesized gold nanoparticles. The Benesi-Hildebrand plots for both Au-BSA and Au-HSA are shown in the bottom [44].

Thus, the absorption and fluorescence studies indicate that the interaction of gold nanoparticles with BSA and HSA are significantly different. This preliminary study on the interaction between sonochemically synthesized nanoparticles and the serum albumins is important since both serum albumins and gold can be used as potential drug carriers [44].

5. SONOCHEMICALLY SYNTHESIZED MACROMOLECULES IN BIOMEDICAL APPLICATIONS

The physical and chemical effects generated during acoustic cavitation have been used for emulsion polymerization [45, 46]. Emulsion droplets of an organic monomer are generated in an aqueous solution due to the shear forces of acoustic cavitation. The primary and secondary radicals formed during cavitation initiate the polymerization process in each of the droplets. Polymer latex particles of less than 100 nm can be synthesized using this method.

Recently, we have developed a simple and efficient technique for the encapsulation of magnetite nanoparticles within host latex particles [47]. During the sonochemical emulsion polymerization of butyl methylmethacrylate (BMA), pre-synthesized magnetite (iron oxide) nanoparticles were added along with the monomer. The magnetite particles were encapsulated within the poly(BMA) latex particles during the polymerization process. Fig. **6** shows that the magnetite loaded latex particles possess a strong magnetic activity. The ultrasound initiated polymerization/incorporation procedure represents an effective means of producing a range of nanocomposites consisting of multiple combinations of different polymers and encapsulation materials. These nanocomposites can, in principle, be used in various biomedical, cosmetics and food processing applications for the controlled release of pharmaceuticals and flavors.

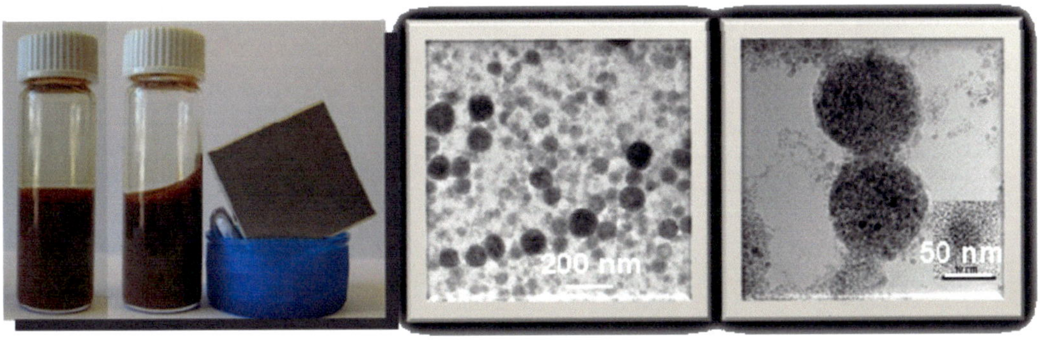

Figure 6: Left: Paramagnetic properties of a magnetite/poly(butymethacrylate) (poly(BMA)) latex nanoparticles composite suspension: the photograph shows that a strong permanent magnet placed next to the fluid containing the latex particles draws the fluid upwards. Middle and right: TEM images of the magnetite nanoparticles in poly(BMA) latex nanoparticles at different degrees of magnification 47].

6. SONOCHEMICALLY SYNTHESIZED PROTEIN MICROSPHERES IN BIOMEDICAL APPLICATIONS

In addition to cross-linking conventional monomers to synthesize functional polymers, bioactive macromolecules can also be sonochemically cross-linked [48-

51] to generate micro-encapsulation agents. In recent years, many researchers focus their attention on the synthesis of microbubbles and microstructures for biomedical applications [7, 50-56]. The design of drug loaded microbubbles for ultrasound triggered delivery applications has been highlighted by Lentacker *et al.* [7]. Cavalieri *et al.* [50] have reviewed various methods of preparation of multifunctional microbubbles for ultrasound image-guided cancer therapy applications. This review highlights the importance of the preparation methods in controlling the biofunctional properties of microbubbles.

The sonochemical synthesis of air-filled and oil-filled microspheres using a bioactive enzyme, lysozyme, as the shell material has been achieved recently [51]. In this synthetic process, the physical forces generated during acoustic cavitation were used to generate microbubbles or oil droplets coated with cross-linked enzyme molecules. The oxidative radicals generated during cavitation were used to generate inter-molecular disulfide bonds leading to the crosslinking of the enzyme molecules. This ultimately led to a stable suspension of air bubbles or oil droplets stabilized by the cross-linked enzyme shell. Fig. **7** shows the experimental procedure followed to synthesize oil-filled and air-filled microspheres [56]. For air-filled microbubbles, the horn was positioned at the aqueous solution-air interface. For oil-filled microspheres, the horn was positioned at the aqueous solution-oil interface.

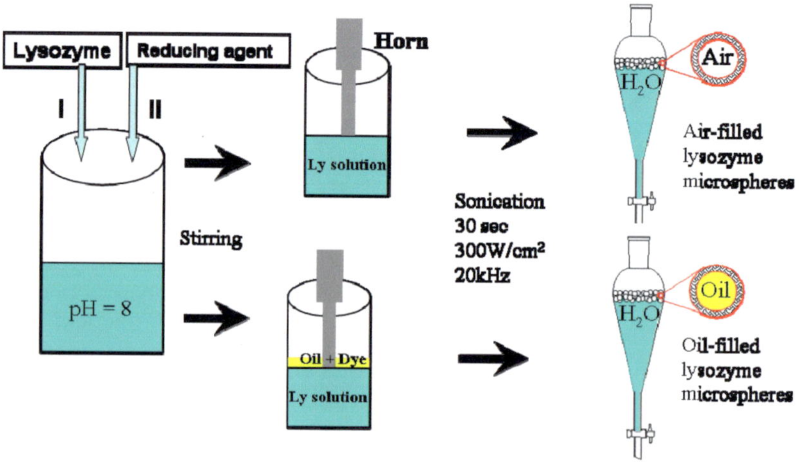

Figure 7: Experimental details of the ultrasonic synthesis of air-filled and oil-filled protein microspheres.

The efficiency of the formation of microbubbles requires careful tailoring of the surface properties of the protein molecules. Most of the proteins are globular in nature where the functional moieties and hydrophobic chains are well hidden within the globule. The primary requirement for the formation of microbubbles is the adsorption of the protein at the bubble/solution interface. The proteins in their globular form are not surface active. In fact, when we attempted to synthesize protein microspheres in their native form, we were not successful. The first step that was required to successfully synthesize protein microspheres was to partially denature the proteins.

The partial denaturation process leads to two changes. The first change is the breaking of intramolecular disulfide bonds leading to the unfolding of the proteins which then aggregate due to the hydrophobic interactions between them. These small aggregates then migrate to the bubble interface as shown in the Scheme (Fig. **8**). In fact, the denaturation of the protein has to be carried out in a controlled manner. Conventionally, the denaturation of the proteins can be achieved by heating (thermal denaturation). However, this leads to the complete denaturation of the proteins which then cannot be cross-linked to form stable microspheres. This is in addition to the loss of the functional properties of the protein. Once the protein nano-aggregates are generated, the oxidative radicals generated during acoustic cavitation could be used to generate inter-molecular disulfide bonds leading to the formation of stable microbubbles as shown in Fig. **8**.

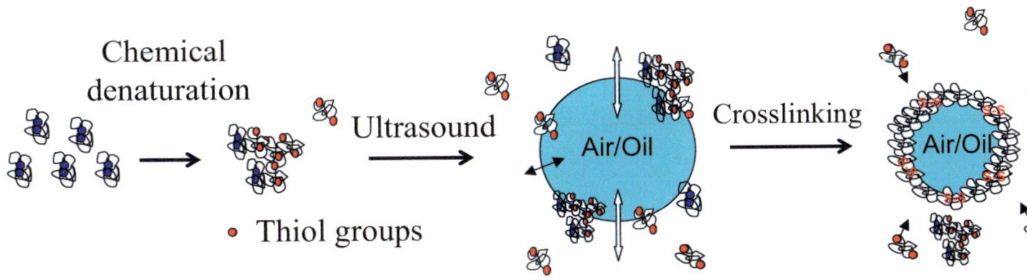

Figure 8: The schematic representation of the mechanism of ultrasonic microbubble formation [51].

Fig. **9** shows the uniform size distribution of the lysozyme microbubbles, synthesized by the ultrasonic method. These microbubbles can be used as ultrasound contrast agents in diagnostic medicine and drug delivery vehicles in therapeutic medicine. There is also a possibility that these microbubbles can be used to encapsulate high value food ingredients, such as food flavors and vitamins.

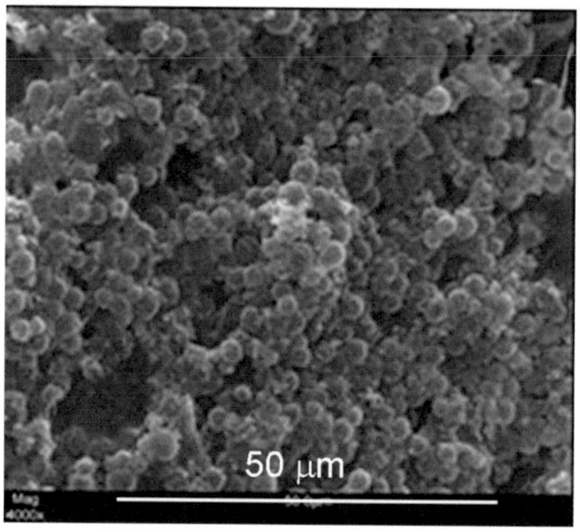

Figure 9: Scanning electron micrograph (SEM) of the ultrasonically synthesized lysozyme microbubbles.

It is well known that lysozyme is a potential bactericidal agent against both gram-negative and gram-positive bacteria [51]. We have shown by the turbidometric micrococcus leuteus test that the enzymatic activity of lysozyme is preserved after microbubble formation. The decrease in the absorption shown in Fig. **10** corresponds to the deactivation of the bacteria. As can be seen in Fig. **10a**, lysozyme molecules that are present on the surface of the microbubbles show a relatively slower activity. However, when the microbubbles are broken by high frequency ultrasound (Fig. **10b**), the deactivation rate is found to be similar to that of native lysozyme. The increased activity is due to the exposure of lysozyme molecules present in the inner side of the shell. This test clearly indicates that the sonochemical crosslinking of the proteins does not affect the functional properties of the proteins. Majority of the protein molecules are preserved in their native forms during the microbubble formation process.

We have also observed that the lysozyme-coated microbubbles are stable for several weeks and retain the enzymatic (antimicrobial) activity of the protein. The stability of these microbubbles after several weeks of preparation can be studied by a new sonoluminescence (SL) based technique [50]. In degassed water, SL can not be observed due to the absence of gas nuclei even after a few minutes of sonication.

The microbubbles, stored in water for several weeks can then be added to degassed water and the initial growth of SL can be restored. What this experiment indicates is that the microbubbles are broken in the high frequency ultrasonic field and the gas bubbles released from these microbubbles act as cavitation nuclei.

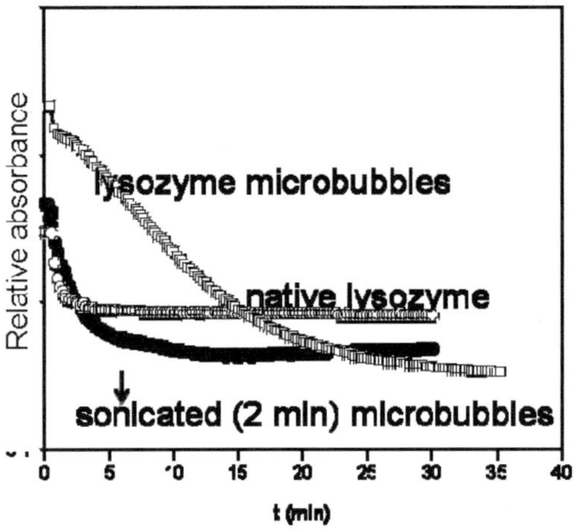

(a)

(b)

Figure 10: (a) Turbidometric micrococcus leuteus test for enzymatic activity [51]; (b) SEM photograph of air-filled lysozyme microbubbles after 2 minutes sonication at 363 kHz.

There is also another aspect to this experiment. If the microbubbles are loaded with drugs, then they can be selectively broken at specific sites and hence can be used in targeted drug delivery applications. While the synthetic procedure shown in Fig. **7** was used for encapsulating simple oil materials, a similar procedure is valid for encapsulating bioactive materials. In order to prove this concept, oil-filled lysozyme microspheres were synthesized using 3 different oils (sunflower oil, tetradecane and perfluorohexane) [57]. Fig. **11a** shows the perflurohexane-filled lysozyme microspheres. The average size of these microspheres is about 2 μm and the size distribution is relatively narrow. In order to explore the possibility of using these oil-filled microspheres for drug encapsulation and delivery applications, tetradecane containing an organic dye (nile red) was encapsulated within the microspheres. In Fig. **11b**, the fluorescence spectrum of nile red encapsulated within the lysozyme microspheres is shown.

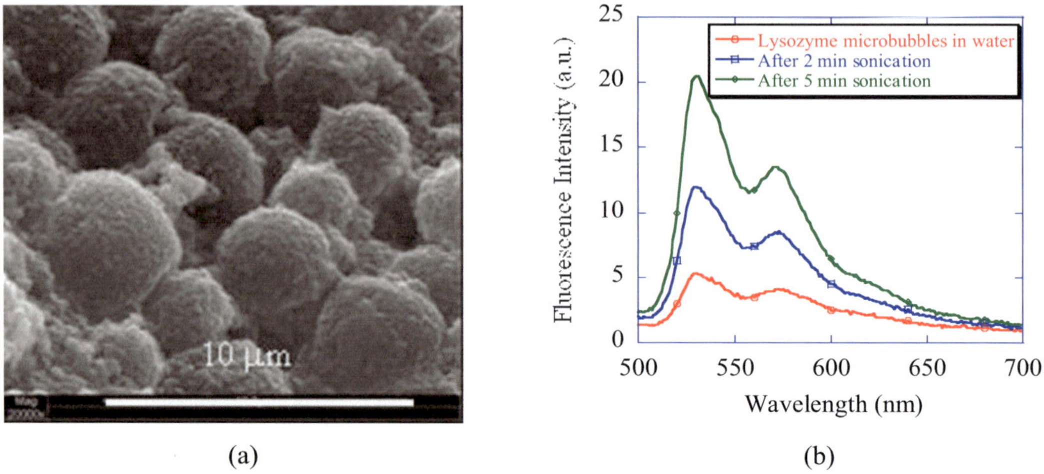

Figure 11: (a) SEM photograph of perfluorohexane-filled lysozyme microbubbles. (b) Fluorescence spectra of nile red/tetradecane encapsulated microspheres before and after sonication. The "lysozyme microbubbles in water" spectrum was recorded using washed microspheres dispersed in water. This solution was then sonicated at 363 kHz to release nile red/tetradecane.

It can be seen that there are two emission bands for nile red at wavelength maxima of 532 nm and 573 nm. The intensity ratio between these two bands is about 1.2 when nile red is dissolved in water. This ratio changes to about 1.5 in pure tetradecane. The fluorescence spectrum observed with washed microspheres in water shows a ratio of about 1.2 between the two emission bands observed (Fig. **11b**). However, this ratio changes to 1.5 when the dye/oil is released into the bulk solution by ultrasonic breaking of the microspheres. In addition to this observation, we have also noticed that the intensity of fluorescence also increases with an increase in the sonication time indicating the breakage of more microspheres with increasing sonication time. This is a clear indication that the lysozyme microspheres could be used as potential drug delivery agents.

7. CONCLUSIONS

The acoustic cavitation generated physical and chemical effects have been shown to generate functional materials that are useful in biomedical applications. The sonochemically generated gold nanoparticles interact with bioactive molecules such as DNA and serum albumins in addition to possessing antimicrobial properties. Both conventional and cross linked bio polymers are shown to be useful in biomedical applications such as drug delivery, therapeutic and imaging. The fundamental understanding and control of the acoustic cavitation processes have provided the opportunities for such developments. The research work discussed in this chapter is only preliminary in nature. However, the experimental data discussed provide valuable information that can be extended to synthesize a variety of tailor-made materials for biomedical applications.

ACKNOWLEDGEMENTS

Declared none.

CONFLICT OF INTEREST

The author(s) confirm that this chapter content has no conflict of interest.

REFERENCES

[1] Davies PK, Jacobson AJ, Torardi CC, Vanderah TA. High pressure synthesis of solid state materials. Solid-State Chem Inorg Mater 1997; 453: 141-6.
[2] Yang GW. Laser ablation in liquids: applications in the synthesis of nanocrystals. Prog Mater Sci 2007; 52: 648-98.
[3] Le Codec Y, Solozhenko V, Klotz S, Mezouar M. Synthesis of advanced materials under extreme conditions: *et al.* studies by X-ray diffraction with synchrotron radiation. Actualite Chimique 2008; 322: 27-34.
[4] Mishra P, Yadava RS, Pandey AC. Starch assisted sonochemical synthesis of flower-like ZnO nanostructure. Digest J Nanomater Biostr 2009; 4: 193-8.
[5] Tavakoli A, Sohrabi M, Kargari A. A review of methods for synthesis of nanostructured metals with emphasis on iron compounds. Chem Pap 2007; 61: 151-70.
[6] Applerot G, Abu-Muck R, Irzh A, *et al.* Decorating parylene-coated glass with ZnO nanoparticles for antibacterial applications: a comparative study of sonochemical, microwave and microwave-plasma coating routes. ACS Appl Mater & Interfaces 2010; 2: 1052-9.
[7] Lentacker I, De Smedt C, Sanders NN. Drug loaded microbubble design for ultrasound triggered delivery. Soft Matter 2009; 5: 2162-70.
[8] Paradossi G, Cavalieri F, Chiessi E, Ponassi V, Martorana V. Tailoring of physical and chemical properties of macro- and microhydrogels based on telechelic PVA. Biomacromol 2002; 3: 1255-62.
[9] Grinstaff MW, Suslick KS. Air-filed proteneous microbubbles-synthesis of an echo-contrast agent. Proc Nat Ac Sci USA 1991; 88: 7708-10.
[10] Wang H, Li W. Selective ultrasonic foaming of polymer for biomedical applications. J Manufact Sci & Eng Trans ASME 2008; 130: 021004.
[11] Suslick KS, Price GJ. Applications of ultrasound to materials chemistry. Ann Rev Mater Sci 1999; 29: 295-326.
[12] Ashokkumar M, Mason T. Sonochemistry. Kirk-Othmer Encyclopedia of Chemical Technology, John Wiley & Sons, 2007, DOI: 10.1002/0471238961. 1915141519211912.a01.pub2, Article Online Posting Date: October 19, 2007.

[13] Theerdhata S, Vitta S, Bahdur D. Magnetic nanoparticles through sonochemistry. Mater Techn 2008; 23: 88-93.

[14] Saez V, Gonzalez-Garcia J, Marken F. Active catalysts of sonoelectrochemically prepared iorn metal nanoparticles for the electroreduction of chloroacetates. Proc Int Cong Ultrasonics 2010; 3: 105-9

[15] Bang JH, Suslick KS. Applications of ultrasound to the synthesis of nanostructured materials. Adv Mater 2010; 22: 1039-59.

[16] Ashokkumar M, Grieser F. Ultrasound assisted chemical processes. Rev Chem Eng 1999; 15: 41-83.

[17] Sonawane S, Shirsath R, Khanna K, *et al.* An innovative method for effective micro-mixing of CO_2 gas during synthesis of nano-calcite crystal using sonochemical carbonization. Chem Eng J 2008; 143: 308-13.

[18] Zuo J, Knoerzer K, Mawson R, *et al.* The pasting properties of waxy rice starch suspensions. Ultrason Sonochem 2009; 16: 462-8.

[19] Maeda Y, Okitsu K, Inoue H, *et al.* Preparation of nanoparticles by reducing intermediate radicals formed in sonolytical pyrolysis of surfactants. Res Chem Intermed 2004; 30: 775-784.

[20] Huang W, Liu S, Tang M, *et al.* Interaction between Ti-gel and silver, copper, as well as nickel compounds under ultrasound irradiation. Res Chem Intermed 2004; 30: 793-806.

[21] Ohayon E, Gedanken A. The application of ultrasound radiation of the synthesis of nanocrystalline metal oxide in a non-aqueous solvent. Ultrason Sonochem 2010; 17: 173-8.

[22] Ashokkumar M. Sonochemical synthesis of inorganic nanoparticles. Chapter 4: Advanced Wet-Chemical Synthetic Approaches to Inorganic Nanostructures, Ed: Cozzoli PD. Transworld Research Network 2008; p. 107-31.

[23] Cintas P, Luche JL. Green Chemistry-The sonochemical approach. Green Chem 1999; 1: 115-25.

[24] Current Trends in Sonochemistry. E. Price GJ. Royal Society of Chemistry, Cambridge, U.K. 1992; p. 34–49.

[25] Ashokkumar M, Grieser F. Sonochemical preparation of colloids. Encyclopedia of Surface and Colloid Science, Ed. Hubbard A. Marcel Dekker NY 2002: p. 4760-74.

[26] Ultrasound contrast agents: Basic principles and clinical applications. 2nd ed. Eds. Goldberg BB, Raichlen JS, Forsberg F. Martin Duntiz Ltd, UK 2001.

[27] Mason TJ, Peters D. Practical Sonochemistry, Power ultrasound uses and applications, 2nd ed., Ellis Horwood Publishers, Chichester, 2002.

[28] Suslick KS. Ultrasound: Its Chemical, Physical, and Biological Effects, VCH Publishers, New York, 1988.

[29] Balachandran S, Kentish E, Mawson R, Ashokkumar M, The use of ultrasound to enhance the supercritical extraction of ginger. Ultrason Sonochem 2006; 13: 471-9.

[30] Kentish SE, Wooster T, Ashokkumar M, *et al.* The use of ultrasonics for nano-emulsion preparation. Innov Food Sci Emerg Technol 2008; 9: 170-5.

[31] Muthukumaran S, Kentish S, Lalchandani S, *et al.* The optimization of ultrasonic cleaning procedures for dairy fouled ultrafiltration membranes. Ultrason Sonochem 2005; 12: 29-35.

[32] He Y, Vinodgopal K, Ashokkumar M, Grieser F. Sonochemical synthesis of ruthenium nanoparticles. Res Chem Intermed 2006; 32: 709-15.

[33] Teo BM, Prescott SW, Ashokkumar M, Grieser F. Ultrasound initiated miniemulsion polymerization of methacrylate monomers. Ultrason Sonochem 2008; 15: 89-94.

[34] Neppolian B, Doronila A, Grieser F, Ashokkumar M. Simple and efficient sonochemical method for the oxidation of arsenic(III) to arsenic(V). Env Sci Techn 2009; 43: 6793-98.

[35] Singla R, Grieser F, Ashokkumar M. Sonochemical degradation of martius yellow dye in aqueous solution. Ultrason Sonochem 2009; 16: 28-34.

[36] Ashokkumar M, Grieser F. The effect of surface active solutes on bubbles in an acoustic field. Phys Chem Chem Phys 2007; 9: 5631-43.

[37] Ashokkumar M, Grieser F. Single bubble sonoluminescence – A Chemists' overview. Chem Phys Chem 2004; 5: 439-48.

[38] Ashokkumar M, Lee J, Iida Y, *et al.* Spatial distribution of acoustic cavitation bubbles at different ultrasound frequencies. Chem Phys Chem 2010; 11: 1680-84.

[39] Okitsu K, Ashokkumar M, Grieser F. Sonochemical synthesis of gold nanoparticles in water: effects of ultrasound frequency. J Phys Chem B 2005; 109: 20673-5.

[40] Okitsu K, Teo BM, Ashokkumar M, Grieser F. Controlled growth of sonochemically synthesized gold seed particles in aqueous solutions containing surfactants. Aus J Chem 2005; 58: 667-70.

[41] Vinodgopal K, He Y, Ashokkumar M, Grieser F. Sonochemically prepared platinum-ruthenium bimetallic nanoparticles. J Phys Chem B 2006; 110: 3849-52.

[42] Anandan S, Grieser F, Ashokkumar M. Sonochemical synthesis of Au-Ag core-shell bimetallic nanoparticles. J Phys Chem C 2008; 112: 15102-5.

[43] Anandan S, Oh SD, Yoon M, Ashokkumar, M. Photoluminescence properties of Sonochemically synthesized gold nanoparticles for DNA biosensing. Spectrochim. Acta, A: Molecular and Biomolecular Spectroscopy 2010; 76: 191-6.

[44] Naveenraj S, Anandan S, Kathiravan A, *et al.* The interaction of sonochemically synthesized gold nanoparticles with serum albumins. J Pharm Biomed Anal 2010; 53: 804-810.

[45] Teo BM, Ashokkumar M, Grieser F. High Intensity Ultrasound initiated polymerization of butyl methacrylate in mini- and microemulsions. Macromolecules 2009; 42: 4479-83.

[46] Teo BM, Ashokkumar M, Grieser F. Microemulsion polymerization *et al.* high frequency ultrasound irradiation. J Phys Chem C 2008; 112: 5265-7.

[47] Teo BM, Chen F, Hatton TA, Grieser F, Ashokkumar M. A novel one-pot synthesis of magnetite latex nanoparticles by ultrasound irradiation. Langmuir 2009; 25, 2593-5.

[48] Suslick KS, Grinstaff MW, Kolbeck KJ, *et al.* Characterization of sonochemically prepared proteinaceous microspheres. Ultrason Sonochem 1994; 1: S65-8.

[49] Gedanken A. Preparation and properties of proteinaceous microspheres made sonochemically. Chem – A Eur J 2008; 14: 3840-53.

[50] Cavalieri F, Zhou M, Ashokkumar M. The design of multifunctional microbubbles for ultrasound image-guided caner therapy. Curr Top Med Chem 2010; 10: (in press).

[51] Cavalieri F, Ashokkumar M, Grieser F, Caruso F. Ultrasonic synthesis of stable, functional lysozyme microbubbles. Langmuir 2008; 24: 10078-83.

[52] Grigoriev D, Miller R, Shchuin D, Mohwald H. Interfacial assembly of partially hydrophobic silica nanoparticles induced by ultrasonic treatment. Small 2007; 3: 665-71.

[53] Krafft MP, Riess JG. Highly fluorinated amphiphiles and colloidal systems and their applications in the biomedical field. A contribution. Biochimie 1998; 80: 489-514.

[54] Tzvetkov G, Paradossi G, Tortora M, *et al.* Water-dispersible PVA-based dry microballoons with potential for biomedical applications. Mater Sci & Eng C-Mater Biodmed. App 2010; 30: 412-6.

[55] Lee SE, Choi KJ, Menon GK, *et al.* Penetrating pathways induced by low-frequency sonophoresis with physical and chemical enhancers: iron oxide nanoparticles *versus* lanthanum nitrates. J Inv Dermatology 2010; 130: 1063-72.

[56] Bawa P, Pillay V, Choonara YE, *et al.* Stimuli-responsive polymers and their applications in drug delivery. Biomed Mater 2009; 4: 022001.

[57] Zhou M, Leong T, Melino S, *et al.* Sonochemical synthesis of liquid-encapsulated lysozyme microspheres. Ultrason Sonochem 2010; 17: 1075-81.

CHAPTER 4

Vibrational Spectroscopies for Surface Characterization of Biomaterials

José Manuel Delgado-López[1,*], Jaime Gómez-Morales[1] and Michele Iafisco[2,3]

[1]Laboratorio de Estudios Cristalográficos (IACT, CSIC-UGR). Avda. Las Palmeras 4. E-18100 Armilla (Granada), Spain; [2]Dipartimento di Chimica "G. Ciamician", Alma Mater Studiorum, Università di Bologna via Selmi 2. 40126 Bologna, Italy and [3]Dipartimento di Scienze della Salute, Università del Piemonte Orientale via Solaroli 4, 28100 Novara, Italy

Abstract: This chapter describes some of the most important spectroscopic methods to study the materials surface. Special IR and Raman configurations providing information from the surface will be reviewed highlighting the main advantages and/or drawbacks for their applications in the study of biomaterials. Additionally, some examples are presented to demonstrate their usefulness for the surface characterization of materials for biomedical applications as well as to understand the complexity of the chemical interactions between biomaterial surface and biological molecules.

Keywords: Infrared spectroscopy, raman spectroscopy, organic-inorganic interaction, calcium phosphates, surface characterization.

1. INTRODUCTION

In the biomaterials field it is well established that knowledge of the chemical nature of the surface is essential to understand the possible physico-chemical interactions of the surface with living tissues of biological systems. Surface interaction of biomaterials with proteins forming the living systems represents the first event when they are implanted within the body and must integrate into the host tissue affecting mainly the cell attachment and activation. These findings allow material scientists to design new biomedical devices. Biomaterials surfaces are difficult to characterize because they can be easily contaminated and often exhibit great mobility and reactivity.

***Address correspondence to José Manuel Delgado-López:** Laboratorio de Estudios Cristalográficos (IACT, CSIC-UGR), Avda. Las Palmeras, 4. E-18100 Armilla, Spain; Tel: +34 958 23 00 00; Fax: +34 958 55 26 20; E-mail: jmdl@lec.csic.es

Lia Rimondini, Claudia Bianchi and Enrica Vernè (Eds)

Table 1: Most common analytical methods for surface characterization of biomaterials

Diffractometric	Electron diffraction	Selected Area Electron Diffraction (SAED)	
	X-Ray diffraction (XRD)	Grazing incidence XRD	
Spectroscopic	Vibrational spectroscopies	Infrared (FTIR)	Attenuated Total Reflection (ATR)
			Microspectroscopy
		Raman scattering	Confocal Raman microscopy
			FT-Raman spectroscopy
	Electron Spectroscopies	X-Ray Photoelectron Spectroscopy (XPS)	
		Auger electron spectroscopy (AES)	
	X-Ray Spectroscopies	X-Ray emission spectroscopy	
		X-Ray absorption spectroscopy	X-Ray Absorption Near Edge Spectroscopy (XANES)
			Extended X-Ray Absorption Fine Structure Spectroscopy (EXAFS)
Microscopic	Scanning Probe Microscopy (SPM)	Scanning Tunnelling Microscopy (STM)	
		Atomic Force Microscopy (AFM)	
	Electronic Microscopy	Scanning Electron Microscopy (SEM)	
		Transmission Electron Microscopy (TEM)	

Most common analytical methods properly measure the bulk but not the surface properties. However, there are several techniques which provide valuable information related to the surface; the most common are summarized in Table **1**. Vibrational spectroscopies (Infrared (IR) and Raman) are a good choice to analyze physico-chemical and structural properties of different materials such as polymers, semiconductors, pharmaceuticals [1] as well as biosamples in the fields of biochemistry, biomedicine and biomaterials science [2, 3]. Additionally, these techniques are non-invasive and surface sensitive when the detected radiation is derived from the outermost layers of these samples. By combining both IR and Raman spectroscopies, complementary identification of the functional groups forming a sample can be achieved.

This chapter is devoted to briefly describe the most useful IR and Raman configurations, commonly available in conventional laboratories, to analyze the biomaterials surfaces. In addition, several examples will be selected to demonstrate how these tools help us to understand the complex chemical interactions between biomaterials and biological macromolecules.

2. INFRARED SPECTROSCOPY

Monochromatic radiation can be absorbed, scattered or reflected after interaction with the matter. IR radiation will be absorbed if its frequency matches the vibrational frequency of the molecule and consequently causes a change in the amplitude of the molecular vibration [1]. The infrared spectrum of a sample is recorded after passing the beam of infrared light through the sample. It displays the IR wavelengths at which the sample absorbs the radiation, which reveals details about its molecular structure. This spectrum can be collected either by an energy-dispersive spectrometer equipped with a monochromatic beam that changes in wavelength over time, or also, by a Fourier transform infrared (FTIR) spectrometer which measures all wavelengths at once.

2.1. Fourier Transform Infrared Spectroscopy

FTIR spectrometers, based on the Michelson principle, collect the intensities as a function of the optical path difference and the interferogram produced is related to the frequency of the incoming radiation by a Fourier transformation [4, 5]. Therefore, the main advantage of using these systems is that the whole spectrum is recorded for each one of the scans that the interferometer makes, thus reducing strongly the total collection time with respect to a dispersive spectrometer [4, 5]. This is the reason why dispersive spectrometers have been totally substituted by FTIR systems in all laboratories.

As commented above, surface interaction between biomaterials and surrounding proteins is the first event taking play when they are implanted within the body and integrate into the host tissue. FTIR spectroscopy, through the study of the amide I and amide II absorption bands, has been proven to be a valuable tool to characterize the secondary structure of the proteins. The amide I broad band (between 1600 and 1700 cm^{-1}) is mainly associated with the C=O stretching whereas amide II (between 1500-1600 cm^{-1}) results from the combination of the N-H bending (40-60%) and the C-N stretching (18-40%) vibrations of the peptide bonds. The amide I band strongly depends on the secondary structure of the proteins (basically the percentage of α-helix, β-sheet and β-turn) whereas the amide II band has been related to the amount of adsorbed proteins [6]. Usually, each broad amide band is composed by overlapped bands. Therefore, spectra deconvolution followed by curve fitting can be applied to estimate quantitatively the area of the component band representing each type of secondary structure. Table **2** summarizes the frequencies of the secondary structure components bands in the amide I region.

Table 2: Typical IR spectral regions of the main components of the secondary structure of proteins in the amide I region [7, 8].

Conformation	Wavenumber (cm^{-1})
β-turn	1660-1690
α-helix	1645-1655
β-sheet	1630-1645
	1685-1700

FTIR has been employed to obtain information on protein-surface interactions in experiments *in vitro* as reported by Iafisco *et al.* [9]. These authors analyzed the adsorption of serum proteins on carbonate-hydroxyapatite nanoparticles (CHA) by FTIR. Indeed, they investigated the secondary structure of human serum albumin (HSA) adsorbed onto the CHA surfaces in the amide I region. Fig. **1** shows the deconvoluted spectra of HSA in such spectral region before and after interaction with plate- and needle-like nanocrystals. The appearance of new IR bands in the spectrum collected after adsorption of this protein on both CHA surfaces (spectra B and C, respectively) suggests that HSA has undergone a structural modification. By analyzing the integrated areas of the secondary structure components bands before (A) and after adsorption on CHA nanocrystals (B and C), the authors interpreted that a significant reduction of its helicity and β-structures contents took place as a consequence of the adsorption on both CHA surfaces.

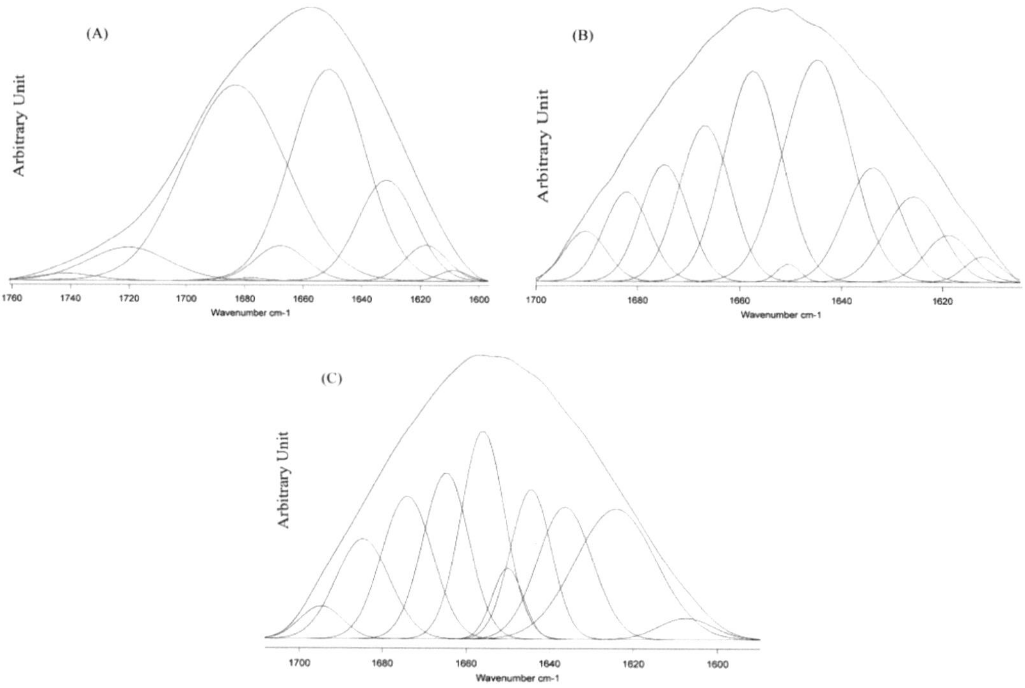

Figure 1: Fourier self-deconvoluted spectra of HSA before adsorption (A) and after adsorption either on plate-shaped CHA (B) or needle-shaped CHA (C). Reprinted from Iafisco *et al.* Colloids and Surface B 2010; 81: 274-84, with permission from Elsevier.

Dasgupta *et al.* performed a similar work [10]. They synthesized HA nanoparticles, undoped or doped with Mg^{2+} and Zn^{2+} and used FTIR to characterize their interaction with bovine serum albumin (BSA) proteins. The presence of adsorbed BSA was proved by the changes of deconvoluted amide I and II bands observed as compared with the pure protein spectrum. The results show a decrease in α-helix content in the adsorbed BSA proteins as compared to pure BSA. Moreover, this decrease was more noticeable in the cases of doped HA, probably due to distortion along the c-axis caused by the substitution of Ca^{2+} by either Mg^{2+} or Zn^{2+} in the crystalline structure of HA.

These two examples demonstrate the usefulness of FTIR to analyze structural changes of proteins after interaction with biomaterial surfaces. Nevertheless, it has to be considered that for proteins in aqueous solutions the amide I spectral zone is normally overlapped with the intense HOH bending mode band of water molecules appearing at *ca.* 1650 cm^{-1}. This is one of the main disadvantages of FTIR spectroscopy. As it will be described in the next section, Attenuated Total Reflection (ATR) is an useful configuration to properly subtract the water absorption bands.

2.2. Attenuated Total Reflection

ATR, also known as internal reflection, is a reflectance method which can be used to analyze solid, liquid or gas samples without further preparations [5, 11, 12]. Under this set up, a sample is placed on a prism (as infrared transparent windows) which has a quite larger refractive index than that of the sample [12]. Normally Ge, Si, ZnSe or more recently diamond are used as IR-transparent window [12]. The incident beam going through the prism is totally reflected at the interface between the prism and the sample only when the incident angle θ_i is larger than the critical angle (θ_c) defined by the Snell law as [12]:

$$\theta_c = \sin^{-1}\left(\frac{n_2}{n_1}\right) \qquad \textbf{(1)}$$

In this equation n_1 and n_2 are the refractive indices of both the prism and the sample, respectively. Upon undergoing total reflection, an electromagnetic wave propagates parallel to the surface in the plane of incidence creating an electromagnetic field in the less dense medium whose intensity exponentially decrease with the distance to the interface (plot in Fig. **2**). This electromagnetic field is termed evanescent wave (Fig. **2**). The intensity of the evanescent wave, and therefore of the reflected beam, is attenuated at wavelengths at which the sample absorbs. This phenomenon, which is the basis of the ATR method, provides the possibility to extract spectroscopic information from the layer next to the interface. The *penetration depth* (d_p) of the evanescent wave depends on the wavelength of the incident beam (λ), the incident angle and the refractive indices of both the prism (n_1) and the sample (n_2), according to the following equation [12]:

$$d_p = \frac{\frac{\lambda}{n_1}}{2\pi\left(\sin^2\theta_i - \left(\frac{n_2}{n_1}\right)^2\right)^{\frac{1}{2}}} \qquad \textbf{(2)}$$

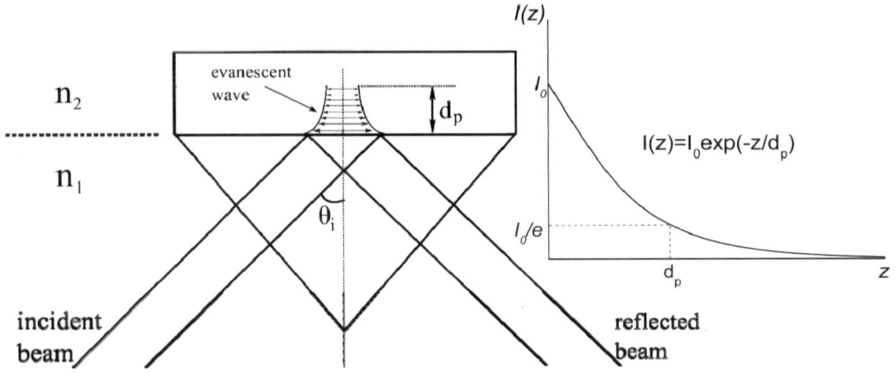

Figure 2: Total internal reflection at the interface between media with refractive indices n_1 and n_2. The graph on the right shows the exponential dependence of the intensity (I(z)) of the evanescent wave on the distance (z) away from the interface. Note: figure not shown to scale.

The intensity of the spectrum can be enhanced by multiple reflections of the beam if the ATR window is a parallel-sided plate or cylinder [5, 11, 12]. An advantage of the ATR configuration related with the limited path length crossing through the sample is the much lower attenuation (compared for example with transmission experiments) of the IR signal in highly absorbing media, such as occurs in aqueous solutions [13]. Since ATR technique facilitates the subtraction of water, it can be successfully applied for the study of the structure of proteins in aqueous solutions [6, 14].

One simple and practical method to obtain a clean protein spectrum consists on the subtraction of pure water spectrum to that collected for protein in water solution [6, 14] but using a scale factor in order to get a flat baseline in the region between 1900 and 1800 cm^{-1} [6, 14]. This method is also suitable to acquire the clean spectrum of the adsorbed protein, but in this case both spectra should be collected in the presence of the biomaterial deposited onto the ATR window.

Other possibility that ATR offers is the quantification of adsorbed protein on a given surface (with the prior knowledge of the molar absorptivity of the protein). Indeed, this method has been successfully applied to estimate the adsorbed amount of monoclonal antibodies, *i.e. Immunoglobulin G* (*IgG*) on coated silicon substrates [15], albumin adsorbed on polyurethane surfaces [16, 17] and Langmuir-Blodgett monolayers of fatty acids deposited on calcium fluoride (which is used as ATR window as well) [18].

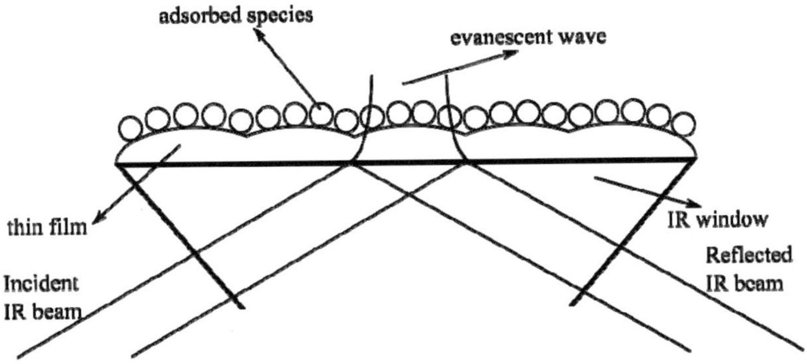

Figure 3: ATR configuration in which the sample is deposited as a thin film on the IR window. In principle, if the d_p is a bit larger than the thickness of the thin film it could be possible to obtain spectroscopic information coming mainly from the molecules interacting with the surface. Note: figure not shown to scale.

In addition, the ATR set-up can provide *in situ* spectroscopic information from the surface when the sample is deposited on the IR window as a thin film (scheme in Fig. **3**). According to equation 2, the penetration depth mainly depends on the wavelength and incident angle of the IR beam as well as the refractive indices of the media. Thus, controlling these parameters together with the thickness of the deposited thin film, the spectroscopic information can be derived mainly from species interacting with the surface (Fig. **3**). Actually, ultrathin films of hydroxyapatite and metal alloys have been deposited on Ge prism surfaces to study interactions between these materials and biofilms [19].

Ong *et al.* [6] demonstrated that ATR is a suitable method to characterize *in situ* protein adsorption onto bioactive amorphous calcium phosphate (CaP) coatings. These authors deposited CaP films by ion beam sputtering on Ge ATR crystals with a thickness lower than 250 Å. ATR spectra collected for coatings immersed in the physiological solutions Alpha Modification of Eagle's Medium (α-MEM) showed the disappearance of the band at *ca.* 945 cm^{-1} and the decrease of the intensity of the 1005 cm^{-1} band, both related to phosphate groups [6]. This finding suggested that CaP coatings were progressively dissolved in α-MEM solutions. The addition of proteins (fibronectin and albumin) to the α-MEM solution leads to differences in the CaP dissolution. In the presence of fibronectin the rate of dissolution was significantly slower than in the presence of albumin. This behaviour was ascribed to the faster adsorption rate of fibronectin on the CaP surfaces. In this work, the ATR method also provided *in situ* information about the interaction between the two proteins and the CaP surface from the deconvolution of the amide I and amide II vibrational bands. Conformational changes associated to the adsorption of both proteins on CaP coatings suggest some denaturation of the proteins. In the case of fibronectin, the analysis of the amide I band suggests a greater loss of β-sheet conformer on the CaP coatings. In the case of albumin, the α-helix to β-turn ratio strongly decrease after adsorption on CaP coatings suggesting a decrease in the amount of α-helix component [6].

An excellent review on ATR-FTIR describing how this technique can be employed for characterization of biomaterials surfaces and events occurring at these surfaces (mainly protein adsorption) is presented in the reference [14].

2.3. FTIR Microspectroscopy

FTIR microspectroscopy is a method that results from the coupling of FTIR spectrometer with a microscope. This setup, in conjunction with Raman confocal spectroscopy, that will be described hereafter, has been proved to be among the most powerful analytical tools for surface characterization [20]. It provides not only the visual inspection of the material surface but also the chemical information of this part of the sample under examination in the form of infrared spectrum. Even biological samples with microscopic size can be non-destructively analyzed with a FTIR microscope [21, 22]. Additional capabilities such as a temperature-controlled stage, polarized radiation, and *in situ* biological sample-holder are available as well.

In addition, FTIR microspectroscopy allows for spatially resolved mapping of the chemical composition and/or distribution on the surface. Usually, this mapping is obtained after collection of the spectra at different points on the surface with a standard microscope equipped with a computer-controlled motorized XY stage. The absorbance of the selected band is represented as a surface map. Unfortunately the main disadvantage of this task is the high time consuming even to map a few square millimetres in size. Moreover, the maximum achievable spatial resolution which is often limited by the geometrical aperture size necessary to achieve an acceptable signal-to-noise ratio (S/N) has to be considered.

Conventional thermal IR sources encounter S/N limitation when apertures confine the infrared beam to an area lower than 20–30 μm in diameter. On the contrary, high brightness synchrotron infrared sources provide lower sized infrared beams that allow smaller regions to be tested with good S/N. Thus, in this latter case, the spatial resolution is no longer controlled by the geometrical aperture size but rather by the numerical aperture of the optical micro-FTIR system as well as by the wavelength of the light [23]. Reference [24] offers detailed information about Synchrotron and Free Electron Laser Sources of Infrared Radiation.

In the field of biomaterials, FTIR micro spectroscopy starts to be widely used to map the topographic distribution of the surfaces. Particularly, this setup is being very useful to study the topographic distribution of hydroxyapatite nanocrystals deposited on Ti substrates. The coated metallic implants are employed as prosthesis for bone substitution [25]. S. Manara *et al.* [26] reported a nice example about the use of FTIR microspectroscopy to characterize these biomaterials. They mapped HA/collagen distribution during the coating of Ti surfaces with biomimetic collagen/apatite nanocrystals by an electrochemically-assisted process. The FTIR maps collected at electrodeposition times of 30 s (F1), 5 min (F2) and 30 min (F3) exhibited heterogeneous radiation absorption (Fig. **4**). They show areas of low (1, blue), medium (2, green) and high absorption (3, red).

The FTIR spectra collected in these areas (not shown) indicate that at lower electrodeposition times (30 s) the Ti surface was covered mainly by separated poorly crystalline HA (areas 1 and 2) and uncalcified collagen phases (area 3). Nonetheless, increasing electrodeposition time to 5 min (F2) and 30 min (F3) lead to the formation of calcified collagen as pointed out by the shift at lower wavenumbers of the collagenic COO^- symmetric stretching band from $1347 cm^{-1}$ to $1337 cm^{-1}$, which evidences a chemical interaction of the protein with the apatite phase. At 5 min (F2) the FTIR spectrum collected in area 3 shows the highest amount of collagen and hydroxyapatite, revealing a preferential nucleation and growth of nanocrystals on calcified collagen fibrils, whereas in areas 1 and 2 the lower intensities of the absorption bands indicate fewer amounts of protein and inorganic phase. At 30 min (F3), Ti surface is totally covered by the HA/collagen composite and the different coloured areas (different absorption intensities) are produced by the different thickness of the composite layer in each area.

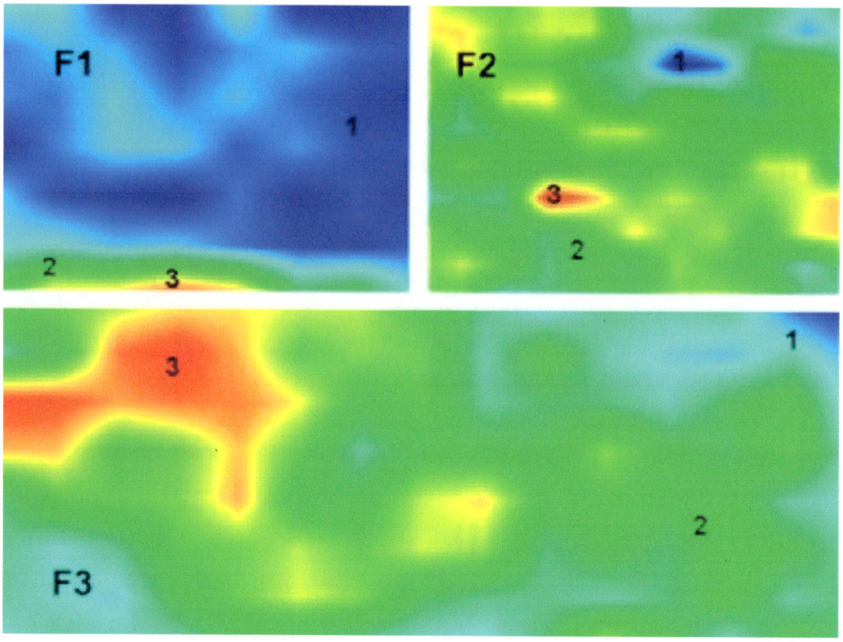

Figure 4: FTIR Chemical maps relative to: collagen fibrils–hydroxyapatite nanocrystals composite coating obtained by electro-deposition at different time (30 s (F1), 5 min (F2) and 30 min (F3)) in electrolytic solutions of $Ca(NO_3)_2$ and $NH_4H_2PO_4$ containing helical type I collagen. The colours code is (1, blue) low intensity, (2, green) medium intensity and (3, red) high intensity. Reprinted from S. Manara *et al.* Inorganica Chimica Acta 2008; 361:1634-45 with permission from Elsevier.

3. RAMAN SPECTROSCOPY

Raman Spectroscopy provides complementary information to that of FTIR concerning the structural and compositional properties of the molecules from their vibrational transitions. However, when using Raman spectroscopy, unlike IR absorption, this information is given by scattered photons of different wavelengths after the interaction between the incident visible light and the sample. Particularly, the possible scattering processes are schematically illustrated in Fig. **5**. When the matter is illuminated with monochromatic light of frequency v_0, the majority of the photons are scattered without energy exchange (Rayleigh scattering). From a quantum mechanical point of view, the molecule is excited to an unstable state (g_2) with energy hv_0 above the ground state (g_0), from which it decays back to the ground state, g_0. Conversely, if the excited molecule decays from the unstable state (g_2) to the first vibrational state (g_1), a photon with energy equal to $hv_0 - hv_v$

will be emitted. Thus, the scattered light will show a Raman peak (*i.e.*, Stokes band) at the frequency $v_0 - v_v$. In addition, the reverse process may also take place. If the collision brings the molecule from the first vibrational state (g_1) to the unstable state (g'_2, with energy hv_0 above the first vibrational state, g_1) and then it decays to the ground state (g_0), the emitted photon will have an energy equal to $hv_0 + hv_v$ and, thus, the emitted light will exhibit an "anti-Stokes" Raman band at the frequency $v_0 + v_v$. Both processes are also schematized in Fig. **5**. By

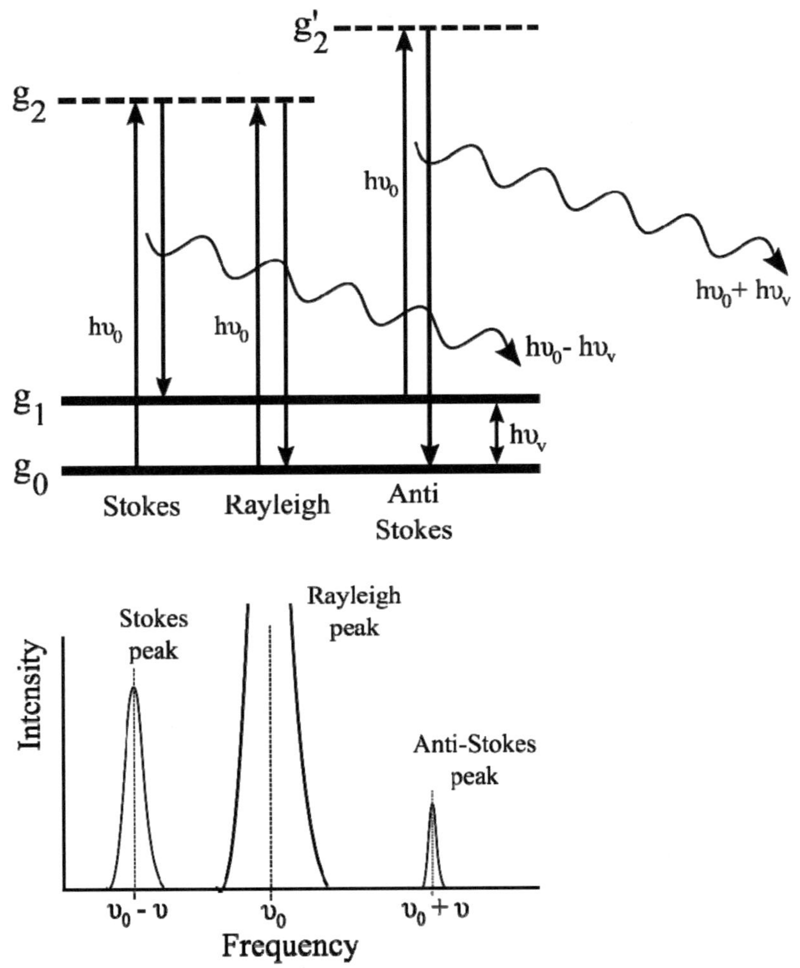

Figure 5: Energy-level diagrams for scattering (Rayleigh and Raman) processes occurring when matter is illuminated with a monochromatic beam. At the bottom, the relationship between the intensities of Stokes, Rayleigh and Anti-Stokes bands in typical Raman spectra are schematized.

far, anti-Stokes peaks will have much lower intensity than Stokes ones, since the number of vibrationally excited molecules is normally much lower. For that reason only the Stokes bands are commonly represented in Raman spectra. Spectrometers are often set up to read the negative shift in frequency with respect to the exciting frequency (v_0) and, therefore the frequency of the peaks corresponds to that of the vibrational transitions.

Unlike IR instruments, where dispersive spectrometers have totally disappeared, dispersive and FT-Raman spectrometers are equally extended. On one hand, the former use mainly visible lasers as excitation source (usually emitting at 514, 663 or 785 nm), notch filters to avoid the Rayleigh radiation and charge couple device (CCD) detectors. On the other hand, the latter employ a near-infrared laser (emitting at 1064 nm) for excitation and a Michelson-based interferometer (similar to that used by FTIR systems commented previously). For more detailed description of the instrumentation of both systems references [24, 27] are strongly recommended.

The election of one or another system largely depends on the type of experiments to be performed as well as on the properties of the sample to be analyzed. One of the main drawbacks in Raman spectroscopy is the fluorescence phenomena. This problem is much relevant when using laser emitting at the visible region than when emitting either in the UV or in the NIR regions. Therefore, the use of FT spectrometers equipped with a NIR laser, usually a neodymium-doped yttrium aluminium garnet (Nd^{3+}:YAG) solid state lasers emitting at 1064 nm, will largely reduce the fluorescence emission with respect to visible excitation (*e.g.*, 514.5 nm) systems. Another problem for Raman characterization is the heating destruction of the sample after long-term laser irradiation [22], especially important for biological samples. To obtain an enhanced Raman spectrum of a sample, a prolonged exposure of the specimen to the laser beam is necessary. However, the heat generated by the laser may alter or destroy the biosamples after a long-term exposure [22]. To overcome the heating destruction of the sample, it is recommended to shorter the irradiation time with a repeated number of scans. Finally, unlike FTIR, the big advantage of Raman spectroscopy is the weak interference of water since water molecules are strong absorbers of infrared radiation but weak scatterers in Raman spectroscopy, which makes the technique particularly suitable for samples in aqueous solutions [2].

3.1. Surface Enhanced Raman Scattering (SERS)

The use of Raman spectroscopy for surface characterization was hugely expanded when in 1974 Fleischmann *et al.* [28] discovered the significant enhancement of the Raman scattering signal coming from molecules adsorbed on rough metal surfaces, an effect known as Surface-enhanced Raman Scattering (SERS) [29]. This effect which enhances the spectra of adsorbates by factors of up to 100 times occurs on the surface of a few metallic particles (notably Cu, Au and Ag) [29]. Iafisco *et al.* [30] have used SERS spectroscopy to investigate the modification of adsorbed myoglobin (Mb) on hydroxyapatite nanocrystals. The myoglobin and HA-myoglobin SERS spectra collected in the presence of silver colloids are reported in Fig. **6**. SERS effect due to the silver particles allows a 1014-fold magnification of the myoglobin heme group vibrational bands, making it possible to reach the limit of single-molecule detection [31]. Raman bands in the region of 1300-1700 cm^{-1} are sensitive to the dimensions of the porphyrin core which is influenced mainly by the coordination, spin, and oxidation state of the heme iron. The vibration of the C-N group at 1360-1370 cm^{-1} is sensitive to the oxidation state of the heme iron: ferric heme shifts this band to higher wavenumbers. In the 1500-1630 cm^{-1} region (vibrations of C-C of the porphyrin ring) bands at *ca.* 1595 and 1565 cm^{-1} are sensitive to the spin state of the heme iron whereas the band at 1620 cm^{-1} (stretching of vinylic C=C) is sensitive to both the spin and the oxidation state [32]. The relative intensity of these latter bands increases if the content of low-spin iron increases.

After interaction with HA, the 1596 cm^{-1} band increases in intensity, which is interpreted as a higher content of low-spin iron. This is also proved by the appearance of another marker band of low-spin heme at 1507 cm^{-1} [33] after the interaction with HA. The presence of the bands at 1596 and 1568 cm^{-1} in the pure myoglobin sample indicates that the powder is a mixture of two spin states (maybe caused by the freeze-drying process). In conclusion, spectra highlight the fact that protein changes its conformation upon adsorption on the inorganic substrate. In particular, the interaction with HA alters the coordination state of the iron in the heme through the formation of a hexa-coordinated low-spin myoglobin heme, possibly involving the distal histidine [30].

Review of reference [34] is highly recommended for extensive information on SERS and its applications for biomedical materials.

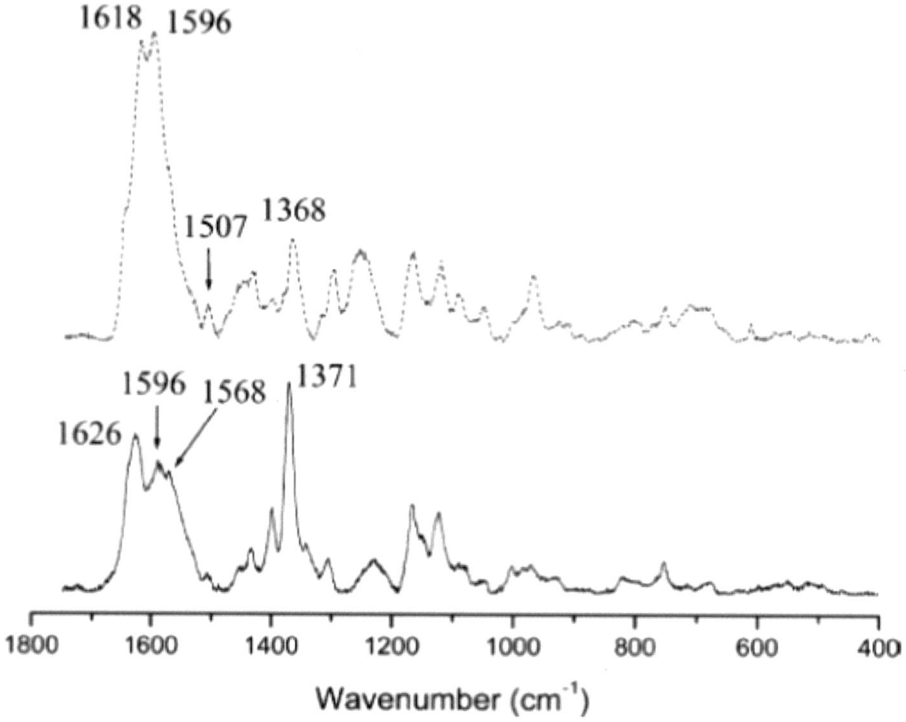

Figure 6: SERS (hydroxylamine-silver colloid) spectra of solid myoglobin (-) and solid myoglobin-hydroxyapatite (---). Reprinted from Iafisco *et al.* Langmuir 2008;24:4924-30 with permission from American Chemical Society.

3.2. Raman Microspectroscopy

As in the case of FTIR, dispersive Raman spectrometers can also be coupled to a microscope, but in this case the microscope can also be set up confocally. Such a system isolates the scattered light originated from a small region of the sample coincident with the illuminated spot, and efficiently eliminates the contributions from out-of-focus zones improving the rejection of stray light and the z-spatial resolution. The high spatial resolution, and the use of automated stages, enables the performance of two-dimensional (2D) or three-dimensional (3D) selective maps (mapping) of the surface of the sample. This type of experiments (also described in the section 2.3 for FTIR micro spectroscopy), which allows the characterization of the surface composition with spatial resolution of many materials, has been repeatedly apply for mapping the surface of hydroxyapatite [35], bone [36, 37] and other biological tissues [22]. In principle, FT-Raman spectrometer could be also coupled to a microscope but the high power density at the focal point reached with NIR lasers can cause easily the thermal degradation of the sample [27].

4. COMPLEMENTARY CHARACTERIZATION BY FTIR AND RAMAN

In Raman spectroscopy, as in IR absorption, not all electronics vibrations are permitted. Specifically, a vibration is Raman active if the molecule changes its polarizability (*i.e.* tendency of the electron cloud of the molecule to be distorted by an external electric field). However, as discussed above, the selection rules for IR establish that the dynamic dipole moment must change during the vibration. Thus, some vibrational modes will be active in both spectroscopies, but others will be only Raman or IR active (Table **3**). Therefore, the complete vibrational behaviour of the molecules forming the sample will be only determined by using both spectroscopies.

Table 3: Infrared and Raman spectroscopic band assignments for bone extracted from ref. [38].

Assignment	Wavenumber range (cm^{-1})	Observed in
$v_2PO_4^{3-}$	422-454	Raman
$v_4PO_4^{3-}$	578-617	Raman/IR
C-C stretch	815-921	Raman
$v_2CO_3^{2-}$	860-890	IR
$v_1PO_4^{3-}$	957-962	Raman
$v_3HPO_4^{2-}$	1003-1005	Raman
$v_3PO_4^{3-}$	1006-1055	Raman
$v_1CO_3^{2-}$	1065-1071	Raman
Amide III	1243-1269	Raman
CH_2 wag	1447-1452	Raman/IR
Amide II	1540-1580	IR
Amide I	1595-1720	Raman/IR
CH_2 stretch	2840-2986	Raman
OH stretch	3572-3575	Raman

A large number of references can be found on the use of IR and Raman spectroscopies to achieve the whole characterization of bone and other biological tissues [2, 22, 38-40]. Main IR and Raman band assignments for bone are summarized in Table **3**. Bala *et al.* [41] studied the mineralization process of ewes bone by FTIR. The spectra for ewes bone with different mineralization times are represented in Fig. 7. From these spectra, several indices concerning bone properties can be estimated: i) *crystallinity index*, given by the full width at half maximum (FWHM) of the 604 cm^{-1} band (ν_4PO_4):

This index corresponds to both crystalline domain size and how well the ions of the crystal are ordered in the unit cells [41]; ii) *mineral maturity* calculated as the area ratio of 1030 cm^{-1} peak (ν_3PO_4, apatitic phosphate) over 1110 cm^{-1} peak (ν_3PO_4, non-apatitic phosphate). This index informs about the transformation of immature precursors into mature apatite [41]; iii) *mineralization index*, which provides the ratio of the mineral phase over the organic phase (collagenic and non-collagenic proteins) and is given by the area ratio between 1030 cm^{-1} (ν_3PO_4) and 1660 cm^{-1} (amide I) bands [41] and iv) *carbonation degree*, measured as the area ratio of the ν_2CO_3 (894–862 cm^{-1}) over ν_3PO_4 (1184–910 cm^{-1}) regions and representing the amount of carbonate groups incorporated into the bone mineral lattice.

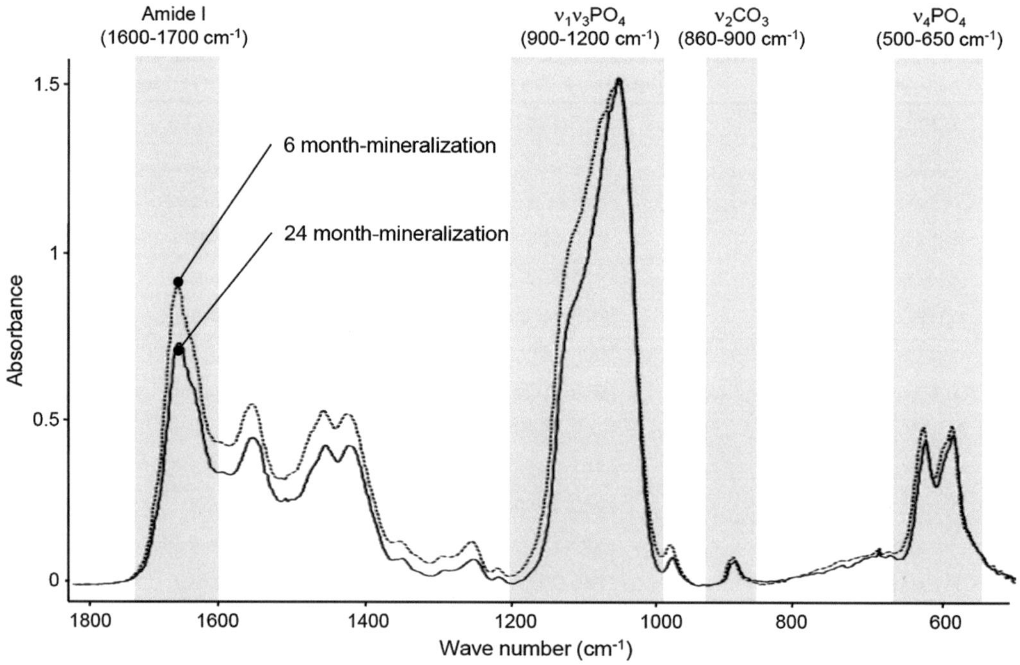

Figure 7: FTIR spectra of ewe bone tissue mineralized during 6 and 24 months (dotted and solid spectra, respectively). Reprinted from Bala *et al.* Bone 2010;46:1204-12 with permission from Elsevier.

By FTIR spectroscopy is possible to determine not only the quantity of carbonate incorporated into the crystal lattice but also the carbonate environment in bone apatite [42]. When carbonate ions are replacing PO_4 ions sites (known as B-type carbonation) infrared spectrum shows bands at *ca.* 1465, 1413 and 873 cm^{-1} [42, 43]. However, in the case of A-type substitutions in which carbonate are replacing OH ions, IR spectrum displays peaks at *ca.* 1546, 1456 (shoulder) and 880 cm^{-1} [42, 43].

Raman spectroscopy is also used to analyze simultaneously the organic and mineral component of bone and other biological tissues [38, 39]. Additionally, when focusing with Raman confocal microscope to the surface of the sample, the scattered Raman photons only will provide information from the surface but not from the bulk [37, 44]. Raman spectrum collected at the surface of a femoral bone from a healthy human female is represented in Fig. **8**. Bands associated with the mineral component mainly correspond to the PO_4^{3-} (430, 590 and 960 cm^{-1}) and CO_3^{2-} (1070 cm^{-1}) vibrations [37] (see assignments in Table **3**). The frequency of this latter band is also related to the carbonate substitution type [45]. Unlike FTIR, the most intense band of the mineral component corresponds with the $\nu_1PO_4^{3-}$ vibration appearing at *ca.* 960 cm^{-1}. From the ratio between the intensity of this band at 960 cm^{-1} ($\nu_1PO_4^{3-}$) and that at 1070 cm^{-1} ($\nu_3CO_3^{2-}$) the carbonation degree can be also estimated [44]. Regarding the organic component, the band at 2940 cm^{-1} is dependent on proteins present in bone. The area ratio of the 960 cm^{-1} over 2940 cm^{-1} bands characterizes the mineral/protein proportion of the bone [46]. Analysis of collagen by Raman spectroscopy is mainly based on the amide vibrations: amide I at about 1640–1670 cm^{-1} (νC=O (α helix)) and amide III, displaying a doublet at 1245–1270 cm^{-1} (δ NH) [37].

The combination of both spectroscopies has also been largely utilized to study the physico-chemical and structural properties of synthesized materials with biomedical applications [47, 48]. Recently Liu *et al.* [49] studied the interaction between functionalized Ti surfaces and calcium phosphate by ATR-FTIR and Raman spectroscopies. They incorporated self-assembled monolayers (SAM) containing different terminal groups such as -OH, PO_4H_2 and –COOH to the Ti surface in order to verify the CaP nucleating abilities of these functional groups. Fig. **9** shows both ATR and Raman spectra of the CaP films deposited on oxidized and functionalized Ti surfaces.

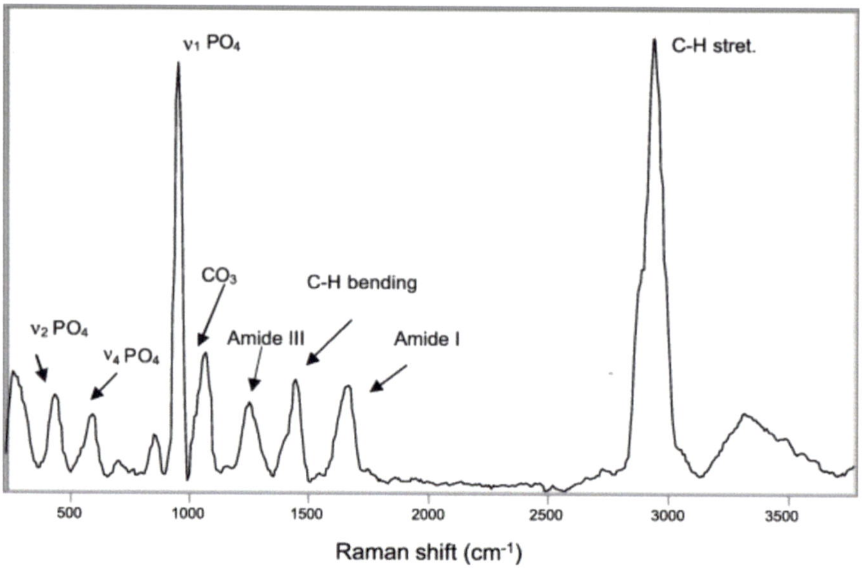

Figure 8: Raman spectrum collected at the surface of femoral human bone. Reprinted from Kazanci *et al.* Journal of Structural Biology 2006;156:489–96 with permission from Elsevier.

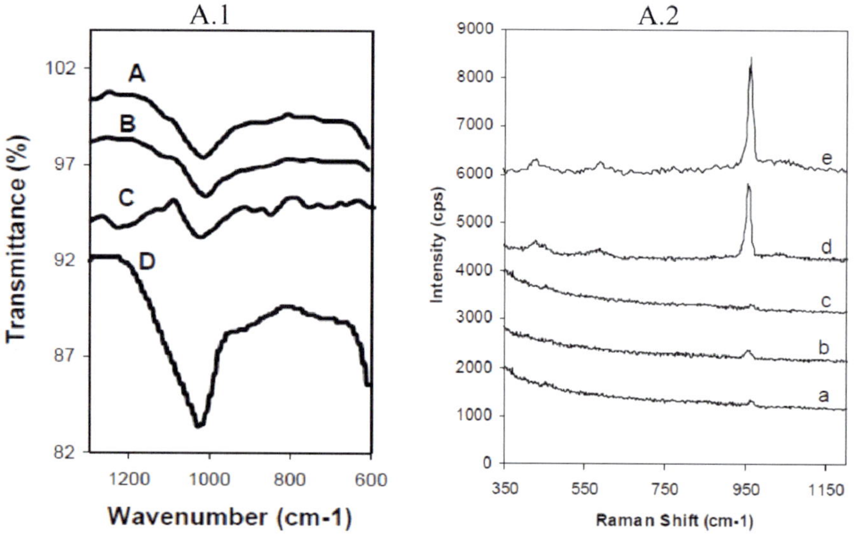

Figure 9: A.1) ATR-FTIR spectra of the CaP coatings showing P-O absorption bands at 1040 cm^{1}. (A) CaP deposited on oxidized Ti; (B) CaP deposited on Ti with phosphorylated SAM; (C) CaP deposited on Ti with carboxylated SAM; (D) biomimetic apatite coating on Ti with pre-deposited CaP on phosphorylated SAM. A.2) Raman spectra of CaP coatings on: (a) Oxidized Ti samples; (b) Ti samples with phosphorylated SAM; (c) Ti samples with carboxylated SAM; (d) sample b after immersion in simulated body fluid (SBF) for 18 h; (e) CaP precipitates collected from the suspension in the CaP deposition experiment. Reprinted from Liu *et al.* Biomaterials 2002;23:3103-11 with permission from Elsevier.

In this study, spectroscopic data confirmed the presence of CaP layers on the surface of the oxidized, carboxylated and phosphorylated-treated Ti samples by the presence of an intense band related to P-O stretching band at *ca.* 1040 cm^{-1} (IR) and 960 cm^{-1} (Raman). However, these latter bands appeared neither for clean Ti surfaces nor for OH-functionalized Ti surfaces. After immersion in SBF, all samples with pre-deposited CaP increased notably the peak intensities in IR and Raman at 1040 cm^{-1} and 960 cm^{-1}, respectively, indicating the formation of new apatite (spectra D in Fig. **9**).

5. CONCLUSIONS

In this chapter we have described the fundamentals of vibrational spectroscopies (IR and Raman) and the applications of special configurations such as ATR, FTIR microspectroscopy, SERS and Raman microspectroscopy commonly used to characterize the physico-chemical and structural properties of biomaterial surfaces as well as to study the *in situ* conformational changes of proteins undergone during adsorption on the biomaterials surfaces. Since some vibrational modes are active in both spectroscopies but others are only Raman or IR active, the use of both spectroscopies is strongly recommended to achieve a complete vibrational characterization of biomaterials.

ACKNOWLEDGEMENTS

JMDL and JGM acknowledge Spanish MINECO for support (Factoría de Cristalización (Consolider Ingenio), Spanish-Italian Integrated Action IT2009-00028, and CRYSFUNBIO ref. MAT2011-28543). JMDL also thanks to Consejo Superior de Investigaciones Científicas (CSIC) for his JAE-Doc postdoctoral research contract. MI is recipient of a fellowship from Regione Piemonte.

CONFLICT OF INTEREST

The author(s) confirm that this chapter content has no conflict of interest.

REFERENCES

[1] Colthup N.B., Daly L.H., Wibereley S.E.; Introduction to IR and Raman Spectroscopy. 3ª ed. Academic Press Inc. New York 1990.

[2] Krafft C. Bioanalytical applications of Raman spectroscopy. Anal Bional Chem 2004;378:60-2.
[3] Yan B., Gremlich H.U.; Infrared and Raman spectroscopy of biological materials. Marcel Dekker. New York 2001.
[4] Gauglitz G., Vo-Dinh T.; Handbook of Spectroscopy. WILEY-VCH. Weinheim 2003.
[5] Griffiths P.R., de Haseth J.; Fourier Transform Infrared Spectrometry. Wiley. New York 1986.
[6] Ong JL, Chittur KK, Lucas LC. Dissolution/reprecipitation and protein adsorption studies of calcium phosphate coatings by FT-IR/ATR techniques. J Biomed Mater Res 1994;28:1337-46.
[7] Nevskaya NA, Chirgadze Y. Infrared spectra and resonance interactions of amide-I and II vibrations of α–helix. Biopolymers 1976;15:637-48.
[8] Chirgadze Y, Nevskaya NA. Infrared spectra and resonance interaction of amide-I vibration of the parallel-chain pleated sheet. Biopolymers 1976;15:627-36.
[9] Iafisco M, Sabatino P, Lesci IG, Prat M, Rimondini L, Roveri N. Conformational modifications of serum albumins adsorbed on different kinds of biomimetic hydroxyapatite nanocrystals. Colloids Surf B 2010;81:274-84.
[10] Dasgupta S, Banerjee SS, Bandyopadhyay A, Bose S. Zn- and Mg-Doped Hydroxyapatite Nanoparticles for Controlled Release of Protein. Langmuir 2010;26:4958-64.
[11] Willis H.A., van der Maas J.H., Miller R.G.J., Eds.; Laboratory Methods in Vibrational Spectroscopy. 3^a ed. John Wiley & Sons. New York 1987.
[12] Harrick N.J.; Internal Reflection Spectroscopy. Harrick Scientific Corporation. New York 1987.
[13] Bordo V.G., Rubahn H.G.; Optics and Spectroscopy at Interface and Surface. WILEY-VCH. Weinheim 2005.
[14] Chittur KK. FTIR/ATR for protein adsorption to biomaterial surfaces. Biomaterials 1998;19:357-69.
[15] Buijs J, Norde W, Lichtenbelt JWT. Changes in the Secondary Structure of Adsorbed IgG and $F(ab')_2$ Studied by FTIR Spectroscopy. Langmuir 1996;12:1605-13.
[16] Jeon JS, Sperline RP, Raghavan S. Quantitative Analysis of Adsorbed Serum Albumin on Segmented Polyurethane Using FT-IR/ATR Spectroscopy. Appl Spectrosc 1992;46:1644-8.
[17] Pitt WG, Cooper SL. Albumin adsorption on alkyl chain derivatized polyurethanes: I. The effect of C-18 alkylation. J Biomed Mater Res 1988;22:359-82.
[18] Jang WH, Miller JD. Verification of the internal reflection spectroscopy adsorption density equation by Fourier transform infrared spectroscopy analysis of transferred Langmuir-Blodgett films. Langmuir 1993;9:3159-65.
[19] Suci PA, Siedlecki KJ, Palmer RJ, Jr., White DC, Geesey GG. Combined Light Microscopy and Attenuated Total Reflection Fourier Transform Infrared Spectroscopy for Integration of Biofilm Structure, Distribution, and Chemistry at Solid-Liquid Interfaces. Appl Environ Microbiol 1997;63:4600-3.
[20] Tolstoy V.P., Chernyshova I.V., Skryshevsky V.A.; Handbook of Infrared Spectroscopy of Ultrathin Films. John Wiley & Sons. New Jersey 2003.
[21] Ellis DI, Goodacre R. Metabolic fingerprinting in disease diagnosis: biomedical applications of infrared and Raman spectroscopy. Analyst 2006;131:875-85.
[22] Krafft C, Sergo V. Biomedical applications of Raman and infrared spectroscopy to diagnose tissues. Spectroscopy 2006;20:195-218.
[23] Carr GL. High-resolution microspectroscopy and sub-nanosecond time-resolved spectroscopy with the synchrotron infrared source. Vib Spectrosc 1999;19:53-60.

[24] Chalmers J.M., Griffiths P.R., Eds.; Handbook of Vibrational Spectroscopy: Theory and Instrumentation, Vol. 1. John Wiley & Sons. Chichester 2002.

[25] Bradt JH, Mertig M, Teresiak A, Pompe W. Biomimetic Mineralization of Collagen by Combined Fibril Assembly and Calcium Phosphate Formation. Chem Mater 1999;11:2694-701.

[26] Manara S, Paolucci F, Palazzo B, Marcaccio M, Foresti E, Tosi G, *et al.* Electrochemically-assisted deposition of biomimetic hydroxyapatite-collagen coatings on titanium plate. Inorg Chim Acta 2008;361:1634-45.

[27] Smith E., Dent G.; Modern Raman Spectroscopy. A practical approach. John Wiley & Sons. Chichester 2005.

[28] Fleischmann M, Hendra PJ, McQuillan AJ. Raman spectra of pyridine adsorbed at a silver electrode. Chem Phys Lett 1974;26:163-6.

[29] Tian ZQ, Ren B, Wu DY. Surface-Enhanced Raman Scattering: from Noble to Transition Metals and from Rough Surfaces to Ordered Nanostructures. J Phys Chem B 2002;106:9463-83.

[30] Iafisco M, Palazzo B, Falini G, Di Foggia M, Bonora S, Nicolis S, *et al.* Adsorption and Conformational Change of Myoglobin on Biomimetic Hydroxyapatite Nanocrystals Functionalized with Alendronate. Langmuir 2008;24:4924-30.

[31] Bizzarri AR, Cannistraro S. Surface-Enhanced Resonance Raman Spectroscopy Signals from Single Myoglobin Molecules. Appl Spectrosc 2002;56:1531-7.

[32] Feng M, Tachikawa H. Raman Spectroscopic and Electrochemical Characterization of Myoglobin Thin Film: Implication of the Role of Histidine 64 for Fast Heterogeneous Electron Transfer. J Am Chem Soc 2001;123:3013-20.

[33] Droghetti E, Smulevich G. Effect of sol-gel encapsulation on the unfolding of ferric horse heart cytochrome c. J Biol Inorg Chem 2005;10:696-703.

[34] Abalde-Cela S, Aldeanueva-Potel P, Mateo-Mateo C, Rodríguez-Lorenzo L, Alvarez-Puebla RA, Liz-Marzán LM. Surface-enhanced Raman scattering biomedical applications of plasmonic colloidal particles. J R Soc Interface 2010;7:S435-S450.

[35] Seah RKH, Garland M, Loo JSC, Widjaja E. Use of Raman Microscopy and Multivariate Data Analysis to Observe the Biomimetic Growth of Carbonated Hydroxyapatite on Bioactive Glass. Anal Chem 2009;81:1442-9.

[36] Sahar ND, Hong SI, Kohn DH. Micro- and nano-structural analyses of damage in bone. Micron 2010;36:617-29.

[37] Kazanci M, Roschger P, Paschalis EP, Klaushofer K, Fratzl P. Bone osteonal tissues by Raman spectral mapping: Orientation-composition. J Struct Biol 2006;156:489-96.

[38] Carden A, Morris MD. Application of vibrational spectroscopy to the study of mineralized tissues (review). J Biomed Opt 2000;5:259-68.

[39] Lin SY, Li MJ, Cheng WT. FT-IR and Raman vibrational microspectroscopies used for spectral biodiagnosis of human tissues. Spectroscopy 2007;21:1-30.

[40] Morris MD, Finney WF. Recent developments in Raman and infrared spectroscopy and imaging of bone tissue. Spectroscopy-An International Journal 2004;18:155-9.

[41] Bala Y, Farlay D, Delmas PD, Meunier PJ, Boivin G. Time sequence of secondary mineralization and microhardness in cortical and cancellous bone from ewes. Bone 2010;46:1204-12.

[42] Rey C, Collins B, Goehl T, Dickson I, Glimcher M. The carbonate environment in bone mineral: A resolution-enhanced fourier transform infrared spectroscopy study. Calcif Tissue Int 1989;45:157-64.

[43] Xu G, Aksay IA, Groves JT. Continuous Crystalline Carbonate Apatite Thin Films. A Biomimetic Approach. J Am Chem Soc 2001;123:2196-203.

[44] Le Blond S, Guilminot E, Lemoine G, Huet N, Mevellec JY. FT-Raman spectroscopy: A positive means of evaluating the impact of whale bone preservation treatment. Vib Spectrosc 2009;51:156-61.

[45] Penel G, Leroy G, Rey C, Bres E. MicroRaman Spectral Study of the PO4 and CO3 Vibrational Modes in Synthetic and Biological Apatites. Calcif Tissue Int 1998;63:475-81.

[46] Penel G, Delfosse C, Descamps M, Leroy G. Composition of bone and apatitic biomaterials as revealed by intravital Raman microspectroscopy. Bone 2005;36:893-901.

[47] Koutsopoulos S. Synthesis and characterization of hydroxyapatite crystals: A review study on the analytical methods. J Biomed Mater Res 2002;62:600-12.

[48] Dorozhkin SV, Epple M. Biological and medical significance of calcium phosphates. Angew Chem Int Ed 2002;41:3130-46.

[49] Liu Q, Ding J, Mante FK, Wunder SL, Baran GR. The role of surface functional groups in calcium phosphate nucleation on titanium foil: a self-assembled monolayer technique. Biomaterials 2002;23:3103-11.

CHAPTER 5

Surface Characterization of Inorganic Nanobiomaterials at a Molecular Scale: Use of Vibrational Spectroscopies

Gianmario Martra*, Valentina Aina, Yuriy Sakhno, Luca Bertinetti and Claudio Morterra

Dipartimento di Chimica & Interdipartimental Centre of Excellence "Nanostructured Interfaces and Surfaces - NIS", University of Torino, via P. Giuria 7, 10125 Torino - Italy

Abstract: In this chapter, the possibilities to highlight atomic/molecular details of the surface of biomaterials by IR and Raman analysis of their interaction with molecules in controlled environments are reviewed. For this aim, information on the experimental setups and procedures is also provided, as the use of vibrational spectroscopies in such conditions is rather new in the field of biomaterials. Particular attention has been devoted to the use of probe molecules. Such molecules can be quite different from those in interaction with biomaterials in actual functional conditions, but, when adsorbed in controlled conditions, become useful "molecular tools", because of the sensitivity of their vibrational features to the physical-chemical properties of surface sites.

Keywords: Infra-red spectroscopy, raman spectroscopy, hydroxyapatite, molecules adsorption, molecules desorption.

1. INTRODUCTION

Many of the most relevant improvements and innovations achieved in the last decades in the field of biomaterials resulted from the emerging awareness of both multiplicity and complementary character of the aspects to be considered, and then of the disciplines that should converge to successfully face tasks and challenges.

Among these aspects there are the physico-chemical features of the surfaces, *i.e.* of the region of a biomaterial actually in contact with the host biological environment, as clearly established since the pioneering work of L. Hench [1], then systematized by B. Kasemo [2], that proposed to name as "biological surface

***Address correspondence to Gianmario Martra:** Dipartimento di Chimica & Interdipartimental Centre of Excellence "Nanostructured Interfaces and Surfaces - NIS", University of Torino, *via* P. Giuria 7, 10125 Torino – Italy; Tel: +39 - 011 670 7538; Fax: +39- 011 670 7855; E-mail: gianmario.martra@unito.it

Lia Rimondini, Claudia Bianchi and Enrica Vernè (Eds)

science" this kind of studies. A lot of research has been carried out in this respect, and a relevant part of it was devoted to the investigation of average surface properties, such as wettability or charge sign and amount, that are commonly used for (extended) DLVO calculations to assess adhesion [3].

Still, the discovery of the benefits brought by the "nanoscale" also in biomaterials field, and the "dynamic complexity" of systems like bioactive glasses, the surface area of which evolves to an impressive extent during the contact with biological fluids, are stimulating an almost mandatory need of knowledge on a more detailed scale, typically down to the atomic/molecular level. In this respect, many methods and techniques have been developed, and are now largely employed for the investigation of engineered surfaces, even of high complexity, but essentially regular and flat, as those of interest for biosensors based on engineered surfaces of gold electrodes [4].

Conversely, this is still a challenge in the case of nanosized/nanostructured materials, typically bioactive glasses and hydroxyapatites. Actually, on a molecular scale, the surfaces of such materials can be extremely rough and heterogeneous in their physico-chemical features, typically as a consequence of the high abundance of possible surface terminations. For instance, defective terminations like steps, edges and corners from one side can represent a source of complexity and actual difficulty for a controlled engineering but, on the other side, are in several cases responsible for the peculiar features that rule the improved functional behaviour of materials in the "nano" form. At the same time, in many cases, nanosizing corresponds to the exposure of high specific surface areas, and then nanosize biomaterials can enter in the "playroom" of well assessed methodologies for the characterization, at the atomic/molecular level, of the materials that represent the oldest commercial application of nanoscience and nanotechnology, *i.e.*, heterogeneous catalysts [5]. However, in this field, several important and problematic aspects must be still faced like, for instance, the severe limits to the temperatures that can be reached for the "clean-up" (activation) of the surfaces, when probe molecules must to be adsorbed, mainly from the gaseous phase. Notably, probe molecules adsorption, to be carried out in a controlled and reproducible way, is vital to extract information on surface characteristics, on the basis of the vibrational or electronic features resulting from the interaction of the probe with specific surface centres. Another delicate problem is still represented by the need to study the interaction between active surfaces and biomolecules in an aqueous environment. Of course, owing to the complexity of "real" active surfaces and of their interfacial interactions, theoretical modeling is another fundamental tool to be added to the "research kit" in the field.

The ensemble is hard and highly challenging, but significant results are being produced. In particular, here we report on the possibility to investigate at atomic/molecular level the surface structure of inorganic nanostructured/nanosized biomaterials (namely, bioactive glasses and hydroxyapatites), and their interactions with molecules in controlled environments by employing IR and Raman spectroscopies.

2. IR STUDIES

2.1. Investigation of the Biomaterial Surface by Controlled Adsorption/Desorption of Molecules from the Gas/Vapour Phase

2.1.1. Experimental Remarks

It is well known that the easiest sampling method for collecting the IR spectrum of a powdered material is the usual technique of the KBr disk, obtained by pressing a mixture of sample and KBr powders. By selecting a proper dilution ratio, also the most intense absorption bands can be thoroughly observed. Usually, such intense bands are due to species present in the bulk of the particles, as it contains most part of the intrinsic oscillators. An example is reported in Fig. **1A**, showing the spectrum of an hydroxyapatite powder, where essentially the absorptions due to bulk species (labelled with B) are observed. Conversely, signals due to oscillators localized on the surface (in the present case, (mostly) adsorbed water molecules) appear as minor features (see weak components labelled with S). Furthermore, the grains of the sample powder are embedded in the KBr matrix, and this prevents the adsorption/desorption of molecules on/from their surface. An effective method to overcome these limitations is the preparation of self-supporting pellets of the sample powder alone (Fig. **1B**). As a consequence, the maxima of some bands due to intense (bulk) species are now out of scale (label: B), whereas the signals related to surface species (label: S) are now clearly observable. Of course, the intensity of such surface signals depends on the specific surface area (SSA) of the material investigated. Usually, a still acceptable signal to noise ratio is obtained for SSA values not lower than some 5 $m^2 \cdot g^{-1}$.

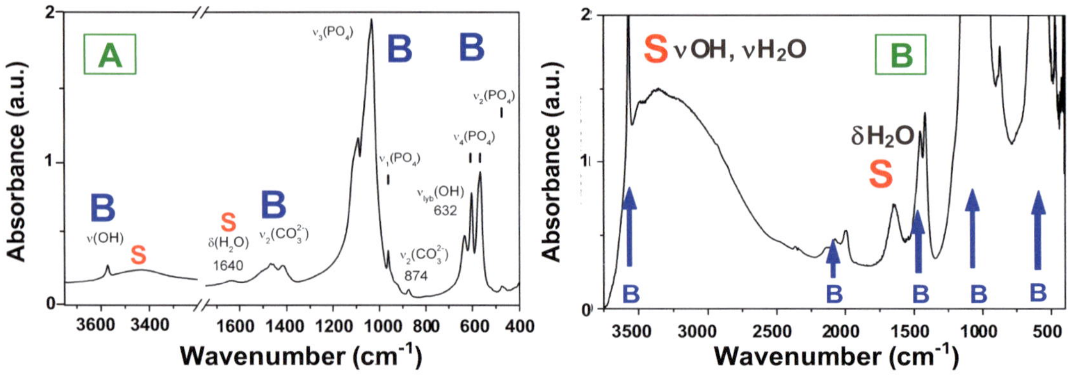

Figure 1: IR spectra of nano-hydroxyapatite (HA). Panel A: spectrum collected with the KBr disk technique (HA: KBr = 1:50); Panel B: spectrum of a self supporting pellet of HA. Letters B and S label bulk and surface IR absorptions, respectively.

Self-supporting pellets can be inserted in properly designed cells (Scheme **1A**), equipped with windows transparent to the IR radiation (KBr, CaF_2), and the cells can be attached to conventional high-vacuum lines allowing adsorption/desorption (and, if required, thermal treatments) to be carried out *in situ*.

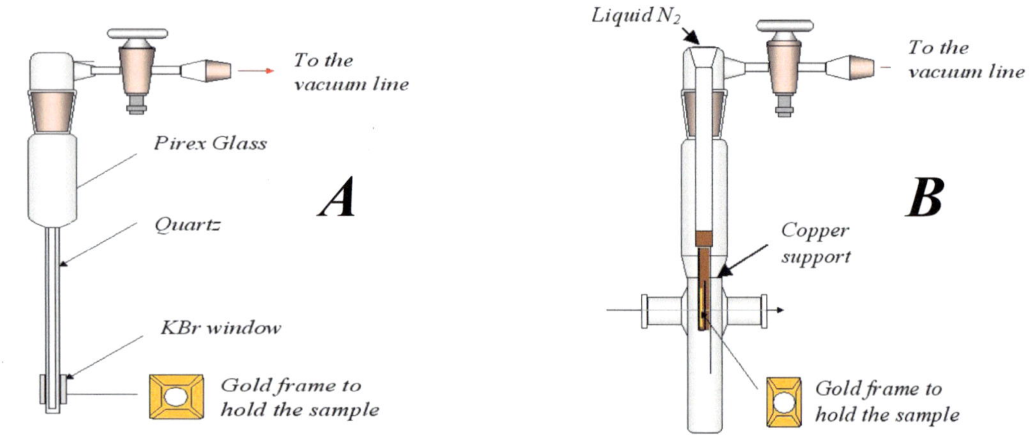

Scheme 1. Schemes of cells for IR measurements in controlled atmosphere at room temperature (A) and at nominal liquid nitrogen temperature (B).

In the case of biomaterials, this possibility is of particular interest for the study of the surface interaction with water at molecular level. Actually, the poor IR transparency of water in infrared region prevents prevents the study of a self-supporting pellet immersed in H_2O (Scheme **2A**). Still, by considering that the interfacial region relevant for the solvent-solid interaction is limited to some molecular layers, self-supporting pellets of biomaterials placed in the IR cell can be preliminary outgassed and then contacted with controlled amounts of H_2O vapour (Scheme 2B, C).

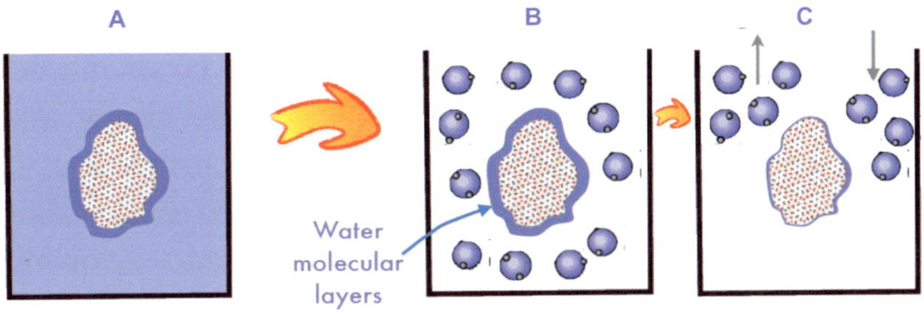

Scheme 2: Schematic drawing of an hydroxyapatite particle in a cell filled with liquid water (A), in equilibrium with H_2O vapour pressure at room temperature (B), in the presence of decreasing/increasing H_2O vapour pressure. The dark blue contour around the particles represent the layer of water molecules adsorbed on the particle surface.

In this way, the spectra of samples with adsorbed "liquid like" water layers (Fig. **2**, scheme and spectrum A), or just retaining the "first hydration layer" constituted by H_2O molecules and hydroxyl groups in direct contact with surface atoms (Fig. **2**, scheme and spectrum B) can be recorded. Furthermore, after outgassing at a proper temperature, the spectrum of the dehydrated material can be obtained (Fig. **2**, Scheme **2** and spectrum C).

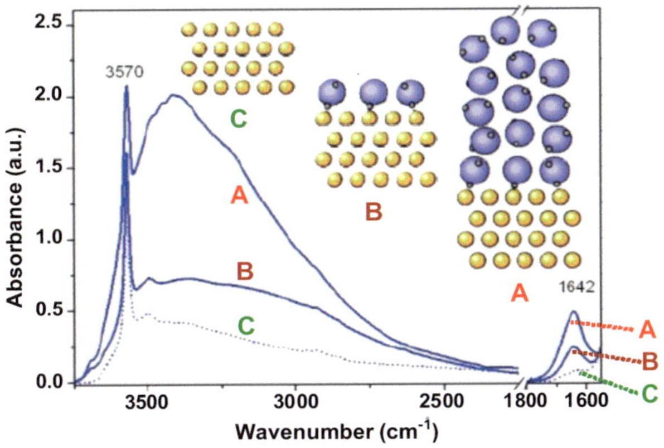

Figure 2: IR spectra of a self-supporting pellet of nano-hydroxyapatite inserted in the IR cell and: (A) left in contact with 20 mbar H_2O vapour, and subsequently outgassed for 60 min at (B) room temperature and (C) 573 K.

In particular, in the case of nano-hydroxyapatites, that are prepared in aqueous media at mild temperature, it is important to proceed with care and assess the limiting outgassing temperature that can be adopted without the occurrence of an extensive, irreversible transformation of the original surface structure. This check can be conveniently carried out by IR spectroscopy, evaluating the capability of the thermally treated sample to re-adsorb water, with respect to the original material [6]. Typically, in the IR spectrum of dehydrated biomaterials, bands due to specific surface oscillators (like, for instance, hydroxyl groups) can be clearly observed, as widely reported in the case of bioactive glasses [7-8]. Conversely, other vibrational modes related to the solid lattice surface terminations are not so easily distinguishable from the corresponding bulk ones. This is the case of the modes involving cationic sites, present at the surface and in the bulk of both in bioactive glasses and hydroxyapatites. However, it is possible to use probe molecules which, when in interaction with surface-localized sites (surface sites), modify their vibrational features in dependence on the physical-chemical features (*i.e.,* oxidation state, coordination, local structure) of the adsorbing centres. This method has been developed and widely adopted in the field of heterogeneous catalysis [9] and many kinds of probe molecules have been selected to this aim. Several papers dealing with the study of biomaterials by this method appeared in the last decades, where the probe molecules considered were H_2O [10, 11], CO_2 [12], CD_3CN [13], CH_3OH [7] and CO [6, 8, 14, 15]. In particular, carbon monoxide plays a special role, because of the simplicity of its IR spectrum, the high sensitivity to the physical/chemical features of adsorbing sites, and the possibility to adsorb on metal sites independently on their electronic configuration. Fundamental applications of CO as vibrational probe molecule have been reviewed by Sheppard *et al.* [16] for the investigation of zero-valent metals, and more recently by Hadjiivanov *et al.* [17] for the investigation of cationic centres. On the basis of the examples reported in the following sections, here, we would just mention that the adsorptive interaction of CO with zero-valent metal sites or with transition metal ions in a low oxidation state results in a C=O stretching wavenumber (νCO) lower than that of CO in gas phase (2143 cm^{-1}), whereas the opposite occurs for CO in interaction with cations in a d^0 electronic configuration.

2.1.2. Biomaterial-Water Interaction: IR Study of H_2O on HA

Since the definition of the "Biological Surface Science" [2], the relevant role of biomaterials-water interaction has been recognized. As an example of the possibility to use IR spectroscopy to investigate such aspect, the results of the study of the interaction of H_2O molecules with nanohydroxyapatite particles exhibiting amorphous (hereafter: HA-am) or crystalline (hereafter: HA-cry) surfaces will be summarized (for more details, see Ref. [15]).

In order to study the structure of the water layers formed at the interface with the materials of interest, increasing amounts of H_2O (from few mbar to the vapour pressure at room temperature, *i.e*, ca. 24 mbar) were admitted on the samples, previously dehydrated by outgassing at 403 K (HA-am) or 433 K (HA-cry). The evolution of the stretching vibrational modes of H_2O (νH_2O) the most sensitive ones to the structure of the H-bond network, was then followed by IR spectroscopy (see Fig. **3A**, **3B**). To extract information related only to the adsorbed water the spectra of the materials preoutgassed at the indicated temperatures (spectra "a" in Fig. **3A**, **3B**) were subtracted as background. The resulting absorption spectra are shown in Fig **3 A'** and **B'** for HA-am and HA-cry, respectively.

At low coverage, where essentially only H_2O molecules in direct contact with the surface are present, a very broad band extended over the 3700-2600 cm^{-1} spectral range and centered at about 3300 cm^{-1} can be observed for both materials. The high similarity between the spectral profile obtained for both HA-am and HA-cry well agrees with the independence of the local physical-chemical features of the surface sites of nanohydroxyapatite on the surface long range order/disorder [15]. As the coverage increases, the band appears to be structured in two components (centered at about 3400 cm^{-1} and 3200 cm^{-1}, respectively). Their relative intensity appeared different in the two cases, and for HA-am the absorption at ca. 3200 cm^{-1} exhibited almost the same intensity of the partner at higher frequency (Fig. **3A'**). As a result, the $\nu(H_2O)$ band appeared significantly broader, indicating the occurrence of stronger H-bonding among water molecule adsorbed in multilayers. Thus, the full set of data suggests the following picture:

i. H_2O molecules in direct contact with the surface, forming the first hydration layer, are apart enough each other to prevent intermolecular interactions/structuring dependent on the long range order/disorder of the surface;

ii. When additional H_2O molecules are adsorbed, they interact with the first hydration layer, exhibiting a pattern dependent on the order/disorder of the underlying surface. Such a pattern affect the H-bonding network progressively built-up, then "transmitting" to the multilayer an "imprinting" dependent on the structure of the hydroxyapatite surface.

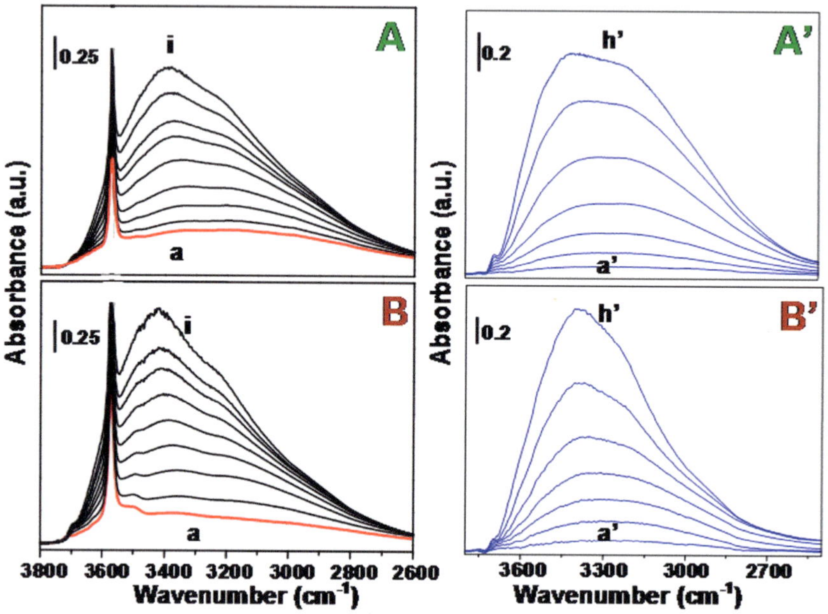

Figure 3: Absorbance IR spectra of the ν(H_2O) region of HA-am (section A) and HA-cry (section B). Curves (a) sample outgassed at 403 K (HA-am) and 433 K (HA-cry); curves (b-i) after admission of increasing amounts (pressures) of water vapour pressure (in the sense of the lettering). Sections A' and B': bands to adsorbed H_2O, obtained after subtraction of the spectra of the materials pre-outgassed at the indicated temperatures (curves (a) in panels A and B) used as background spectra. Adapted from Ref. [14].

2.1.3. CO as a Vibrational Probe for Surface Sites of Biomaterials

Surface Cations of Nanohydroxyapatites

Biomimetism is a leading criterion in addressing the preparation of synthetic hydroxyapatites. In this respect, other than nanosize, the introduction in the structure of both anionic and cationic substitutents of stoichiometric constituents (OH^-, PO_4^{3-}, Ca^{2+}) is important. As for cations, Mg^{2+} is considered to play a relevant role. Its exposure at the surface results in a significant effect towards water adsorption [11], as a consequence of the much higher polarizing power of this cation with respect to Ca^{2+}. IR spectroscopy of adsorbed CO offers the possibility to probe the actual presence of different cationic centers at the surface of Mg^{2+}/Ca^{2+} nanohydroxyapatites (for more details see Ref. [14]). As shown in Fig. **4A**, the spectra of an all-Ca^{2+} HA and a Mg^{2+}/Ca^{2+} HA (Mg content: 8 at%) contacted with carbon monoxide exhibit a common band at 2182 cm^{-1}, due to CO molecules adsorbed on surface Ca^{2+} sites. Notably, in the case of Mg^{2+}/Ca^{2+} HA two weak additional components are present at 2210 and 2200 cm^{-1}, clearly monitoring the exposure at the surface of magnesium cations, also. Indeed, their location at higher frequencies results from the larger polarization of CO on Mg^{2+} ions, that have a higher charge/radius ratio than Ca^{2+} ions. Furthermore, IR spectroscopy of adsorbed CO was employed to probe a possible evolution of surface features induced by samples ageing. This process plays a relevant role for biomaterials, and in the case of multicationic hydroxypatite is believed to be accompanied by cations migration, and, possibly, by Mg^{2+} enrichment at the surface. In this respect, Fig. **4B** compares the spectra of CO adsorbed on

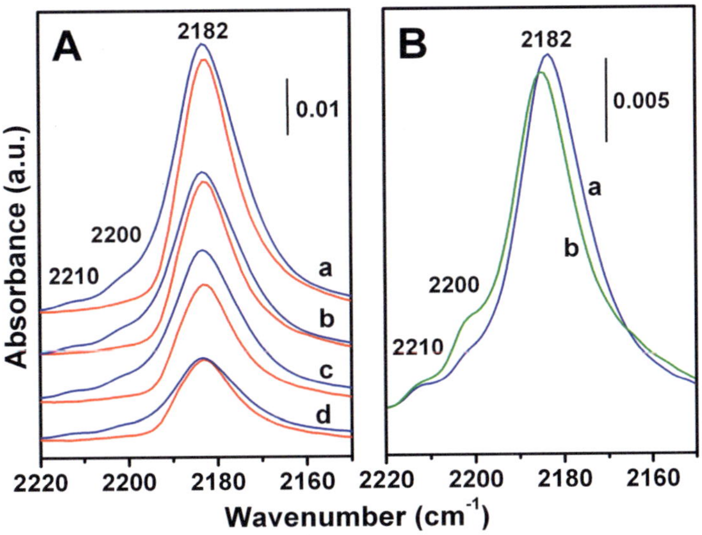

Figure 4: Absorbance IR spectra of CO adsorbed at 77 K on all-Ca^{2+} HA and Mg^{2+}/Ca^{2+} HA (Mg^{2+} content: 8 at %) pre-outgassed at 573 K. Section A: spectra collected at decreasing CO coverage, from 3.0×10^{-2} to 1.0×10^{-3} mbar CO (in the sense of the lettering) for Ca^{2+} HA (red lines) and Mg^{2+}/Ca^{2+} HA (blue curves). Section B: spectra collected in the presence of 1.0×10^{-3} mbar CO for Mg^{2+}/Ca^{2+} HA in the fresh form (a, blue line; the same spectrum as curve d in section A), and aged for 2 months (b, green line).

Mg^{2+}/Ca^{2+} HA in the fresh form (blue line) and after 2 months of ageing (green line). It can be observed that the intensity of the two components at 2210 and 2200 cm^{-1} is somewhat increased in the case of the aged sample, indicating that a migration of Mg^{2+} ions toward the surface occurred. These data well agree with bulk structure investigations carried out by XRD. During aging at room temperature, lattice parameters are found to increase gradually up to the restoration of values typical of pure Ca^{2+} HA, indicating that Mg^{2+} substituting Ca^{2+} is progressively expelled from the bulk of the HA structure.

Au Species at the Surface of Gold-Functionalized Bioactive Glasses

Metallic nanoparticles can be grown in glasses: an excellent example is the gold ruby glass, in which small particles of metallic gold are dispersed in a silicate glass matrix [18]. The use of gold nanoparticles (AuNPs) is becoming increasingly important also in the biological and pharmaceutical fields because of the variety of surface coatings (biofunctionalization) and the great biocompatibility.

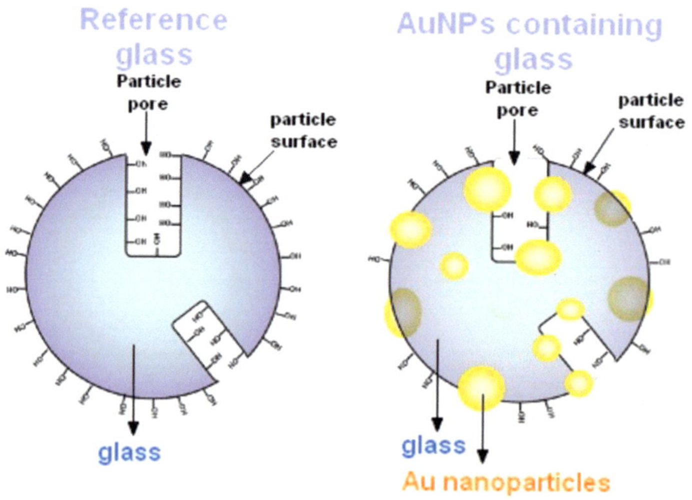

Scheme 3: Schematic representation of reference glass (SG) and AuNPs containing glass (SGAu) Surface (Adapted from Ref. [8]).

Gel glasses of variable $CaO\text{-}SiO_2\text{-}P_2O_5$ composition have been widely investigated as biomaterials because, if a proper ratio among $CaO\text{-}SiO_2\text{-}P_2O_5$ is reached, the systems are likely to develop bioactivity. Bioactive glasses containing AuNPs are interesting biomaterials, as they join the bioactivity of the glass with the possibility of releasing gold nanoparticles in a specific body site. Moreover, the great ability of gold nanoparticles to become functionalized can be advantageous in order to produce bioactive glasses functionalized with specific biomolecules. To this purpose it is very important to determine nature, optical properties, shape, size, and distribution of AuNPs (see for more details Ref. [8]). In Scheme **3** a representation of SG and SGAu samples.

The most common explanation for the variability of the catalytic properties of gold catalysts is the amount of low coordinated gold sites [19]. The concentration of low-coordinated sites is usually related to the size of gold metallic particles, which in most cases is determined by transmission electron microscopy (TEM). When performed accurately, TEM provides a particle size distribution from which a mean size can be derived. But since very small metallic particles (<1 nm) are barely detectable, the method is not ideally applicable for samples containing highly dispersed gold. In these cases, other techniques, such as EXAFS and XANES, can be used to obtain information on gold coordination number of gold. But these alternative measurement techniques are not widely used yet and thus cannot be proposed as routine tests. The use of IR spectroscopy of adsorbed CO allows the determination of gold speciation.

In the case bioactive glasses containing AuNPs, experiments of CO adsorption have been carried out in order to obtaine information about Ca and Au surface sites.

Sections A and B of Fig. **5** report the spectra obtained at (nominally) 77K in the νCO spectral region, by gradually decreasing CO coverage on the sol-gel glass reference system (SG_873, glass without gold thermally treated at 873K) and on the sol-gel glass containing AuNPs (SGAu_873, glass with gold nanoparticles thermally treated at 873K) respectively. The thermal treatment at 873K for 3 hours ensures the elimination of solvents and reagent residues obtaining a gel glasses.

On the basis of the self-explanatory spectral behavior and of literature data, [20] the 2155-2160 cm^{-1} band, present on both systems, can be attributed to CO molecules in weak coordinative interaction with coordinatively unsaturated (*cus*) surface Ca^{2+} species.

In the case of the Au-containing sample (Fig. **5B**), a somewhat more complex situation is present, as the CO band at ~2160 cm^{-1} mentioned above is still present, but also another weaker CO component is clearly evident at lower wavenumbers. The latter νCO component, centered at ~2138 cm^{-1} is most likely to derive from the interaction of CO molecules with $Au^{\delta+}$ sites, in which a tiny partial positive charge would be induced by the interaction with nearby oxygen species, as reported in the case of AuNPs dispersed on SiO_2 or other oxides [19].

This study allows to verify presence and nature of Au metal species at the sample surface. As previously anticipated AuNPs are used for different medical applications and the important role played by the SH group of cysteine in proteins binding to the surface of gold particles has a useful tool to improve the biomaterials biocompatibility.

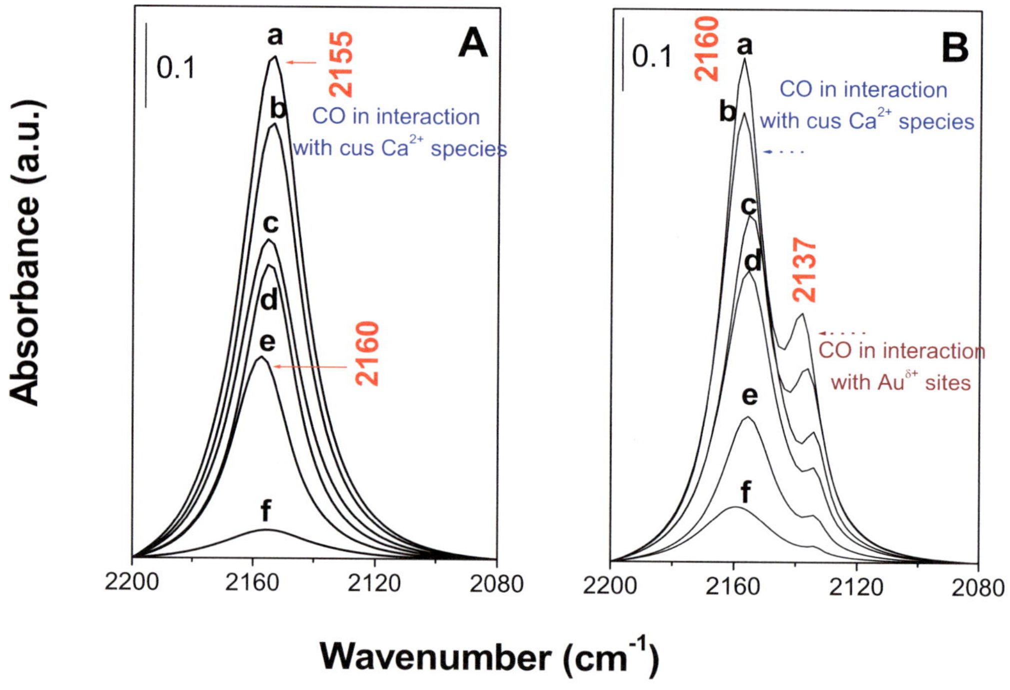

Figure 5: Absorbance differential IR spectra (in the region 2200-2080 cm^{-1}) relative to the adsorption of CO at 873K on SG_873 (A) and SGAu_873 (B) in contact with: a) 50, b) 20, c) 10, d) 5, e) 2.5 and f) 1 Torr CO. (Differential spectra were obtained with respect to the background spectrum of the glass samples outgassed at 478K after a preliminary treatment at 673K in the presence of O_2 and subsequent rehydration with H_2O at r.t.; see for details Ref. [8]).

2.2. Interaction of Biomaterials with Molecules in Aqueous Media: *in situ* IR Study of Proteins Adsorbed on HA

2.2.1. Experimental Remarks

Once biomaterials are implanted, proteins are the first biomolecules that are adsorbed on their surface, giving origin to a hybrid biomaterial/protein interface that is the actual substrate subsequently interacting with cells. For this reason, the adsorption of proteins on biomaterials has been and still is the subject of a number of investigations. Beside the obvious quantitative aspect, a relevant point is also the assessment of the structure of adsorbed proteins, and namely if "native" or partially/totally denatured. In this respect, IR spectroscopy is an effective tool, owing to the sensitivity to the secondary structure of amide I (the $\nu C=O$ vibration) band [21]. In several cases, the observation of such analytical band for adsorbed protein has been carried out on samples dehydrated, mixed with KBr and pressed, so as to obtain pellets for the usual collection of IR spectra in the most common transmission mode. But, this sampling method can heavily affect the protein structure, independently of its interaction with the biomaterial surface. An interesting alternative is offered by the possibility to keep the powder with adsorbed protein in an aqueous, buffered medium and collect the spectra in the ATR mode. Moreover, the use of D_2O instead of H_2O allows to augment the structural sensitivity of the IR investigation due to the resistance to H/D exchange of peptide units contained in hydrophobic pockets. As a consequence, a portion of the amide II ($\delta CNH+\nu CN$) band should be present in the IR spectrum of native proteins dissolved in D_2O-based media. Conversely, in denatured proteins hydrophobic pockets are not longer present, so that the H/D exchange occurs for all units of the polypeptide backbone and the amide II band is depleted [21]. On such a basis, the spectra of proteins irreversibly adsorbed on biomaterial nanoparticles can be effectively obtained by pressing on an ATR crystal the slurry obtained at the end of the centrifugation/washing cycles usually following the incubation of the sample in a protein solution. To limit the amount of sample required, single-reflection diamond ATR crystals are preferable (Scheme **4A**). By properly setting the liquid/solid ratio and the pressure over the ATR crystal, the resulting spectrum should exhibit bands due to the deuterated buffer, the bulk of biomaterials and adsorbed proteins (Scheme **4B**). These bands are the less intense features by far, as expected for molecules retained on the sample surface.

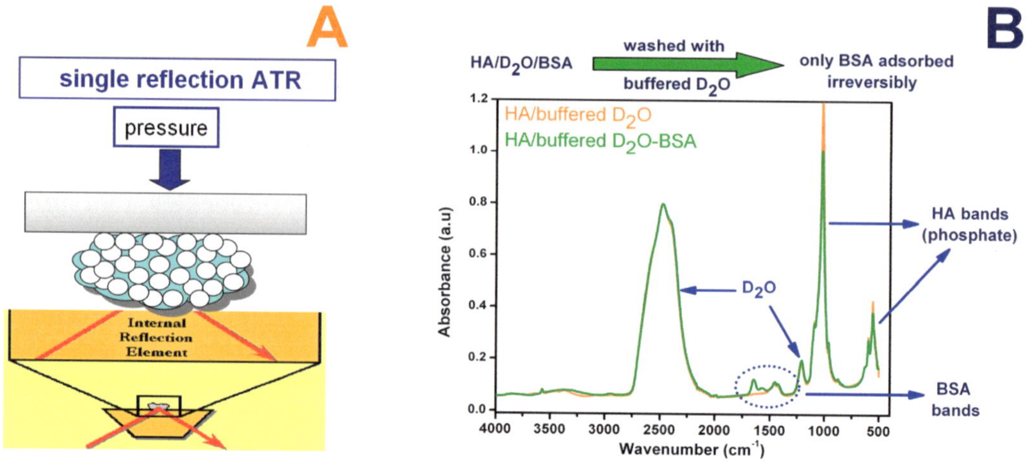

Scheme 4: Panel A: schematic drawing of the experimental set-up for the collection of IR/ATR spectra of proteins absorbed on powdered biomaterials, keeping the sample in an aqueous buffered medium. Panel B: comparative view of the IR/ATR spectrum of a nanohydroxyapatite, HA, without (orange curve) and with irreversibly adsorbed BSA (green curve), in the form of a slurry with buffered D_2O.

2.2.2. Study of the Structure of BSA Adsorbed on Silica and Hydroxyapatite

The method described above was used for a comparative study of the structure of bovine serum albumin (BSA) adsorbed on a commercial pyrogenic silica (Aerosil AOX 50) and a home prepared nano-hydroxyapatite (hereafter HA) [22]. The samples analysed contained only the irreversible fraction of BSA adsorbed from a 5.0 mg·mL protein D_2O solution, buffered at pD= 7.4.

For comparison purposes, the spectra of native BSA in buffered D_2O (pH 7.4) and after thermal denaturation were also collected (Fig. **6A**, curves a and b, respectively). As a consequence of the denaturation, amide I band shifted from 1652 to 1646 cm^{-1}, and the remains of the amide II band at 1547 cm^{-1} disappeared. The spectrum of BSA irreversibly adsorbed on silica AOX 50 (Fig. **6A**, **c**) appeared very similar to that of denatured BSA in solution. Conversely, in the case of BSA adsorbed on HA the amide I peak underwent a downward shift limited to 1648 cm^{-1}, and the amide II band was still present. Because of the weak intensity of the latter spectral feature, confirmative evidence of its presence was obtained by inducing the denaturatioin of the adsorbed protein by heating an equivalent slurry in a closed vessel at ca. 353 K. Curve “b” of Fig. **6B** shows that, in the spectrum of the thermally treated sample, the amide II band was no longer present, and the amide I peak appered shifted to 1645 cm^{-1}.

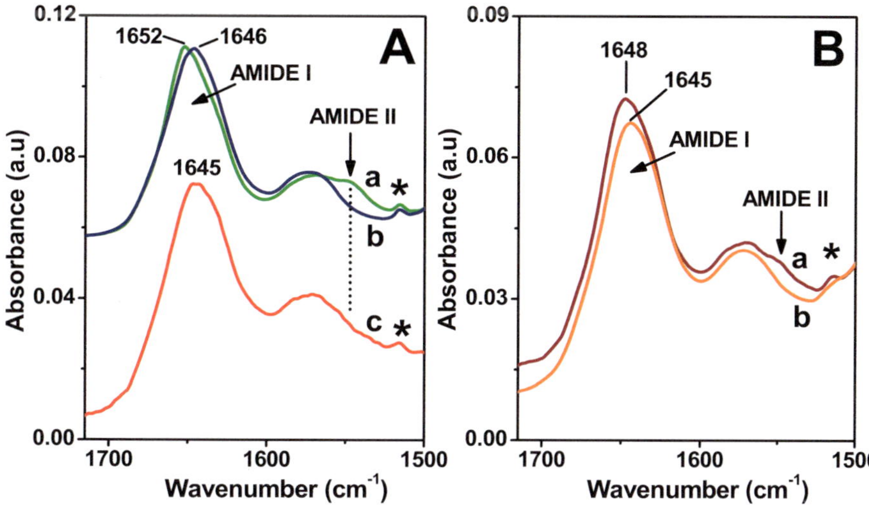

Figure 6: IR/ATR spectra of BSA in solution and irreversibly adsorbed on silica (Aerosil AOX50) and hydroxyapatite powders (at room temperature, from a 5.0 mg·mL buffered protein solution. Spectral data were collected by using the method described in the text, with samples in the form of slurries in buffered D_2O (pD = 7.4). Section A: (a) native BSA in buffered D_2O (20 mg·mL); (b) the same solution, after denaturation by heating at 373 K for 15 min; (c) BSA irreversibly adsorbed on SiO_2. Section B: (a) BSA adsorbed on HA; (b) the same sample, after heating at 353 K for 15 min. After the treatment the sample was dried, and then a small amount of buffered D_2O was added before collecting the spectrum. The weak band labelled with the asterisk is due to tyrosine residues.

As indipendent NIR studies [22] evidenced that protein adsorption resulted in the surface dehydration of both materials, the observed behaviour can be rationalised as follows. In the case of the BSA/A50-silica system, the hydrophobic interaction between surface sites (mainly apolar siloxane bridges) and apolar groups originally in the internal part of the protein prevailed, leading to a significant modification of the structure of the adsorbed protein. Conversely, in the BSA/HA case, the cationic and anionic sites exposed on the saline surface should actually interact with the polar residues exposed by the proteins in a way strong enough to prevent denaturation.

3. CONFOCAL RAMAN STUDIES

The study of the structural evolution of biomaterials during the reaction in simulated body fluids of different complexity is very important in order to understand the initial steps of the reaction mechanism that involve solubilization of the surface layers of the biomaterials.

Confocal Raman should be highly suitable for the investigation of the materials surface in contact with aqueous solutions of biomolecules, due to the essential insensitivity of Raman scattering to the presence of water. The innovation of this technique resides in the new approach to the study of glass bioactivity, in that the modifications undergone by the biomaterial have been followed directly in solution.

Here we briefly report the reactivity of a sol-gel bioactive glass immersed in plain TRIS-buffered water (tris(hydroxymethyl)aminomethane). This simplified procedure has been adopted because the contact of the bioactive glass with TRIS-buffered water is expected to maximize the solubility of the glass (no Ca, P and/or other ions are present in the solution), and to minimize all back-precipitation processes (only chemical species first extracted by dissolution from the solid itself can lead to re-precipitation processes at its surface). In Scheme **5** a pictorial representation of the Raman experimental setup is reported (for more details, see Ref. [23]).

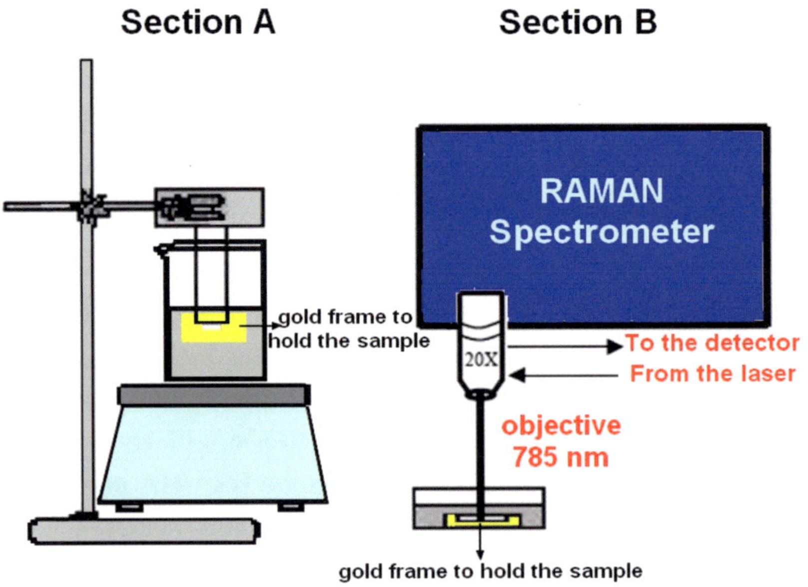

Scheme 5: A pictorial representation of the experimental setup. Section A shows the home-made manifold in which the sample pellet was immersed into a TRIS solution under stirring. Section B represents the capsule used for containing the sample, covered by a fraction of reaction solution, during the Raman measurements (Adapted from Ref. [23]).

In Fig. **7** are reported the Raman spectra carried out on a sol-gel bioactive glass (77S: 77% SiO_2; 14% CaO and 9% P_2O_5, %wt), both as-such and after 7 days reaction in TRIS solution. For comparison purposes, also the spectra of HA (trace A) and type B HCA hydroxyl-carbonate apatite (HCA) (trace B) are also reported.

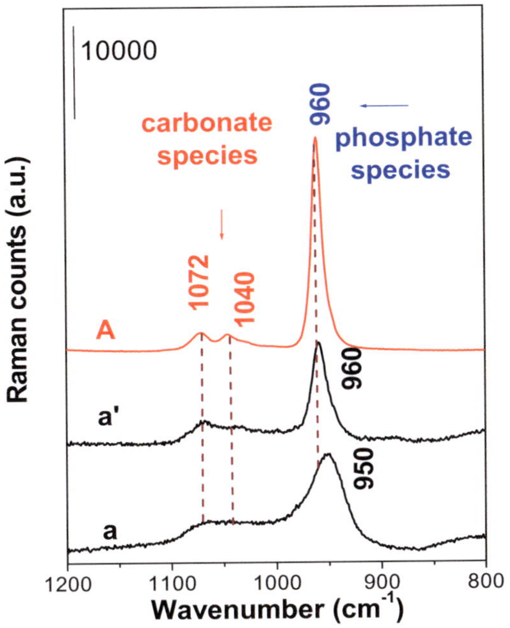

Figure 7: Raman spectra of a 77S sample: a as synthesized and b after 1 week of reaction in TRIS solution. Trace (A) is the spectrum of a type B HCA as reference sample.

In the bioactive glass reacted 7 days (curve a'), the appearance of all the fingerprint spectral features typical of HA-like species confirms that, upon TRIS reaction, phosphate species present on the glass surface underwent a phase transition from amorphous to crystalline. Actually, it is worth noting that we are probably in the presence of HCA and, in particular, of that modification termed type B, as can be confirmed by comparing the vibrational features of the reacted glass sample (curve a' of Fig. **7**) with those of crystalline type B HCA (see curve (A) of Fig. **7**).

Carbonate species present in HCA could derive from CO_2 taken from the atmosphere during the reaction, the (covered) beaker being in contact with air. In fact, in the 'as synthesized' glass sample, no contaminant carbonate surface species seem to be present. Instead, in an aged sample, we expect to observe the main features of carbonate species and, in particular, a sharp signal around 1080 cm^{-1}, that is the most active vibrational mode of ionic carbonate species in Raman spectroscopy (see the broken-line segment in Fig. 7). This observation demonstrates that the just prepared sample we used was quite "clean" (*i.e.*, still virtually free from surface contamination). It has been verified that, after 1 week of reaction, the formation of crystalline HCA can be rather easily recognized by some of its typical Raman features, as proved by the direct vibrational comparison with a reference sample of type B HCA. These simple preliminary results concerning the spectral evolution of chemical species at the surface of a bioactive glass should be considered as the starting point for the vibrational study of the *in vitro* interaction between biosystems (bioactive glasses or apatites) and more complex solutions containing biomolecules, such as amino acids and proteins (for details, see Ref. [23]).

4. CONCLUSIONS

The collection of examples here reported prove that IR and Raman spectroscopies can be effectively employed to investigate the interaction of biomaterials and molecules, leading to unique information on their surface structure and interfacial behaviour, as well as on their possible transformation when in contact with body fluids. On the basis of the experimental and interpretative targets achieved, the study at a molecular level of new and more complex biomaterials might be engaged. A synergistic combination of the results of this kind of investigation with those provided by biochemical studies, cell tests and, possibly, *in vivo* tests, could actually represent a source of innovative knowledge to refine and/or disclose the links among biomaterials design, synthesis/preparation, and expected surface functional behaviour. If successful, this integrated approach could represent a significant step toward the frontier of precisely tailored and highly effective biomaterials.

ACKNOWLEDGEMENTS

The Authors are grateful to Italian Ministry MIUR (PRIN Projects Prot. Numbers 2003032158 and 2006032335) and Regione Piemonte (Project CIPE-2004, Prot. Code D33; NANOMAT-ASP, Docup 2000-2006, Linea 2.4a) for financial support.

The Italian colleagues Dr. E. Verné (Politechnic of Torino), Dr. A. Tampieri (ISTEC-CNR, Faenza), Prof. L. Menabue (University of Modena e Reggio Emilia) and Prof. N. Roveri (University of Bologna) are acknowledged for fruitful collaborations.

CONFLICT OF INTEREST

The author(s) confirm that this chapter content has no conflict of interest.

REFERENCES

[1] Hench LL, Ethridge EC. Biomaterials, an interfacial approach, New York: Academic Press 1982.
[2] Kasemo B. Biol Surf Sci 2002; 500: 656-77.
[3] Hermansson M. The DLVO theory in microbial adhesion. Coll Surf B: Bioint 1999; 14: 105-12.
[4] Scheller FW, Wollenberger U, Warsinke A, Lisdat F. Research and development in biosensors. Curr Opin Biotech 2001; 12: 35-40.
[5] Somorjai A, Borodko YG, Research in Nanosciences. Great Opportunity for Catalysis Science. Catal Letters 2001; 76: 1-5.
[6] Bertinetti L, Tampieri A, Landi E, Ducati C, Midgley P, Coluccia S, Martra G. Surface structure, hydration, and cationic sites of nanohydroxyapatite: UHR-TEM, IR, and microgravimetric studies. J Phys Chem C 2007; 111: 4027–35.
[7] Bolis V, Busco C, Aina V, Morterra C, Ugliengo P. Surface properties of silica-based biomaterials: Ca species at the surface of amorphous silica as model sites. J Phys Chem C 2008; 112: 16879-92.
[8] Lusvardi G, Malavasi G, Aina V, Bertinetti L, Cerrato G, Magnacca G, Morterra C, Menabue L. Bioactive glasses containing Au nanoparticles. Effect of calcination temperature on structure, morphology, and surface properties. Langmuir 2010; 26: 10303-14.
[9] Busca G. The Use of Infrared Spetcroscopic Methods in the Field of Heterogeneous Catalysis by Metal Oxide In: Jackson D, Hargreaves JSJ, Eds. Metal Oxide Catalysis. Weinhein, Wiley-VCH, 2009; pp. 95-176.
[10] Ishikawa T, Wakamura M, Kondo S. Surface characterization of calcium hydroxylapatite by Fourier transform infrared spectroscopy. Langmuir 1989; 5: 140-4.
[11] Bertinetti L, Drouet C, Combes C, Rey C, Tampieri A, Coluccia S, Martra G. Surface Characteristics of Nanocrystalline Apatites: Effect of Mg Surface Enrichment on Morphology, Surface Hydration Species, and Cationic Environments. Langmuir 2009; 25: 5647-54.
[12] Cheng ZH, Yasukawa A, Kandori K, Ishikawa T. FTIR Study of Adsorption of CO_2 on Nonstoichiometric Calcium Hydroxyapatite. Langmuir 1998; 14: 6681-6.
[13] Cerruti M, Morterra C, Ugliengo P. Surface features of P-doped silica explored with CD_3CN adsorption: Can Si atoms act as Lewis centers? Chem Mater 2005; 17: 1416-23.
[14] Bertinetti L, Tampieri A, Landi E, Martra G, Coluccia S. Punctual investigation of surface sites of HA and magnesium–HA. J Eur Cer Soc 2006; *26*: 987-91.
[15] Sakhno Y, Bertinetti L, Iafisco M, Tampieri A, Roveri N, Martra G. Surface hydration and cationic sites of nanohydroxyapatites with amorphous or crystalline surfaces: a comparative study. J Phys Chem *C* 2010; 114: 6640 -6648.
[16] Sheppard N, Nguyen TT. The vibrational spectra of carbon monoxide chemisorbed on the surfaces of metal catalysts a suggested scheme of interpretation. In: Clark RJH, Hester RE Eds. Advances in Infrared and Raman Spectroscopy. London, Heyden, 1978; pp. 67–148.

[17] Hadjiivanov KI, Vayssilov GN. Characterization of Oxide Surfaces and Zeolites by Carbon Monoxide as an IR Probe Molecule. Adv Catal 2002; *47*: 307-511.

[18] Weil W. Colored Glasses; The Society of Glass Technology: Sheffield, UK 1951.

[19] Ilieva L, Sobczak JW, Manzoli M, Su BL, Andreeva D. Reduction behaviour of nanostructured gold catalysts supported on mesoporous titania and zirconia. Appl Catal A 2005; 291: 85-92.

[20] Babaeva MA, Bystrov DS, Kovalgin, AY, Tsyganenko AA. CO interaction with the surface of thermally activated CaO and MgO. J Catal 1990; 123: 396-416.

[21] Maentele FW. Infrared Spectroscopy of proteins. In: Chalmers JM, PR Griffiths Eds. Handbook of Vibrational Spectroscopy. Chichester, Wiley, 2002; pp. 3399-426.

[22] Bertinetti L, Ceschino R, Bollati D, Landi E, Tampieri A, Martra G, Protein adsorption on nanohydroxyapatites: Spectroscopic investigations at molecular level. Key Eng Mat 2008; 361-363: 87-90.

[23] Bonino F, Damin A, Aina V, Miola M, Vernè E, Bretcanu O, Bordiga S, Zecchina A, Morterra C. *Et al.* study to monitor bioactive glasses reactivity. J Raman Spectr 2008; 39: 260-4.

CHAPTER 6

Morphological Surface Characterization of Nano-Structured Inorganic and Polymeric Materials for Biomedical Application

Barbara Palazzo[1,2,*], Ismaela Foltran[2,3] and Dominic Walsh[4]

[1]R&D Medical Device, Ghimas S.p.A. c/o Melab, Cittadella della Ricerca SS7 per Mesagne Km 7+ 300, 72100 Brindisi, Italy; [2]Department of Medical Sciences, Università del Piemonte Orientale "Amedeo Avogadro", via Solaroli 17, 28100 Novara, Italy; [3]Department of Chemistry "G. Ciamician", Alma Mater Studiorum Bologna University, via F. Selmi 2, 40126 Bologna, Italy and [4]Advanced Research Fellow, School of Chemistry, University of Bristol, Cantocks Close, BristolBS8 1TS, UK

Abstract: Control of surface morphology impacts greatly upon the performance of nano-structured devices. This is particularly true in the case of biomaterials and medical devices, because cell responses are highly affected by topography. This chapter describes some of the most up to date microscopy principles and methods, offering a flavor of the broad range in which AFM, SEM and TEM techniques have impacted upon advances in understanding. Additionally, some applied examples of biomedical materials surface characterizations are described in order to point the way to gain useful information about the interaction between biomaterials and biological systems.

Keywords: Biomaterials, nanostructured devices, atomic Force Microscopy, Electron Microscopy, nanofibers, nano-hydroxyapatite, collagen.

1. INTRODUCTION

Hierarchically nanostructured materials exhibit order on multiple length scales, with at least one of which in nanometre dimension. The control of the morphology which governs material performance and manufacturing remains a challenge. The development of novel qualitative and sometimes quantitative characterization techniques is important in order to correlate the underlying morphology to the relevant processing parameters and functional macroscopic behaviours. A further understanding of the relationship between the structure and the resultant

***Address correspondence to Barbara Plazzo:** R&D Medical Device, Ghimas S.p.A. c/o Melab, Cittadella della Ricerca SS7 per Mesagne Km 7+ 300, 72100 Brindisi, Italy; Tel: +39 0321 660673, +39 0321 32221; E-mail: barbarapalazzo@gmail.com

Lia Rimondini, Claudia Bianchi and Enrica Vernè (Eds)

properties is also required to bridge the gap between lab-scale research and functional application, in particular when the materials are intended for Medicine. In fact, an important issue in biocompatibility is how the structural make-up of the material influences the response of a biological environment including protein absorption and functionality of cells and tissues. For devices and materials that do not leach undesirable substances, this phenomenon occurs at the interface.

Surface analysis methods are common in solid state physics and in microelectronic science but are relatively new in biology. These methods have been categorized by Ratner [1] into six classes: Electron Spectroscopies (*e.g.* ESCA-Electron Spectroscopy for Chemical Analysis), Ion Spectroscopies (*e.g.* SIMS –Secondary Ion Mass Spectroscopy), Vibrational Spectroscopies (*e.g.* ATR-IR – Attenuated Total Reflectance Infrared), Contact Angle Methods, Diffraction methods (*e.g.* XPS X-ray Photoelectron Spectroscopy) and Scanning Probe and Electron Microscopies (*e.g.* AFM- Atomic Force Microscopy, STM Scanning Tunneling Microscopy, SECM Scanning Electrochemical Microscopy, NFSOM Near Field Scanning Optical Microscopy, SEM Scanning Electron Microscopy).

This chapter will offer insight into some of the most used surface microscopy technologies in Scanning Probe Microscope (SPM), atomic force, and in scanning and transmission electron microscopies. The topics described are not exhaustive of the area but only aimed to offer a flavor of the broad range in which AFM, SEM and TEM have improved ways to obtain useful information about biomaterials and biological systems.

2. SCANNING PROBE MICROSCOPY

In the last few decades, the improvement of Scanning Probe Microscope has provided a wide range of novel surface analytical techniques useful in the characterization of biomedical interfaces. SPMs are unique compared to other microscopes in because they are based on a fine probe tip which senses the surface of the samples.

Depending on the physico-chemical interaction(s) responsible for the image formation, a whole range of material surface properties are accessible [2].This is reflected in various modes of operation such as Atomic Force Microscopy (AFM, repulsive forces), Scanning Magnetic Microscopy (SMM, magnetic interactions), Friction Force Microscopy (FFM, lateral shear force). Furthermore, Scanning Thermal Microscopy (SThM) [3-5] and Near-Field Scanning Optical Microscopy (NSOM) [6] use heat and light respectively as signal to obtain surface profiles of the materials including pharmaceutical samples.

The AFM is the most widely used SPMs in the field of drug delivery technology [7, 8].

2.1. Atomic Force Microscopy

2.1.1. The Principle of AFM

The main features of the AFM include the ability to *(i)* image non-conductive samples (*e.g.*ceramic, biological and drug particles) and *(ii)* measure the surface topography of the specimens at sub-nanometric resolution. In addition, AFM can work under a range of environmental conditions including in air and in liquid and over a range of temperatures, with minimal sample manipulation and low running costs [9-11].

During AFM imaging, a sharp probe tip raster scans over a specimen surface. The motion is obtained using a piezoelectric scanner.

The main element of an AFM microscope is actually the sensor, a silicon nitride pyramidal tip attached to a 100-400 μm long flexible silicon cantilever [12]. The geometry of the apex of the tip represents one of the key parameters for the image resolution. Highest resolutions can be achieved with a tip that ends in a single atom. The AFM creates three-dimensional representations of the specimen surface by monitoring the forces of interaction between the surface and the above described probe.

The cantilever acts as a soft spring constant that obeys to Hookey's Law:

$$F = -zK \quad \textbf{(1)}$$

where F is the force acting on the cantilever, K is the cantilever spring constant and z is the vertical displacement of the end of the cantilever.

Therefore it is possible to measure the forces acting on the AFM probe by monitoring the vertical displacement of the cantilever (z). In fact when the AFM tip engages the surface, the cantilever is bent due to the forces between tip and the specimen. Most commonly the cantilever deflection is detected by an optical lever; this is comprised of a red laser that reflects from the cantilever surface into a four-quadrant photodiode (Fig. **1**).

Feedback control responds immediately to any changes in tip–sample distance and sends a signal to the z piezoelectric micropositioner. The signal received by the computer is finally converted into an image (reviewed in Takano *et al.* 1999, [13]).

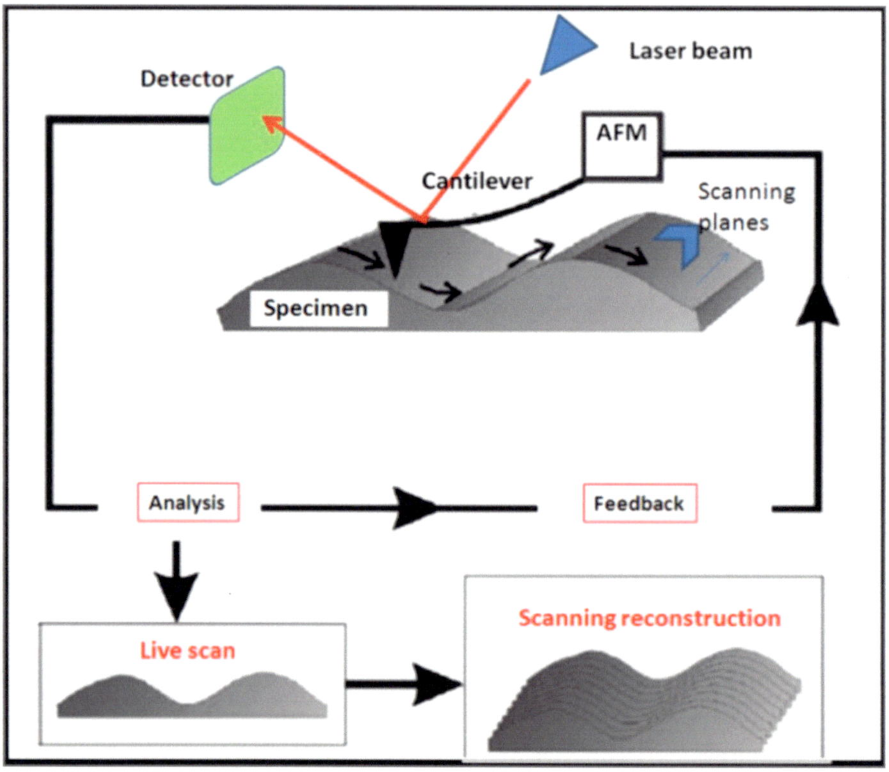

Figure 1: Schematic set-up of an atomic force microscopy(adapted from Sitteberg *et al.* [8]).

2.1.2. Operation Mode

A wide range of different operation modes exist in AFM, such as contact AFM, intermittent contact AFM, non contact AFM, phase imaging and atomic force spectroscopy.

A scheme of some operation mode is shown in Fig. **2 (a)**, **(b)** and **(c).**

In constant height contact mode AFM (Fig. **2 (a)**) the tip is continuously engaged with the sample surface, causing the cantilever to deflection. The degree of the deflection is detected to gain sample surface topography information.

Protruding features on a sample surface undergo high physical stress (pressure in the Gpa range may occur during this scanning process [14]), because the cantilever deflection is proportional to the force applied to the sample.

In order to avoid the damage of the surface in case ofs oft specimen, constant force contact mode has been developed. The desired repulsive force between the probe and the surface is set up by raising the sample and using the piezo-scanner tube until the specimen surface just contacts with the AFM probe tip. By controlling the height of the sample relative to the probe, the repulsive force can be maintained lower than 1nN. If the AFM probe scans over a raised area the repulsive force increases because the electron orbitals are forced to overlap further. A feedback loop then acts to lower the sample and thereby restores the pre-set repulsive force. Three dimensional pictures of constant force are obtained by plotting the sample height against the lateral probe position.

In intermittent contact AFM (Fig. **2 (b)** and **(c)**), the cantilever is forced to vibrate close to its resonance frequency. In this mode, AFM operates by a tapping tip-sample contact which lasts for only a nano-seconds period.

When the probe interacts with the sample, a damping of oscillation can be observed. The damping is used to give an input to the feedback loop which is adjusted at a constant rate by adjusting the sample stage. The specimen topography can be extracted from the up and downwards movements of the AFM stage when the probe scans across the sample [15].

A technique closely related to intermittent contact AFM, is *phase imaging*, which allows the investigation of material surface properties. When the tip interacts with the sample, an energy transfer to its surface takes place. This phenomena is responsible for a loss of oscillation energy and amplitude, and a phase shift in the cantilever oscillation, which in turn is a function of material stiffness or softness [16, 17]. As a consequence of this, the distribution of different materials or phases within the sample can be detected [18].

The AFM creates three-dimensional representations of the sample surface by monitoring the forces of interaction between the sample and the tip probe. Therefore, this force determination can be used not only in a feedback loop for imaging, but also as an indicator for the intensity of the interaction between sample and tip [19-21].

A scheme for a force-distance curve is reported in Fig. **2 (d)**. The benefits of atomic force spectroscopy (AFS) is the high lateral positioning of the probe which allows the observation of the behavior of the sample at nanoscale resolution and the high force sensitivity; in fact forces down to the pN range can be recorded.

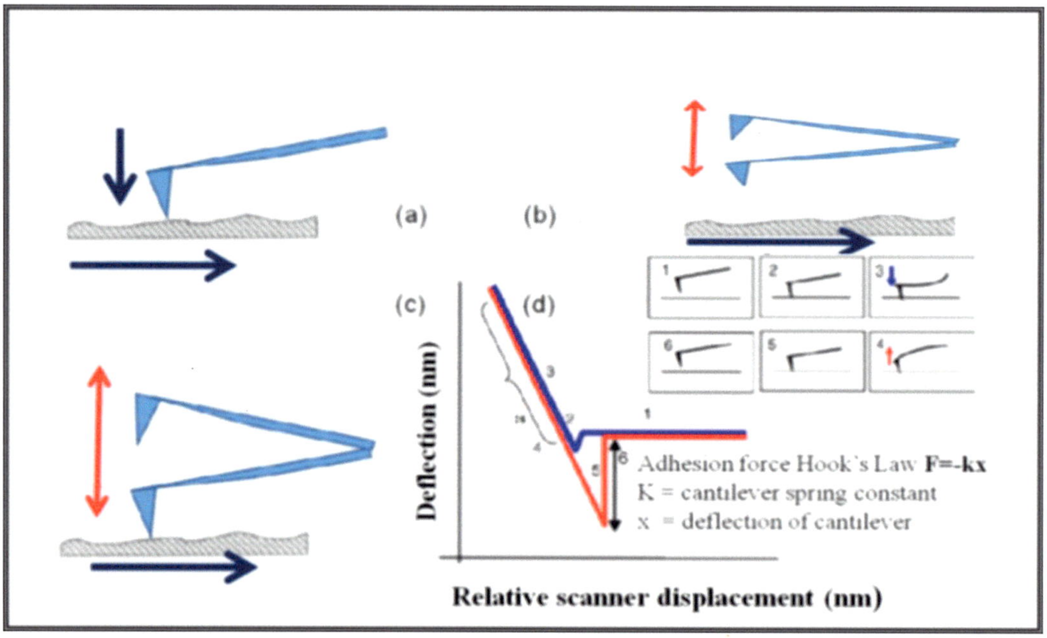

Figure 2: Modes of operation of the AFM (adapted from reference [7(a)]). (a) Contact mode: probe constantly in contact with sample surface. (b) Non contact mode: cantilever modulated at resonance frequency above surface (c) Tapping mode: cantilever modulated at resonance frequency with probe in contact with the surface at low point. (d) Scheme of a force–distance curve (adapted from reference [7(b)]). The blue and red lines are indicative of the approach and retraction of the probe tip from the sample surface respectively.1–6 shows the order at which approach and retraction occur.

Wherever possible, the force measurement is carried out in a liquid. In a basic experiment the forces applied to the cantilever when the tip touches and presses the sample surface, can provide information on sample stiffness. Information on adhesion forces between the surface and the tip may be obtained varying the retraction forces applied to pull off the tip. In the first approach information on the visco-elastic behavior of the specimen during the deformation can be drawn. These kinds of data can be used to investigate the stability and ageing of the material [22]. In retraction experiments, unspecific adhesion between the specimens and the top of the tip can be investigated. The top of the tip may be variously modified and a classical example of this application is the study of the interaction between a receptor and its ligand [23].

2.1.3. Cantilever-Tip System

An appropriate choice of tip in conjunction with the sample and the scope of the experiments are essential. Actually the tip geometry, radius and chemical composition are key factors for high resolution imaging. In particularly the tip apex radius limits the maximum resolution: features in the size range of the tip are visualized in a distorted fashion and broadened, while smaller features cannot be visualized at all.

For very high (atomic) resolution, super-tips (with a radius less than 30 nm) can be prepared. Although usually tips are made of silicon or silicon nitride, diamond, carbon or mineral single crystals have been employed for special applications. For example, controlled growth of carbon filament at the ends of Si_3N_4 tip can be conducted and further surface modification can be applied in order to obtain fine-tuned and surface tailored nano-sensor systems [24, 25].

2.1.4. Artifacts

The shape of an AFM tips can vary considerably in term of both the geometry and the radius. These variations may cause artifacts. Indeed only a truly two dimensional tip could image correctly some particular features. The image contains elements arising from the combination of the tip shape and the true topography of the surface. While long thin tips will clearly best reproduce the topography, their fragility can lead to other artifacts.

Sometimes the damage of tip leads to the occurrence of double or multiple tips. Such tips will generate images in which features will be present as pairs or larger multiple. These artifacts are anyway easy recognizable.

Artifacts may also occur when tip-sample interactions are greater than the mechanical strength theoretically planned at the beginning of the experiment. This may be encountered in the contact mode imaging of a soft surface or when the response time of the feedback loop are incorrectly set. This can happen also when there are rapid changes in the specimen profile while the feedback loop has been set on a flat part of the surface.

Finally, biological samples scanned in air are prone to artifacts arising from the presence of buffer in the solution used to prepare the specimens. On drying the buffer can crystallize, yielding dendritic patterns [26].

3. AFM IN BIOMATERIAL AND BIOMEDICAL INTERFACE CHARACTERIZATION

The surface chemistry of biomaterials plays a major role in determining host response and therefore its *in vivo* performance. Protein adsorption and cellular adhesion phenomena when materials are exposed to body fluids are dependent on to the surface energy and morphology of the materials [27, 28].

Some of the key problems in biomaterial performance relate to surface interactions. For instance, encrustation of urinary catheters is mediated through inorganic deposits forming on an adsorbed protein layer from urine. Another example is the development of biofilms upon implants, this process also depends on surface structure and is responsible for implant-associated infections [29].

Numerous AFM studies on *biomaterials surface morphology* have been reported in the literature (Kajiyama *et al.* and references therein [30]) including inorganic coating [31] (Fig. **3**), single polymer crystals, mechanically oriented thin coatings, epitaxially grown films, polymer blends and polymerized monolayers.

The major application for biomaterial imaging for AFM lies in the sub-micrometre scale, where resolution of macromolecular assembly and organization at the surface is the key target. Typical example includes images resolving crystallographic features separated by only 0.9 nm, attributable to chain endings or folded structures.

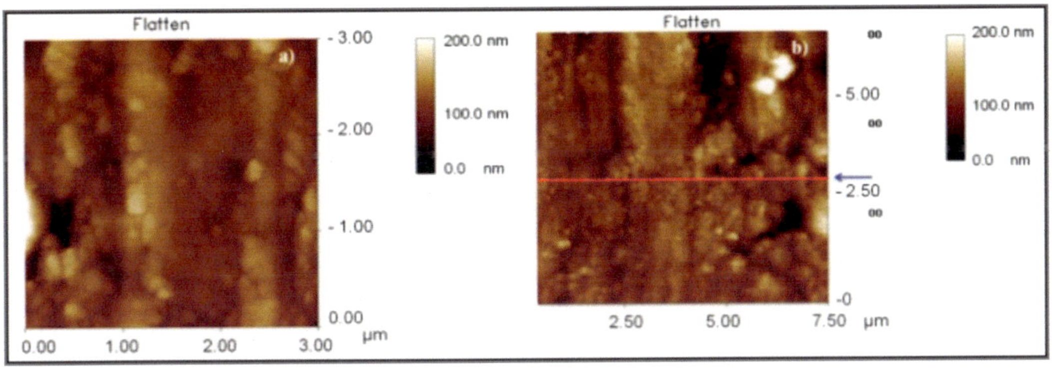

Figure 3: AFM images of a Ti plate before (a) and after (b) a hydroxyapatite electrochemically assisted deposition (30 s): the red line indicates the mineral deposition front (b), scale bar = 200 nm (adapted from reference [30]).

The use of AFM in the acquisition of force measurements, and the resulting frictional and mechanical data allows revealing surface details not detectable with topographical observation alone. For example different mechanical stiffness deriving from specimen hydration levels or from different cross-linking degree can be observed.

Most conventional polymeric materials do not meet the high demands required for their surface properties such as roughness, adhesiveness, surface energy or biocompatibility when used in real applications. An effective approach for developing high performing polymers is to *modify the surface* of these materials which already have excellent (bio) functionality and bulk properties [32]. Widely used surface modification techniques include the addition of surfactants, anti-static agents or lubrificants, coating procedures and controlled oxidation by low-temperature plasma and grafting [33-35].

In AFM studies, chemisorptions at gold surfaces of a number of hydrophilic thiolated polymers such as poly-ethylene-oxide (PEO), dextran and PEO-dextran conjugates by AFM have been studied. Subsequently, the resistance of the formed polymer brushes towards plasma proteins has been investigated. The AFM images indicate that the substitution of dextran by PEO improved the protein resistance and therefore blood compatibility compared to chemisorbed dextran and PEO only surfaces [36].

One of the advantageous features of AFM is the possibility of imaging surfaces within an aqueous or generally a fluidic environment. This ability allows examination of *dynamic processes* such as erosion, hydration and adsorption at the interface.

Degradation Mechanisms of bioresorbable inorganic surfaces and biodegradable polymers can be monitored by AFM techniques. Erosion processes will be clearly influenced by material crystallinity and morphology. Actually, crystalline regions are known to inhibit water ingress and this results in preferential degradation of amorphous regions, where water diffusion is enhanced.This phenomenon has been observed in the rapid degradation an dsubsequent loss of amorphous material during AFM imaging of melt-crystallised PSA (poly-(sebacic-anhydride)) films in buffer, resulting in the exposure of crystalline fibers that are less susceptible to degradation [37]. Polymer blend composition may also have a significant impact on the degradation kinetics. AFMstudies revealed differential degradation rates for blends of PSA and poly-lactic acid (PLA), so demonstrating the rapid erosion of the PSA component of the surface which left the porous PLA network intact [38].

Adsorption Phenomena. The conditioning of biomaterial surfaces is a key stage in determining the biocompatibility of medical implants or the development of novel drug delivery systems. Few studies have illustrated that AFM is capable of providing dynamic data on the local nature of adsorption processes. The influence of time and concentration on the formation of a monolayer of albumin, fibrinogen and IgG on polystyrene surfaces has been studied by Green *et al.* in buffer at molecular resolution level using AFM [39].The surface of a wide range of commercial soft contact lenses of different chemical composition was examined under physiological conditions using AFM by Baguet *et al.* [40, 41]. The surface roughness was investigated in combination withthe lachrymal components accumulation on the soft contact lens surfaces after various periods of continuous wear. Using electrophoresis and immunoblotting, the extracted deposits were identified as IgA, lactoferrin, tear lipocalin and lysozyme. This is an example of the combination of different technologies to obtain a very comprehensive evaluation of biocompatibility.

Oberleithner and coworkers [42] investigated the calcification process of porin bio-prosthesis implanted into the human heart by combining scanning electron microscopy and AFM. In view of the occurrence of calcified deposits in the vicinity of the surface defects on the valve tissue and the frequent observation of calcification spots and crystals in the subsurface structures just beneath the defective valve surface, their results proved that alteration of the valve surface is an important factor for calcification occurring in the implanted bioprosthetic heart valves.

Hydration. It has been noted that the hydration of a biomaterial is likely to result in the considerable restructuring of the interfacial chemistry and this naturally will have a pronounced effect on the surface morphology and as shown for gelatin the surface mechanical properties. Of particular relevance is a report by Haustad *et al.* [43] which showed the possibility to distinguish two phases in the dry gelatin films adsorbed on mica by revealing different frictional characteristics. The influence of hydration of gelatin films was investigated as well by other authors [44, 45]. Different mechanical stiffness was observed where increasing hydration led to a reduced elastic modulus of the surface. As would be anticipated, in the presence of cross linking agents hardening of the interface was seen.

AFMhas also been applied to monitor the changes in electroactive polymers where changes in sample height allowed the swelling of the material to be monitored [46].

It is evident that such a strategy may be extremely valuable to study of biomaterial interfaces at nanoscale, in the fields in which hydration is a key element for performance *e.g.*; coating technology, responsive drug delivery systems and hydrogels.

Cells. With the introduction of the AFM techniques, many cell structural surface features as well as dynamic behaviour of living and fixed cells(such as red and white blood cells, bacteria, platelets, cardiac myocytes, epithelial cells and glial cells) or their membranes in both native or in reconstituted liquids have been investigated [47-49]. For example, optimized methods were applied to topographically study MR-1 cells bearing flagella in their native environment [50]. The topographic image shown in Fig. **4** demonstrates that the ends of these elongated cells are thicker than the central portion (Fig. **4a**). Additional information and structural details are shown by phase imaging mode (Fig. **4b**).

AFM has been used not only to measure cell surface topography but also cell compliance, Young's modulus [51] and cell volume [52], to carry out cytological [53] and histological analyses [54], and as a robot for nano-manipulation [55].

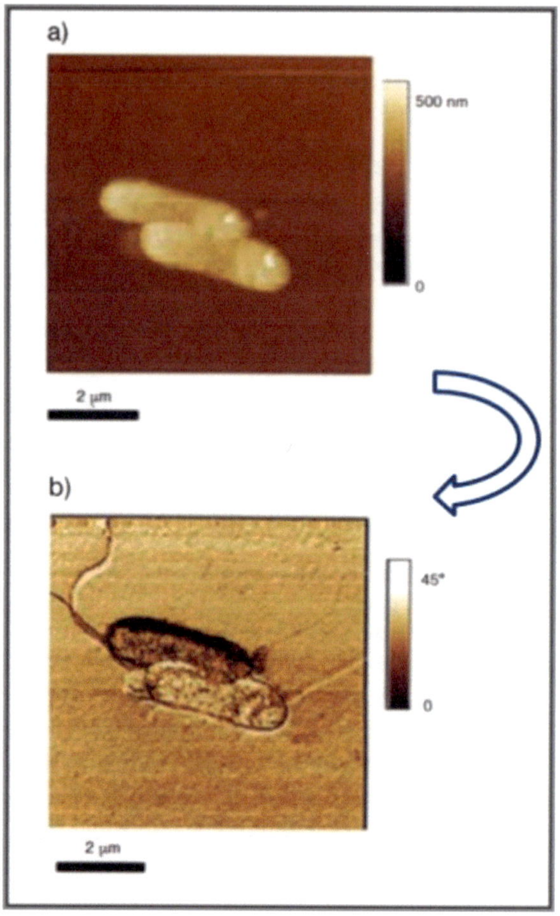

Figure 4: AFM tapping mode images of a pair of bacterial cells on a poly-L-lysine surface: topographic height image (a) and phase image(b) (adapted from reference [50]).

A fascinating AFM application in biomaterial science is the investigation of cellular interaction and mechanical properties. Essential for multi-cellular development are cell-cell adhesions mediated by specific cell surface molecules. By combining single molecule force spectroscopy with genetic manipulation and by controlling the interaction between single cells it is possible to quantify de-adhesion forces at the resolution of an individual cell. An early case study of measurements of adhesion forces between individual cells are the experiments of Benoit *et al.* [56] on a glycoprotein, contact site A, expressed in aggregating cells of Dictiosteliumdiscoideum, which are engaged in development of a multi-cellular organism proteins. Using the light microscope of a custom-made force spectrometer for guidance, the authors picked up a D. discoideum cell with a tip less AFM cantilever, the end of which had been covalently functionalized with alectin, resulting in firm attachment of the cell to the cantilever. A target cell at the bottom of a Petri dish was positioned underneath the cantilever-mounted cell, and was approached until a predefined repulsive contact force was established (Fig. **5**). This contact force was held constant for a predefined time interval to allow establishment of cell adhesion. Upon retraction of the cantilever, the authors recorded the force as a function of the distance that the cantilever was moved until contact between the cells was broken [56].

Conceptually the experiments could be reproduced, with same modification in order to estimate the adhesion force between a cell and a nanostructured biomaterial.

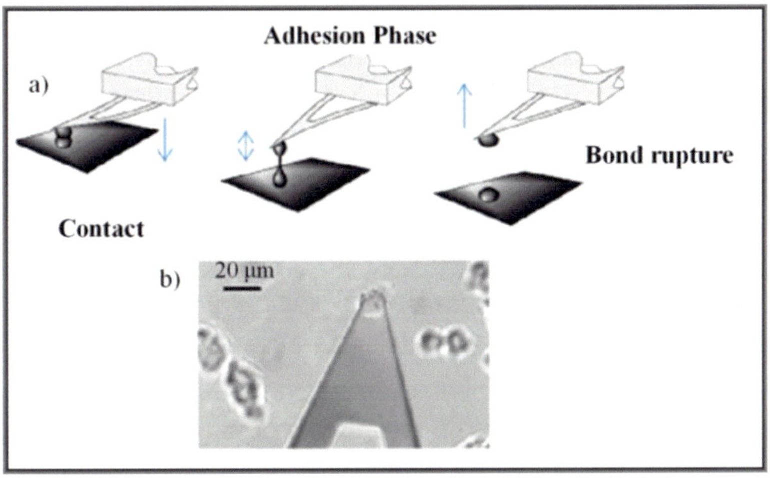

Figure 5: (a) Experimental procedure for cell-cell adhesion; (b) light-microscopic image of a cantilever-mounted cell before being brought into contact with another cell (adapted from reference [56]).

4. SCANNING AND TRANSMISSION ELECTRON MICROSCOPY

Microscopy relies on the interference of electromagnetic radiation with the object. In electron microscopy, the source of electromagnetic radiation consists of electrons accelerated by high voltage in a vacuum. The velocity of these electrons is close to the speed of light, and at this high relativistic speed, electrons can be considered as waves with wavelengths of a few picometres. In fact, electron microscopes use electrons instead of photons, because electrons have a shorter wavelength than photons. This property allows you to observe matter at atomic resolution.

There are two general types of electron microscopes: the Scanning Electron Microscope (SEM) that scans an electron beam over the surface of an object and measures how many electrons are displaced and emitted or scattered back, and the Transmission Electron Microscope (TEM) that shoots electrons through the sample and measures how the electron beam interacts with the sample [57, 58].

Electron waves interfere with an object in several modes. In other words, accelerated electrons in an electric microscope carry significant amounts of kinetic energy. This energy is dissipated and generates a variety of signals that derive from the electron-sample interactions (Fig. **6**).

Most of the incoming electrons pass through a thin object without interfering. Secondary electrons and backscattered electrons are commonly used for imaging samples in SEM: secondary electrons are suitable for showing morphology and topography of the samples and backscattered electrons are most interesting for illustrating contrasts in composition in multiphase samples (*i.e.* for rapid phase discrimination).

Some of the incoming electrons are scattered inelastically and deposit energy on the sample. X-ray generation is produced by inelastic collisions of the incident electrons with electrons in discrete orbitals of atoms in the sample. As the excited electrons return to lower energy states, they yield X-rays that are of a fixed wavelength (that is related to the difference in energy levels of electrons in different orbitals for a given element). This is known as Energy-Dispersive X-ray Analysis (EDXA) and gives information on the elements present in the sample and a rough indication of the proportion of the elements present relative to each other.

Transmission electron microscopy uses the portion of electron which are transmitted through the sample, together with the diffracted backscattered electrons (that are used to determine crystal structures and orientations of minerals), photons (characteristic X-rays that are used for elemental analysis and continuum X-rays). A certain fraction of electrons is elastically scattered by small angles and interfere with un-scattered beam. These electrons contribute to phase contrast, which is the main source of contrast in transmission EM of a thin object [59].

Another fraction of electrons is scattered by higher angles and do not reach the back-focal plane and these electron contribute to amplitude contrast.

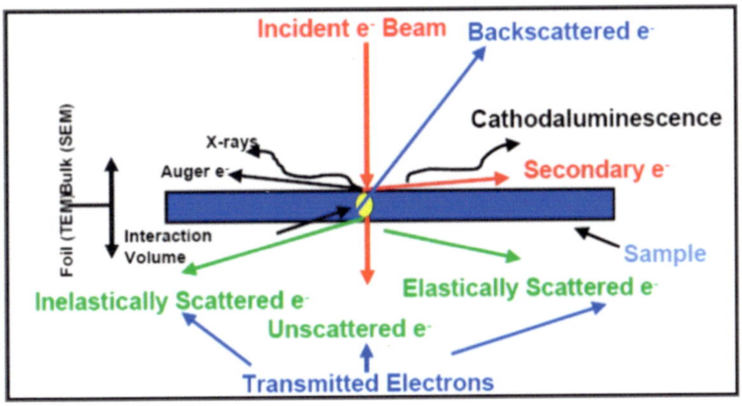

Figure 6: Interaction of electrons with the object in EM. Phase contrast, which is the main source of contrast for thin unstained specimens composed of elements with low atomic numbers is generated by interaction of the elastically scattered electrons with the unscattered electrons.

4.1. Scanning Electron Microscopy

A scheme of a scanning electron microscope is reported in Fig. **7**. The "Virtual Source" at the top represents the electron gun, producing a stream of monochromatic electrons [60]. The stream is condensed by the first condenser lens (usually controlled by the "coarse probe current knob"). This lens is used to both form the beam and limit the amount of current in the beam. It works in conjunction with the condenser aperture to eliminate the high-angle electrons from the beam. The beam is then constricted by the condenser aperture (usually not user selectable), eliminating some high-angle electrons. The second condenser lens forms the electrons into a thin, tight, coherent beam and is usually controlled by the "fine probe current" dial. A user selectable objective aperture further eliminates high-angle electrons from the beam. A set of coils then "scan" or "sweep" the beam in a grid fashion, dwelling on points for a period of time determined by the scan speed (usually in the microsecond range). The final lens, the objective, focuses the scanning beam onto the part of the specimen desired.

When the beam strikes the sample (and dwells for a few microseconds) interactions occur inside the sample and are detected with various instruments.

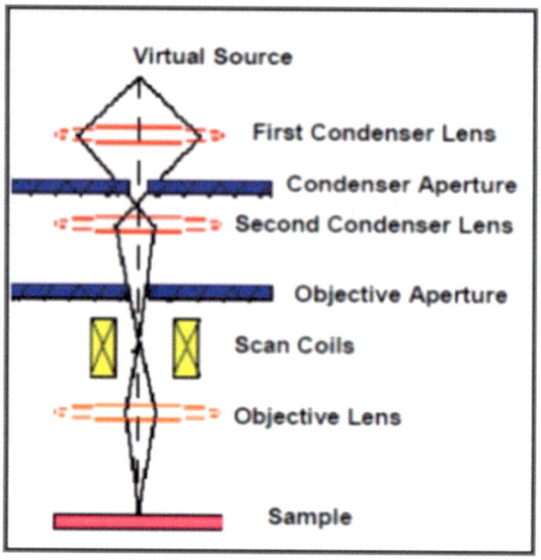

Figure 7: Diagram of a scanning electron microscope.

Before the beam moves to its next dwell point these instruments count the number of interactions and display a pixel on a screen whose intensity is determined by this number (the more reactions the brighter the pixel). This process is repeated until the grid scan is finished and then repeated, the entire pattern can be scanned 30 times/sec.

Typically, SEM imaging requires a high vacuum, 10^{-8}Torr [60]. Few biological specimens tolerate these conditions without rapid collapse [61] and fewer still survive [62]. The environmental SEM [63] uses a series of pressure limiting apertures [64] to allow the electron gun to operate at10^{-7}Torr while preventing gas leakage from the specimen chamber, which can be maintained at 1^{-20}Torr. A gaseous secondary electron detector (GSED) exploits the gas in the specimen chamber for signal amplification. GSED operation produces positive ions that have the added benefit of limiting charging of non-conductive specimens.

4.2. Transmission Electron Microscopy

A transmission electron microscope (Fig. **8**) is constituted of: (1) two or three condenser lenses to focus the electron beam on the sample, (2) an objective lens to form the diffraction in the back focal plane and the image of the sample in the image plane, (3) some intermediate lenses to magnify the image or the diffraction pattern on the screen. If the sample is thin (< 200 nm) and composed of light chemical elements, the image presents a very low contrast when it is focused. To obtain an amplitude contrasted image, an objective diaphragm is inserted in the back focal plane to select the transmitted beam (and possibly few diffracted beam): the crystalline parts in Bragg orientation appear dark and the amorphous or non Bragg oriented parts appear bright. This imaging mode is called *bright field* mode BF. If the diffraction is composed of many diffracting phases, each of them can be differentiated by selecting one of its diffracted beams with the objective diaphragm. To achieve this, the incident beam must be tilted so that the diffracted beam is directed on the objective lens axis such that off axis aberrations are avoided. This mode is called *dark field* mode DF. BF and DF modes are used for imaging materials to nanometer scale [65].

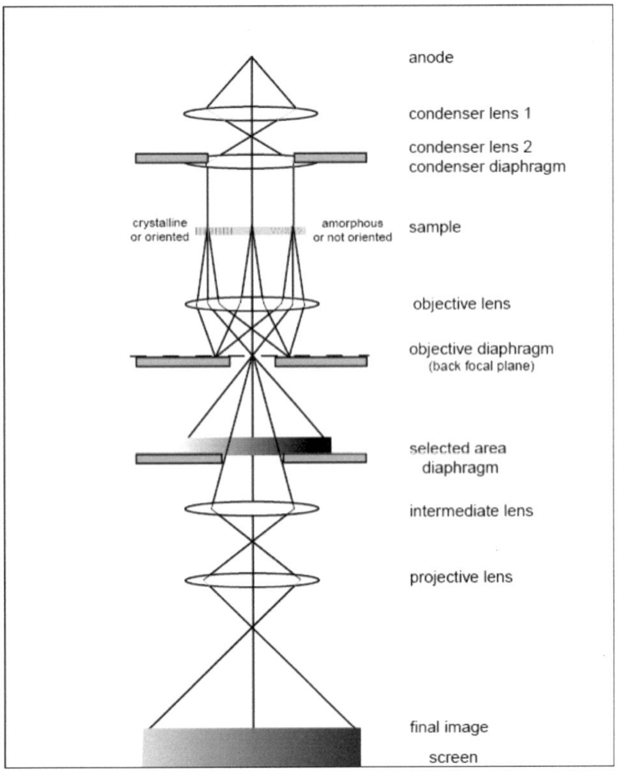

Figure 8: Diagram of a transmission electron microscope in image mode. In diffraction mode, another intermediate lens is inserted to image the diffraction pattern of the back focal plane on the screen.

The selected area diaphragm is used to select only one part of the imaged sample, for example a particular particle or sub-area of interest. This mode is called *selected area diffraction.* The spherical aberrations of the objective lens limit the area of the selected object to few hundred nanometers. Nevertheless, it is possible to obtain diffraction patterns of a smaller object by focusing the electron beam with the projector lenses to obtain a small spot size on the object surface (2-10 nm). The spots of selected area diffraction become disks whose radii depend on the condenser diaphragm. This is called *microdiffraction.*

Selected area diffraction and microdiffraction patterns of a crystal allow calculation of the symmetry of its lattice and its interplanar distances (with the Bragg's law). This is useful to confirm the identification of a phase, after a tentative identification generally based on the literature of the studied system and on chemical analyses.

4.3. High Resolution Electron Microscopy

Conventional TEM uses only the transmitted beams *or* some of the forward scattered beams to create a diffraction contrast image. HREM [66] uses the transmitted *and* the scattered beams to create an interference image. The TEM microscope must have a high performance (low spherical aberrations and high stability of the high tension, of the lens currents and of the energy of the electron beam). Image formation must take into account the two following steps: (1) the propagation of the incident wave through the object, (2) the transfer of the scattered wave by an optical system of the microscope (the objective lens).

4.4. Image Formation in a Nutshell

As we pointed out above, some of the incoming electron are scattered inelastically and deposit energy on the sample which leads to beam damage and consequently to loss of high resolution information. This event is not very highly focalized, so that the inelastically scattered electrons do not contribute to the high resolution information but contain images at low spatial frequencies [59].

The cross section of atoms for scattering electron increases with the atomic number, so that heavy metal compounds play a crucial role as stains in electron microscopy.

Also the cross section for inelastically scattered electrons depends on the atomic number, up to an atomic number of 10 the inelastic cross section is larger than the elastic cross-section meaning that for every elastic scattering event, which contributes to phase or amplitude contrast, some electrons are scattered inelastically and contribute to the beam damage. The problem is particularly crucial in the case of biological or organic polymer samples, which consist mainly of elements with low atomic number.

For example in carbon, for each electron that is elastically scattered, three electrons are inelastically scattered. This phenomenon is responsible of specimens charging which may be evaluated in a range of energy between 10 and 25 KV to the sample. Therefore to recover high resolution information electron doses are limited to 5-10 e A^{-2}. At this small dose images are noisy and many low dose images have to be merged to recover the high resolution information.

Whatever the sample preparation technique is, the result is a relatively thin, but nevertheless three-dimensional (3D) structure, the image is recorded on a two-dimensional (2D) film or CCD camera. The image is only a true projection of the 3D object if all the atoms of the lattice are within the depth of focus D of the microscope. Because this is a stringent prerequisite for all subsequentefforts to reconstruct the 3D structure of a projected object, it is worthwhile to estimate the D required to obtain an anticipated resolution d. Assuming that a particular TEM performs at the diffraction limit, D is given by:

$$D \leqslant \frac{d^2}{0.61\lambda} \qquad \textbf{(2)}$$

where λ is the electron wavelength. The diffraction limit can be approached with modern instruments equipped with field-emission either by a later correction of the contrast transfer function (CTF) or by using a corrector for the spherical aberration of the TEM's objective lens [67]. The depth of focus for a 300 kV instrument allows the study of objects thicker than 1 μm at a resolution of a few nm. A diagram summarizing the meaning of the depth of focus D, the resolution d, the sample thickness T and the tilt angle θ is reported in Fig. **9**. The depth of focus D corresponds to the distance between locations where the diffraction disc and the geometric discs are equal. Scattering centers within the resolution volume indicated in red cannot be resolved. An optical system that collects waves emanating from scattering centers of a tilted 3D object whose projected thickness T/cosα is smaller than D will produce the true 2D projection of the 3D object in the image plane. In Fig. (**9**) the Log–log plot of estimated resolution limit against the depth of focus D, as given by Eq. (2) is shown. The ordinate displays the maximal thickness T of the sample, which would be within the depth of focus D for a given resolution and a given tilt angle. For example, at a tilt of 60°, a resolution of 5 Å is achievable at 300 kV acceleration voltage with a sample thinner than 200 nm. If are solution of 20 Å is to be reached, the sample slab can be even 2 μm thick.

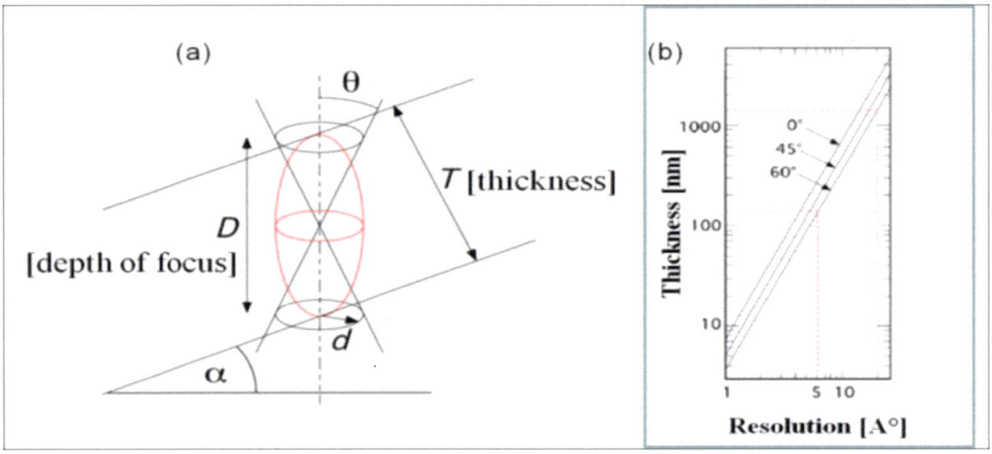

Figure 9: Depth of focus and resolution in a transmission electron microscope (a). Log–log plot of estimated resolution limit against the depth of focus D, as given by Eq. (I) (b) (adapted from Muller *et al.,* reference [68]).

However, multiple scattering concurrent with loss of signal and image blurring, impose further limitations when imaging thick objects such as entire cells [69]. Electrons are scattered elastically by the nuclei of the sample atoms and inelastically by their electron shells. Inelastically scattered electrons carry chemical information of particular interest, but are also directly related to beam-induced specimen damage. The small fractionof inelastically scattered electrons carrying useful chemical information and the beam damage preclude the chemical analysis of biological samples at high resolution. Usually researchers concentrate on the image formation by elastically scattered electrons, since only these provide high resolution information.

The mostly used TEM imaging mode is the axial bright-field mode (BF), where elastically scattered electrons interfere with the unscattered beam to produce a phase contrast image.

The BF mode is described by the nomenclature of Fourier optics [65]. Assuming that the optical system is space invariant, the image is described by the convolution of the geometric image with the impulse response that defines the optical system. Therefore, the transfer of different spatial frequencies characterizing the object is dictated by the CTF, which is the Fourier transform ofthe impulse response. Since electron optical systems exhibit mainly spherical aberration, the information transfer occurs in successive frequency intervals of opposite sign, which requires appropriate subsequent correction.

In theory, a thin object can be considered as a weak phase, weak amplitude object. Interference of the electron wave with the phase object introduces a phase shift χ. This phase shift depends on the defocus Δ, the spatial frequency R and the wavelength of the electron λ. With a perfect lens a pure phase object in focus has no contrast. However, electron lenses deflect electron much more strongly at the edges of their optical axis than in their centre. This causes an aberration which introduces an additional phase shift that depends on the hardware of the electron microscope, such as the spherical aberration of the lens C_s *etc.* The phase shift is given by Eq. (3)

$$X = \pi/2\ (C_s \lambda^3 R^4 - 2\Delta\lambda R^2) \quad \textbf{(3)}$$

In phase contrast, the contrast is proportional to the sine of the phase shift, while amplitude contrast is proportional to the cosine. For an accurate description of the image information the exact ratio of amplitude and total contrast (*a*) has to be known. For most conditions, it is a sufficient approximation to assume that the ratio between amplitude and phase contrast is constant over the spatial frequencies [70].

In this case

$$\mathrm{CTF} = a\cos(\chi) + \sqrt{(1 - a^2 \sin\chi)} \quad \textbf{(4)}$$

4.5. Sample Preparation and Cryo-Electron Microscopy

Sample preparation in EM and particularly in transmission electron microscopy is an important issue and has to overcome three requirements: (*i*) the object has to be reduced to a suitable size, *i.e.* 40-500 nm thick, less than 3 mm in diameter *(ii)*the object has to be stabilized to resist the high vacuum inside the EM *(iii)* the structure has to be preserved to with stand beam damage over a high electron dose. These limitations appear particularly evident in biological specimens and organic polymeric samples. Sample preparation techniques have therefore to ensure that the structure of the specimen is preserved during the preparation process.

Dehydration by drying EM grids in air is suitable for inorganic nanocrystals but when applied to biological samples, drying not only changes their native environment, but also exposes them to disruptive surface tension forces.

Partially embedding aqueous suspensions of macromolecular complexes or membrane fragments in a heavy metal salt solution produces a high contrast due to the strong scattering of the heavy atoms. This technique is called *negative staining*, since what you see in TEM is not the biological material itself but the envelope of heavy metal 'stain'. Negative staining provides useful structural information to a resolution down to about 1.5 nm [71].

Embedding bio-macromolecules in a sugar solution provides a similar buffer against sample collapse and maintains some hydration, but the small difference in the electron scattering power of sugar and protein means that differences in the contrast produced by such samples are low. Consequently, this preparation method is mainly used for electron crystallography where only ordered structure, and not randomly arranged molecules such as sugars contribute to the crystallographic signal that can be enhanced by averaging [72].

The detrimental effect of surface tension can be avoided by freeze-drying and has prompted the development of various cryo sample preparation methods. The simplest form is employed for mass measurement by scanning transmission electron microscopy (STEM). Here the unstained sample is adsorbed to a thin carbon film, plunge-frozen, freeze–dried in the specimen chamber of the STEM and observed without further exposure to the atmosphere.

The technique of flash-freezing, and hence vitrifying thin solution layers containing bio-macromolecules for direct observation in the electron microscope, was developed by Dubochet and colleagues in the 1980s [73, 74]. It was the birth of cryo-EM, which is now a routine method widely used in single particle EM and electron crystallography.

Cryo-electron microscopy incorporates both vitrification, a freezing process that subverts crystalline ice formation and image acquisition under cryo-condition.

Slow cooling leads to the formation of the ice crystals which can destroy the fine structure of the object. Therefore the aim of the preservation is to maintain a state of water that is as close as possible to a liquid, in other words a vitrified amorphous form of water. Vitrification is achieved by high cooling rates, of the order of 3 x $10^6 Ks^{-1}$ in order to avoid the water forming sizeable ice crystals. Cryo-fixation is rapid, of the order of 0.1 ms, which is 10^4 faster than conventional infiltration method and preserves spatial as well as temporal biological states.

In practice vitrified samples are prepared by forming a thin film of suspended particles and plunge freezing it in liquid ethane. The development of an automated, computer controlled vitrification system has been initiated which ultimately culminated in the Vitrobot TM [75], used in conjunction with a temperature and humidity controlled glove box [76].

The low contrast produced by such samples means that stains are also sometimes employed. Although these have yielded images of unprecedented clarity [77] the stain concentration required (generally 16% ammonium molybdate) creates a non-physiological environment that limits the application of this approach.

Today, the emphasis is therefore on the direct observation of the buffered bio-macromolecules. This is facilitated by the use of refined image processing algorithms, which exploit the enhanced signal-to-noise ratio obtained by projection averaging to enhanced signal inherent low image contrast.

The application of flash-freezing is limited by the maximum size of the samples that can be handled; only small or flat cells can be vitrified in this way. High pressure freezing is the method of choice for large cells and tissues. This technique has been developed in several laboratories and generally involves the substitution of water by organic solvents for plastic embedding at low temperature [78, 79]. Although the method has been used very successfully [80-83], the substitution step may induce artifacts also encountered by classical block-embedding techniques. The ultimate approach, again developed by Dubochet and co-workers, is the preparation of thin sections from vitrified ice blocks [84, 85].

Combining cryo-sectioning with cryo-electron tomography in the technique called cryo-electron microscopy of vitreous ice sections or CEMOVIS, promises to give an insight into the native contents of a cellat nm-scale resolution [85] although cutting the sections can be problematic [84].

Common to all these techniques is the fact that only a small fraction of the biological sample is actually visualized. The best method is an embedding/thin sectioning approach, where the biological object can theoretically be imaged in its entirety. All other methodologies involve blotting or washing and blotting, both procedures that remove the bulk of the sample solution from the grid. Hence, only a minute fraction of protein complexes tightly adhering to the grid, or remaining in the residual water film in the case of cryo-EM, remain for imaging.

5. SEM AND TEM IN BIOMATERIAL AND BIOMEDICAL INTERFACE CHARACTERIZATION

Both SEM and TEM techniques are able to give information on material topography (*i.e.* the surface features of an object), morphology (*i.e.* the shape and size of the particles making up the object or of the nanoparticle in the case of a nano-colloid), and crystallographic properties.

The same features can be visualized both onto synthetic nanostructured materials and living tissue and biological samples.

Within the experimental limitations related to beam damage and electron optical distortions, transmission electron microscope scan visualize structures down to the atomic level. In Fig. **10** high resolution images (HRTEM) of bone and dentine-like synthetic HA nanocrystals (panel B and C) are shown as an example [86]. The crystal lattice spacing d(nm) can be observed and measured and the c axis orientation calculated.

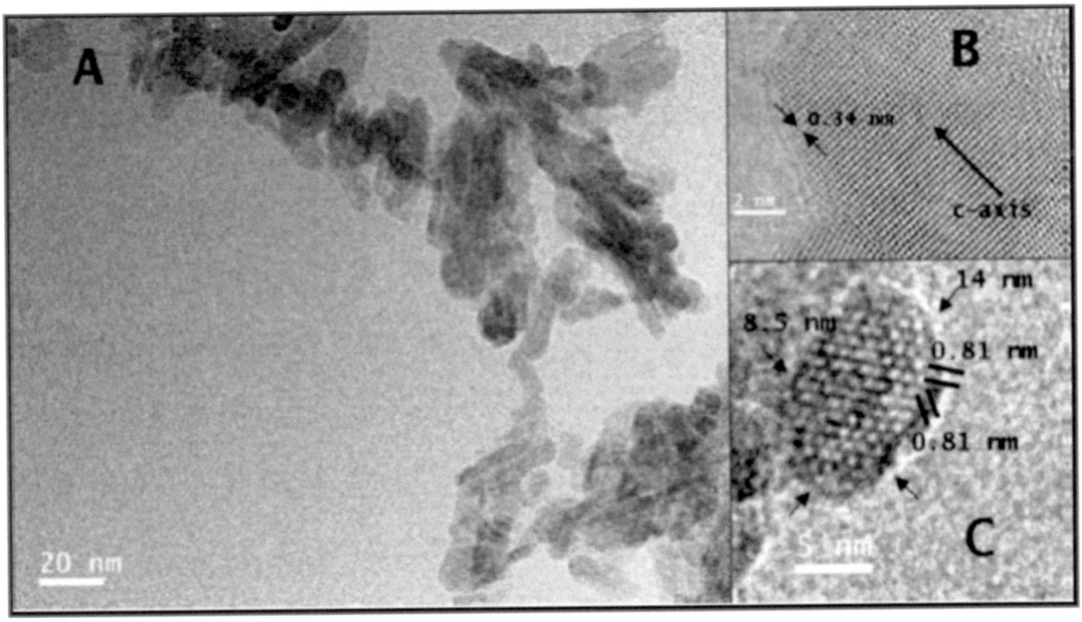

Figure 10: TEM images. (A) General view of bone and dentine-like synthetic HA nanocrystals (original magnification: 100, 000); (B and C) high-resolution images of particles with the c-axis parallel or perpendicular to the image plane, respectively (adapted from reference [86]).

A biomimetic approach towards new biomedical and surgical fibrous synthetic biomaterials can be achieved by a technique known as electrospinning. Here a polymer or a polymer containing suspended nanoparticles is dissolved in a solvent and pumped to the tip of a needle which is separated from an earthed metal collecting plate.

A high voltage potential difference (typically 20-40 kV and at a few mA) is applied between the needle and the plate. A fine polymer fibre in is drawn with from the needle tip and the solvent evaporates en-route. A mat of spun fibres is thus gradually built up at the collection point that can then be peeled away. Novel biologically inspired collagen nanofibers have been reconstituted by electrospinning method in mild conditions, providing an efficient approach towards biomimetic in size, morphology and structure nanofibers (Fig. **11**).

A fascinating example of TEM application for the study of the interface between biomedical materials and biological specimen is shown in Fig. **12**. Herein the possibility to obtain "living" biocomposites made of calcium phosphate-coated mesenchymal stem cells (MSCs) is represented. The cell-nanoparticles composites where synthesized by addition of arginine functionalized hydroxyapatite nanorods to a suspension of immunoselected human MSCs. The amino acid-functionalized nanoparticles were less than100 nm in length and positively charged, and both these factors facilitated adsorption of the particles specifically onto the negatively charged outer layer of the MSC cell membrane. The surface layer of aggregated HAP/R nanoparticles was readily discernible in TEM images of sectioned cells [88].

Figure 11: SEM picture of electrospun collagen fibers (a) and TEM image of a single, electrospun fiber positively stained in order to appreciate the typical collagen banding pattern (b).

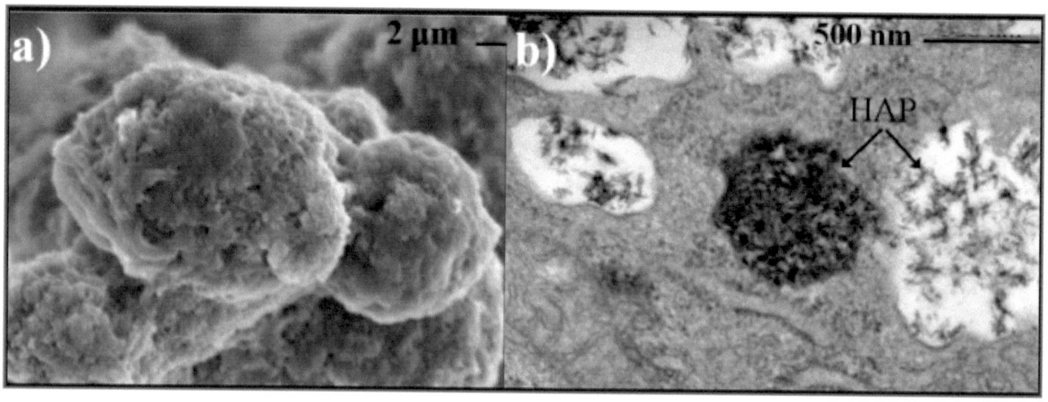

Figure 12: (a) SEM image of arginine-functionalized Hap coated human MSC 24 h after addition of a dispersion of bioinorganic nanoparticles to the suspended cells (bar=1μm); (b) TEM image of a thin section of a MSC showing the cell encased in arginine functionalized Hap nanorods (arrow). The dark region in the centre of the image shows an intracellular aggregate of the nanoparticles (bar=500 nm) (adapted from reference [88]).

Bone is the typical example of an "organic matrix-mediated" mineralization process. Fig. **13** (a) shows the bone collagen fibril structure as a revised quarter-stagger model [89, 90] where molecules are lined up head-to-tail in rows that are staggered of 64 nm along their long axis. Each molecule is transposed by 64 nm with respect to the previous one, and is divided into five sectors, the first four are 64 nm long and the fifth is 25 nm long. This arrangement, stabilized by strong intermolecular crosslinking, produces a regular array of small gaps, 40 nm long and about 5 nm wide, referred to as "hole zones" which are considered to be the *loci* of the nucleation and growth of hydroxyapatite nanocrystals. Adjacent hole zones overlap to form grooves structured in parallel rows along the main axis of the three-dimensional (3D) fibril structure, allowing hydroxyapatite to grow into plate-shaped nano-crystals, oriented such that the crystallographic c-axis is preferentially aligned along the collagen fibril axis (Fig. **13b**) [91].

Recently this mechanism has been further elucidated. In fact combining nanometer scale resolution cryogenic transmission electron microscopy and cryogenic electron tomography [92] with molecular modeling, Sommerdijk and coworkers [93] demonstrated that collagen functions in synergy with inhibitors of hydroxyapatite nucleation to actively control mineralization.The positive net charge close to the C-terminal end of the collagen molecules promotes the infiltration of the fibrils with amorphous calcium phosphate (ACP). Furthermore, the clusters of charged amino acids, both in gap and overlap regions, form nucleation sites controlling the conversion of ACP into a parallel array of oriented apatite crystals. A cryoTEM processing procedure was combined with uranyl acetate staining [94], where uranyl acetate binds to both the negatively and positively charged amino acids in the fibrils, thus significantly increasing the local mass density, and hence the image contrast [95, 96]. The banding pattern along the 67nm repeat matched with the positions of the charged amino acids in the crystal structure of collagen, as shown in Fig. **13c** [96, 97]. After 24 h of mineralization, ACP was observed surrounding and entering the fibril, associated with the a-bands (Fig. **13d**). The infiltration of mineral into the fibril through the a-bandregion is not dependent on the availability of space. Gaps with in the microfibril are present throughout the whole 67nm repeat, both in the gap and overlap regions [93], and could provide entrysites for the mineral phase into the fibril. Therefore, the site-specific localization of mineral infiltration must result from a specific interaction between the amorphous mineral phase and the collagen at this location.

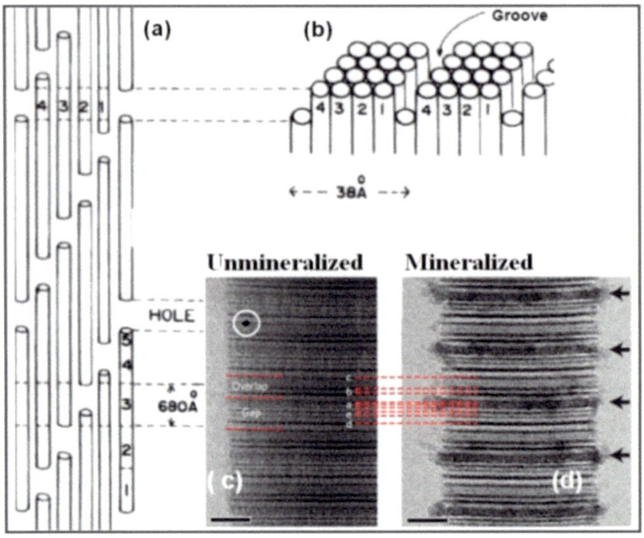

Figure 13: Quarter stagger model of collagen fibers: (a) alignment of collagen gap regions to form groves. In them HA crystals grow with crystallographic c- and a-axes preferentially aligned parallel to the collagen fiber axis and groove directions, respectively (b) (reproduced by permission from Ref. [89]). (c) CryoTEM image of stained, non-mineralized collagen. White circle: 10nm gold marker for electron tomography. (d) CryoTEM image of stained collagen, mineralized for 24 h. Calcium phosphate is associated with the fibril in a regular pattern, following the staining bands (black arrows) (adapted from Ref. [93]).

6. CONCLUSIONS

New technical developments in the last few years have opened new pathways for the characterization of biomaterials by atomic force microscopy (AFM) and electron microscopy (EM). A whole family of microscopy instruments has been developed, extending our sense of reaching to the scale of atoms and molecules. Biomaterials are challenging since the samples to be probed are typically soft and delicate but even more challenging is probing the biomaterials-biological environments interface. Groups all over the world have worked on solutions to image these samples, to determine their local elastic or frictional properties by AFM or to gain insight into the single-molecule mechanics by TEM and HRTEM. In the present chapter a few (and sometimes spectacular) achievements are described, the best current work gives a perspective of what may be possible in the future. The reply to this final question would seem to be a move towards integrated studies using SPM and EM, along with other physical techniques, to probe biological activity in its actual environment and also to determine biological entity responses to biomaterials in the implant location.

ACKNOWLEDGEMENTS

Declared none.

CONFLICT OF INTEREST

The author(s) confirm that this chapter content has no conflict of interest.

REFERENCES

[1] Ratner BD. Surface Analysis of Biomaterials and Biomineralization in "Learning From Nature: How to design new implantable biomaterial". Ruiz R.L. and Weiner S. Eds 2004.

[2] Hansma HG. Varieties of imaging with scanning probe microscope.ProcNatlAcadSci USA 1999; 96 (26) :14678–14680.

[3] Price DM, Reading M, Hammiche A, Pollock MH. Crio-thermal analysis: scanning thermal microscopy and localized thermal analysis. Int J Pharm 1999; 192 (1): 85–96.

[4] Sanders GHW, Roberts CJ, Danesh A, Murray AJ, Prices DM, Davies MC *et al.* Discrimination of polymorphic forms of a drug product by localized thermal analysis. SHORT COMMUNICATION, J Microsc 2000; 198 (2): 77–81.

[5] Ward S, Perkins S, Zhang J, Roberts CJ, Madden CE, Luk SY *et al.* Identifying and mapping surface amorphous domains. Pharm Res 2005; 22 (7): 1195–1202.

[6] Pohl DW, Fischer UC, Durig UT. Scanning near-field optical microscopy (SNOM). J Microsc 1988; 152 :853–861.

[7] (a) Vansteenkiste SO, Davies M, Roberts CJ, Tendler SJB., Scanning probe microscopy of biomedical interfaces. Progr Surf Sci 1998; S7(2):95-136. (b) Turner YTA, Roberts CJ, Davies MC. Scanning probe microscopy in the field of drug delivery. Adv Drug De Rev 2007; 59: 1453–1473.

[8] Sitterberg J, Özcetin A, Ehrhardt C, Bakowsky U. Utilising atomic force microscopy for the characterization of nanoscale drug delivery systems. Eur J Pharm Biopharm 2010; 74: 2–13.

[9] Kasas S, Thomson NH, Smith BL, Hansma PK, Miklossy J, HansmaHG.Biological applications of the AFM: from single molecules to organs. Int J Imag Syst Tech 1997; 8: 151–161.

[10] Hansma HG, Pietrasanta LI, Auerbach ID, Sorenson CD, Golan R, Holden PA. Probing biopolymers with the atomic force microscope: a review. J Biomater Sci Polymer Ed 2000; 11 (7): 675–683.

[11] Alessandrini A, Facci P. Review Article, AFM: a versatile tool in biophysics. Measurement Science and Technology 2005; 16: R65–R92.

[12] Akamine S, Barrett R, Quate.F. Improved Atomic Force Microscope Images Using Microcantilevers with Sharp Tips. ApplPhyLett 1990; 57: 316-318.

[13] Overney RM, Meyer E, Frommer J, Brodbeck D, Luthi R, Howald L, *et al.* Friction measurements on phase-separated thin films with a modified atomic force microscope. Nature 1992; 359: 133-135.

[14] Butt HJ, Cappella B, Kappl M. Force measurements with the atomic force microscope: technique, interpretation and applications. Surf SciRep 2005; 59: 1–152.

[15] Shahin V, Barrera NP. Providing unique insight into cell biology *et al.* atomic force microscopy. Int Rev Cytol 2008; 265: 227–252.

[16] Schmitz I, Schreiner M, Friedbacher G, Grasserbauer M. Phase imaging as an extension to tapping mode AFM for the identification of material properties on humidity-sensitive surfaces. Appl Surf Sci 1997; 115 (2): 190–198.

[17] Winkler RG, Spatz JP, Sheiko, Moller M, Reineker P, Marti O. Imaging material properties by resonant tapping-force microscopy: a model investigation. Phys Rev B Condens Matter1996; 54 (12): 8908–8912.

[18] Simon M, Wittmar M, Bakowsky U, Kissel T. Self-assembling nanoplexes from insulin and water-soluble branched polyesters, poly[(vinyl-3-(diethylamino)-propylcarbamate-co-(vinyl acetate)-co-(vinyl alcohol)]-graft-poly(L-lactic acid): A novel carrier for transmucosal delivery of peptides. Bioconj Chem 2004; 15: 841–849.

[19] Radmacher M, Cleveland JP, Fritz M, Hansma HG, Hansma PK. Mapping interaction forces with the atomic force microscope. Biophys J 1994; 66 (6): 2159–2165.

[20] Quist AP, Bergman AA, Reimann CT, Oscarson SO, Sundqvist BU. Imaging of single antigens, antibodies, and specific immunocomplex formation by scanning force microscopy. Scan Microsc 1995; 9 (2): 395–400.

[21] Florin EL, Moy VT, Gaub HE. Adhesion forces between individual ligand– receptor pairs. Science 1994; 264 (5157): 415–417.

[22] Weiss B, Schaefer UF, Zapp J, Lamprecht A, Stallmach A, Lehr CM. Nanoparticles made of fluorescence labelled poly(L-lactide-co-glycolide):preparation, stability, and biocompatibility. J Nanosci Nanotechnol 2006; 6: 3048–3056.

[23] Hinterdorfer P, Baumgartner W, Gruber HJ, Schilcher K, Schindler H. Detection and localization of individual antibody–antigen recognition events by atomic force microscopy. Proc Natl Acad Sci USA 1996; 93: 3477–3481.

[24] Keller D, Chung CH. Imaging steep, high structures by scanning force microscopy with electron beam deposited tips. Surf Sci 1992; 268: 333–339.

[25] Micic M, Chen A, Leblanc RM, Moy VT. Scanning electron microscopy studies of protein-functionalized atomic force microscopy cantilever tips. Scanning 1999; 21 (6): 394–397.

[26] Lazar AN, Shadgaldian P, Coleman AW. Anion recognition effect in the structuring of bovine serum albumin films.J Supramol Chem 2001; 1: 193–199.

[27] Iafisco M, Palazzo B, Falini G, Di Foggia M, Bonora S, Nicolis S *et al.* Adsorption and conformational change of myoglobin on biomimetic hydroxyapatite nanocrystals functionalized with alendronate. Langmuir 2008; 24 (9): 4924–4930.

[28] Iafisco M, Sabatino P, Lesci IG, Prat M, Rimondini L, Roveri N. Conformational modifications of serum albumins adsorbed on different kinds of biomimetic hydroxyapatite nanocrystals.Colloids and Surf. B: Biointerfaces 2010; 81 (1): 274–284.

[29] Wilcox MH. Microbial BioJilms: Formation and Control. Blackwell Scientific, Oxford 1993.

[30] Kajiyama T, Tanaka K, Ge SR, Takahara A. Morphology and mechanical properties of polymer surfaces *et al.* scanning force microscopy. Prog Surf Sci 1996; 52 (1): 1–52.
[31] Manara S, Paolucci F, Palazzo B, Marcaccio M, Foresti E, Tosi G *et al.*Biomimetic hydroxyapatite-collagen coating electrochemically deposited on titanium plate. Inorg Chim Acta 2008; 361 (6) :1634–1645.
[32] IkadaY.Surface modification of polymers for medical applications.Biomaterials 1994; 15: 725–736.
[33] Coen MC, Dietler G, Kasas S, Groening P. AFM Measurements of the topography and the roughness of ECR plasma treated polypropylene. Appl Surf Sci 1996; 103: 27–34.
[34] Hopkins J, Badyal JPS. Plasma Modification of Poly(ether sulfone). Macromol 1994; 27: 5498–5593.
[35] Ryan ME., Badyal JPS. Surface Texturing of PTFE Film Using Non-Equilibrium Plasmas.Macromol 1995; 28: 1377–1382.
[36] Frazier RA. Ph.D.-thesis, University of Nottingham 1996.
[37] Shakesheff KM, Davies MC., Dumb A, Roberts CJ, Tendler SJB, Williams PM. *Et al.* Atomic Force Microscopy Visualization of the Degradation of Melt-Crystallized Poly(sebacic anhydride). Macromol 1995; 28: 1108 –1114.
[38] Shakesheff KM, Davies MC, Roberts CJ, Tendler SJB, Shard AJ, Dumb A. *Et al.* Atomic Force Microscopy Imaging of Polymer Degradation in an Aqueous Environment. Langmuir 1994; 10 (12): 4417–4419.
[39] Green RJ, Davies J, Davies MC, Roberts CJ, Tendler SJB. Surface plasmon resonance for real time *et al.* analysis of protein adsorption to polymer surfaces. Biomaterials 1997; 18: 405–413.
[40] Baguet J, Sommer F, Duc TM. Imaging surfaces of hydrophilic contact lenses with the atomic force microscope. Biomaterials 1993; 14: 279–284.
[41] Baguet J, Sommer F., Claudon-Eyl V, Duc TM. Characterization of lachrymal component accumulation on worn soft contact lens surfaces by atomic force microscopy.Biomaterials 1995; 16: 3–9.
[42] Oberleithner H, Brinckmann E., Biebisch G, Geibel J. Visualizing life on biomembranes by atomic force microscopy. Kidney Intern 1995; 48: 923–929.
[43] Haugstad G, Gladfelter WL. Force-displacement measurements in a beam-reflection scanning force microscope.Calibration issues Ultramicro 1994; 54: 31–40.
[44] Haugstad G, Gladfelter WL, Weberg EB. Scanning force microscopy of gelatin films in the dry, swollen and redried states.Mat SciEng C - Biomimetic Mat SensSyst 1995; 3: 85–89.
[45] Radmacher M, Fritz M, Hansma PK. Imaging soft samples with the atomic force microscope: Gelatin in water and propanol. Biophys J 1995; 69: 264–270.
[46] Nyffenegger R, Ammann E, Siegenthaler R, Kotz R, Hass O. *In situ* scanning probe microscopy for the measurement of thickness changes in an electroactive polymer.ElectrochimActa 1995; 40: 1411–1415.
[47] Hoh JH, Sosinsky GE, Revel JP, Hansma PK. Structure of the extracellular surface of the gap junction by atomic force microscopy.BiophysJ 1993; 65: 149–163.
[48] Henderson E. Imaging of living cells by atomic force microscopy. Prog Surf Sci 1994; 46: 39–60.
[49] Hoh JH, Hansma PK. Atomic force microscopy for high-resolution imaging in cell biology. Trend Cell Biol 1992; 2: 208–213.

[50] Micic M, Hu D, Suh YD, Newton G., Romine M, Lu HP. Correlated atomic force microscopy and fluorescence lifetime imaging of live bacterials cells. Colloids Surf B: Biointerfaces 2004; 34 (4): 205–212.
[51] Radmacher M, Tillmann RW, Fritz M, Gaub HE. From molecules to cells: imaging soft samples with the atomic force microscope.Science 1992; 257: 1900–1905.
[52] Schneider SW, YanoY, Sumpio BE, Jena BP, Geibel JP, Gekle M *et al.* Rapid aldosterone-induced cell volume increase of endothelial cells measured by the atomic force microscope. Cell BiolInt 1997; 21: 759–768.
[53] Ross B, Motherby H, Saurenbach F, Frohn J, Kube M, Bocking A. Atomic force microscopy in effusion cytology. Anal Quant CytoHistol 1998; 20: 97–104.
[54] Osada T, Arakawa H, Ichikawa M, Ikai A. Atomic force microscopy of histological sections using a new electron beam etching method. J Microsc 1998; 189: 43–49.
[55] Falvo MR, Washburn S, Superfine R, Finch M, Brooks Jr. FP, Chi V *et al.* Manipulation of individual viruses: friction and mechanical properties.Biophys J 1997; 72: 1396–1403.
[56] Beinoit M, Gabriel D, Gerish G, Gaub HE. Discrete interaction in cell adhesion measured by single-molecule force spectroscopy. Nat Cell Biol 2000; 2 (6): 313–317.
[57] Gabor D. The electron microscope.Electronic Engineering Technical Monographs.Hulton Press 1945.
[58] Haguenau F, Hawkes PW, Hutchinson JL, Satiat-Jeunemaıtre B, Simon GT, Williams DB. Key events in the history of electron microscopy.Microscopy and Microanalysis 2003; 9: 96–138.
[59] Langmore J.P., Smith MF. Quantitative energy-filtered electron microscopy of biological molecules in ice.Ultramicroscopy 1992; 46: 349–373.
[60] Stewart ADG. The origins and development of scanning electron microscopy. J Microsc 1985; 139 :121–127.
[61] Heslop-Harrison Y. Scanning electron microscopy of fresh leaves of Pinguicula.Science 1970; 167: 172–174.
[62] Read ND, Lord KM. Examination of living fungal spores by scanning electron microscopy. ExpMycol 1991; 15: 132–139.
[63] Stokes DJ, Donald A.M. *Et al.* mechanical testing of dry and hydrated breadcrumb in the environmental scanning electron microscope (ESEM). J MaterSci 2000; 35: 599–607.
[64] Muscariello L, Rosso F, Marino G, Giordano A, Barbarisi M, Cafiero G, Barbarisi A. A critical overview of ESEM applications in the biological field. J Cell Physiol2005; 205: 328–334.
[65] Goodman J. Introduction to Fourier Optics. McGraw-Hill Book Company, New York 1968.
[66] Williams DB. Electron Microscopy Transmission in Encyclopedia of Materials: Science and Technology Eds: Buschow KHJ, Cahn RW, Flemings MC, Ilschner B. 2008; 2577–2584.
[67] Freitag B, Kujawa S, Mul PM, Ringnalda J, Tiemeijer PC. Breaking the spherical and chromatic aberration barrier in transmission electron microscopy.Ultramicroscopy 2005; 102: 209–214.
[68] Müller SA, AebiU, Engel A. What transmission electron microscopes can visualize now and in the future. J StructBiol 2008; 163: 235–245.
[69] Koster AJ, Grimm R, Typke D, Hegerl R, Stoschek A, Walz J, Baumeister W. Perspectives of molecular and cellular electron tomography. J StructBiol 1997: 120: 276–308.

[70] Wade RH. A brief look at imaging and contrast transfer.Ultramicroscopy 1992; 46: 145–156.

[71] Bremer A, Henn C, Engel A, Baumeister W, Aebi U. Has negative staining still a place in biomacromolecular electron microscopy? Ultramicroscopy 1992; 46: 85–111.

[72] Fujiyoshi Y. The structural study of membrane proteins by electron crystallography.Adv Biophys 1998; 35: 25–80.

[73] Dubochet J, Adrian M., Chang JJ, Homo JC, Lepault J, McDowall AW, Schultz P. Cryo-electron microscopy of vitrified specimens. Q Rev Biophys 1988; 21: 129–228.

[74] Adrian M, Dubochet J, Lepault J, McDowall AW. Cryo-electron microscopy of viruses.Nature 1984; 308: 32–36.

[75] Frederik P, Bomans P, Braet F, Wisse E. Cells of Hepatic Sinusoid eds.Wisse E., Knook DL, Balabaud C.Kuppferr Cell foundation, Leiden 1997: 476–478.

[76] Frederik P, Hubert DH., Cryelectron microscopy of liposomes. Methods Enzymology 2005; 391: 431–448.

[77] Adrian M, Dubochet J., Fuller SD, Harris, JR. Cryo-negative staining.Micron 1998; 29: 145–160.

[78] Hess MW. Cryopreparation methodology for plant cell biology. Methods Cell Biol 2007; 79: 57–100.

[79] McDonald K. Cryopreparation methods for electron microscopy of selected model systems. Methods Cell Biol 2007; 79: 23–56.

[80] Buser C, Walther P, Mertens T, Michel D. Cytomegalovirus primary envelopment occurs at large unfolding of the inner nuclear membrane. J Virol 2007; 81: 3042–3048.

[81] Giddings Jr. TH, O'Toole ET, Morphew M, Mastronarde DN, McIntosh JR, Winey M. Using rapid freeze and freeze-substitution for the preparation of yeast cells for electron microscopy and three-dimensional analysis. Methods Cell Biol 2001; 67 (377): 27–42.

[82] He W, Kivork C, Machinani S, Morphew MK, Gail AM, Tesar DB, *et al.* A freeze substitution fixation-based gold enlarging technique for EM studies of endocytosednanogold-labeled molecules. J StructBiol 2007; 160: 103–113.

[83] Robertson CE. Electron microscopy of Archaea. Methods Cell Biol 2007; 79: 169–191.

[84] Al-Amoudi A, Studer D., Dubochet J. Cutting artefacts and cutting process in vitreous sections for cryo-electron microscopy. J StructBiol 2005; 150: 109–121.

[85] Bouchet-Marquis C, Zuber B, Glynn AM, Eltsov M, Grabenbauer M, Goldie KN *et al.* Visualization of cell microtubules in their native state.Biol Cell 2007; 99: 45–53.

[86] Palazzo B, Walsh D, Iafisco M, Foresti E., Bertinetti L, Martra G *et al.* Amino acid synergetic effect on structure, morphology and surface properties of biomimetic apatite nanocrystals.ActaBiomater 2009; 5(4): 1241–1252.

[87] FoltranI, Foresti E, Parma B, Sabatino P. Novel biologically inspired collagen nanofibers reconstituted by electrospinning method. MacromolSymp 2008; 269 (1): 111–118.

[88] Gonzalez-McQuire R, Green DW, Partridge KA, Oreffo ROC, Mann S, Davies SA. Coating of human mesenchymal cells in 3D culture with bioinorganic nanoparticles promotes osteoblastic differentiation and gene transfection. Adv Mater 2007; 19: 2236–2240.

[89] Lowestan HA, Weiner S. On biomineralization. OxfordUniversity Press, New York, 1989.

[90] Collagen: Biochemistry, Biomechanics, Biotechnology. NimniME. (Ed), CRC Press, Boca RatonFL, 1988.

[91] Weiner S, Wagner HD. The Material bone:structure-mechanical function relations. Ann Rev MatSci 1998; 28: 271–298.

[92] Pouget EM, Bomans PHH, Goos JACM, Frederik PM, De With G, Sommerdijk NAJM. The initial stages of template-controlled $CaCO_3$ formation revealed by cryo-TEM. Science 2009; 323: 1455–1458.

[93] Nudelman F, Pieterse K, George A, Bomans PHH, Friedrich H, Brylka LJ, *et al.* The role of collagen in bone apatite formation in the presence of hydroxyapatite nucleation inhibitors. Nat Mater2009; 9 (12): 1004–1009.

[94] Beniash E, Traub W, Veis A, Weiner S. A transmission electron microscope study using vitrified ice sections of predentin: Structural changes in the dentin collagenous matrix prior to mineralization. J StructBiol 2000; 132: 212–225.

[95] Hodge AJ, Schmitt FO. The charge profile of the tropocollagen macromolecule and the packing arrangement in native-type collagen fibrils, ProcNatlAcadSci USA 1960; 46: 186–197.

[96] Chapman JA, Tzaphlidou M, Meek KM, Kadler KE. The collagen fibril-a model system for studying the staining and fixation of a protein.Electron Microsc Rev 1990; 3: 143–182.

[97] Orgel JPRO, Irving, TC, Miller A, Wess, TJ. Microfibrillar structure of type I collagen *in situ*. ProcNatlAcadSci USA 2006; 103: 9001–9005.

CHAPTER 7

Biological Characterization of Biomaterials: *In vitro* Tests

Neelam Gurav*, Borvornwut Buranawat and Lucy Di Silvio

King's College London, Biomaterials, Biomimetics & Biophotonics Group, Dental Insitute, at Guy's, King's College and St Thomas' Hospitals, Floor 17, Guy's Tower Wing, Guy's Hospital, St Thomas Street, London SE1 9RT

Abstract: There has been a transition in the definition of biocompatibility from a material that was once required to be inert and simply act as 'filler' or scaffold, to one which is dynamic, interactive and has a beneficial effect. In addition to the definition changing, the actual role of the biomaterial has also changed. More is demanded from biomaterials and also a greater understanding about the way cells in the body interact with biomaterials on a three dimensional level. Biomaterials are expected to be used in many different forms and devices in medicine; blocks, fillers, scaffolds that provide structure and shape and if possible be completely resorbable and remodeled. A biomaterial is required not only to be non-toxic but it should also allow the cells in contact with it to benefit from the interaction. The testing of all these functions is quite difficult *in vivo* but can be studied in detail with a sensitive panel of *in vitro* tests. There is an increasing pressure from society to reduce animal experimentation which has led to more sensitive and sophisticated methods and today *in vitro* testing is an important and major part of testing biomaterials and in some cases has been able to replace *in vivo* testing completely. In this chapter, we will discuss these tests and how they can be used to evaluate a range of materials used for clinical application.

Keywords: Scaffold, resorption, *in vitro* test, biomaterial.

BIOCOMPATIBILITY

The background concept of biocompatibility starts from the term "toxicity" which is defined as the release of leachable product from materials in a quantity sufficient to kill cells. *In vitro* experiments have been used as initial screening methods to determine if biomaterials exhibit cytotoxicity or compatibility with living cells for more than three decades. Biomaterials were once considered biocompatible if they didn't elicit a rejection response once they were placed in

***Address correspondence to Neelam Gurav:** King's College London, Biomaterials, Biomimetics & Biophotonics Group, Dental Insitute, at Guy's, King's College and St Thomas' Hospitals, Floor 17, Guy's Tower Wing, Guy's Hospital, St Thomas Street, London SE1 9RT; Tel: +44 (0) 20 7188 1789; Fax: +44 (0) 207188 1823; E-mail: neelam.gurav@kcl.ac.uk

Lia Rimondini, Claudia Bianchi and Enrica Vernè (Eds)

the body. The expectations however of biomaterials as medical devices have changed dramatically over the last decade. Now the requirement deals with two principal components. Firstly, "biosafety" that is to exclude damaging effects of the biomaterial to the organism itself and secondly "biofunctionality". Biofunctionality addresses the need for the material to elicit a beneficial host-response for optimal functioning of the medical device [1, 2]. When a new material has been developed, the assessment of its biocompatibility is essential with several methods available that have been validated for repeatability and reproducibility for both short and long term study. The Medical Devices directive states that a material to be used for medical devices should comply with toxicity tests [3-6]. The International Organisation for Standardisation (ISO) also incorporates the "Biological evaluation of medical devices" in ISO 10993. These define a series of guidelines that can help in the selection of the most appropriate tests for biocompatibility screening. Particularly, *in vitro* tests according to ISO 10993 part 4; selection of tests for interactions with blood, and ISO 10993 part 5 tests for *in vitro* cytotoxicity, the standard offers the opportunity to test early in product development the materials that are being used for the medical device. The ISO standards 10993 provide a guideline of specific tests that should be performed with recommendations relating to sample size, number, types of cells, medium, controls and incubation times. Potential toxicity can be tested and the material can be eliminated at an early stage in the screening process therefore avoiding expensive and time consuming developmental work. Ultimately the biocompatibility of a device is determined by *in vitro tests*, animal models, *in vivo* tests and then clinical tests. This chapter aims to give an overview of the *in vitro* tests and the types of materials currently being tested.

Biocompatible, in the *in vitro* situation implies that the material or any leachable products from it, do not cause cell death or impair cellular functions. Implant biomaterials should be biocompatible both in the short term and also the long term during degradation of the material and the release of degradation products. Assessment of bioresorbable materials is more complex due to the degradation of the materials during testing and the need to artificially age the materials in order to assess longer term effects. Degradable biomaterials are constantly changing with time as the biomaterial undergoes bulk and surface degradation which leads to changes in the environment around the material. Non degradable materials also undergo changes in their chemical and structural nature, but this change is not as dramatic and the time involved for physical and chemical changes to occur, is longer.

To begin testing a biomaterial for its biocompatibility the first thing that needs to be determined is the final application of the implant. This will allow the determination of the duration of time the implant and the cells will be in direct contact. *In vitro* tests are not confirmatory of the performance of a biomaterial but do provide essential quantitative data on cell viability and proliferation as the material is placed in a biological environment over time. A material however, can perform very badly in an *in vitro* test, but go on to perform very well in an *in vivo* situation. This is usually due to the accumulation of toxic products in the *in vitro* situation which in an *in vivo* environment would be readily removed by the body. Interpretation of data from *in vitro* tests can be quite subjective more so than *in vivo* data in certain circumstances and hence, it is important to know what these are before allowing the material to progress onto the next stage of development. More sophisticated tests that utilise perfusion and continuous flow of cultures and cells in co-culture are being developed and should be incorporated into *in vitro* testing protocols [7-11].

IN VIVO VERSUS IN VITRO TESTING

There is no doubt that properly regulated animal research is absolutely essential to the discovery of new treatments as well as to the assignment of safety and efficacy of medicines and implants. There is, however, increasing pressure in today's society to reduce animal experimentation and the last 20 years has seen a trend towards a decrease in the number of animals used for scientific research. In the field of musculoskeletal tissue engineering for example, *in vivo* testing is widely performed, where a large number of animals are used, usually involving numerous experiments, some of which are uncontrolled and produce limited or qualitative information with an objective that may or may not be achieved. These studies usually address only one particular tissue type or one particular response, or the population is too small thereby limiting the amount of significant information that can be acquired. Furthermore, information can only be obtained once the animals have been killed and short-term studies are costly and also increase animal suffering. The final translation of any tissue-engineered construct to the patient necessitates the use of animal tests, usually involving large mammals, in significant numbers, at a very high cost and animal suffering. However, final animal testing should only be performed when the material has satisfied all other criteria of biosafety and biofunctionality *in vitro*.

The use of *in vitro* tests to predict the outcome of a material once it has been implanted *in vivo* can sometimes be contradictory [12], Biocompatibility *in vitro* doesn't always lead to biocompatibility *in vivo* and *vice versa*. By direct seeding of cells on materials it is possible to measure the numbers of cells attaching, the extent of cell spread, cell proliferation over time, cell differentiation, cytotoxicity assays and the changes occurring in the medium following contact of the cells with the materials. This however does not give an indication of how the material will actually perform once it's implanted *in vivo* and allowed to interact with the local environment a much more complex system.

Reducing the use of animal experimentation has been an objective of the European Union for the last two decades. However, attaining relevant *in vitro* information has been hindered by the lack of standardised and reproducible *in-vitro* models which are able to mimic the natural environment of the tissues. In order to achieve such sophisticated and relevant tests, it is necessary to use appropriate cell types representative of the final application. Hence, characterized *in vitro* models which realistically simulate the physiological environment and reproduce the architectural complexity of the natural tissues *in vivo* are required. Optimization of relevant models will allow and facilitate biocompatibility, biofunctionality and mechanical testing of the developed material in its entirety addressing all possible aspects. Thus, by performing the necessary multiparametric analysis of all the factors involved in the cell-material interactions, it is possible to establish a standardised pre-screening test protocol. That will reduce the need for animal testing during the initial testing period of exiting and novel materials for clinical use. Whilst it is not envisaged that large animal testing can be completely eliminated, establishing an *in vitro* system to be employed in the initial stages of material development any further Tissue Engineered (TE) constructs will permit only fully optimised materials, devices or scaffolds to enter into *in vivo* animal studies, thus considerably, reducing animal numbers, refining test methods thus reducing animal suffering, and where appropriate replacing existing *in vivo* testing.

In previously cited papers on 'biocompatibility testing' for example, early 1970's cells were simply placed directly on the material surfaces. Testing in its simplest form involved assessing cell death, adhesion and proliferation over a short period of time. More recent papers, however, use a whole battery of tests pre and post cell exposure to measure changes in the surface structure. This change has occurred over the years due to the advances in technology and also as a direct result of the change in the types of materials and devices developed. In the 1960s and 1970s biomaterials were expected to be inert *i.e.* not to exert any response, by the early 1980s there was a shift from the inert response to the bioactive response [13]. This new generation of biomaterials is aimed at eliciting a bioactive response in a specific and targeted manner [14]. This can be done by altering biomaterial structure and surface chemistry. Specific materials can be utilised to elicit a particular response depending on the area of body to be remodelled or repaired. Very defined responses can be obtained from cells just by altering the surface of the biomaterial [15].

As the field of tissue engineering develops, there is a significant increase in the types of materials and scaffolds being developed. In the main, these are biodegradable and are based on mimicking natural tissues. Hence, factors such as degradation products, release kinetics of incorporated factors, composite structures with varying degrees of degradation and change in local environment such as pH for example all have to be considered under the test of biocompatibility. Such progress would not be possible without the aid of *in vitro* test methods.

IN VITRO TEST METHODS

In vitro testing of biomaterials shouldn't begin with cellular biocompatibility but the determination of the exact behaviour of the materials in bodily fluids [16]. The changes in mass, the release of leachable and degradation products and the swelling properties, then cellular interaction can be considered taking into account the physical and chemical limitation of the material. Frequently a cell is brought into contact with biomaterials and the cells perish due to an increase or a decrease in pH. Release of toxic products may occur as a result of the swelling of the materials and subsequent degradation or solubilisation of products. All these can be determined before complex *in vivo* cellular tests are carried out. There are now many different methods available to test different aspects of the cell interaction with the materials *in vitro*. These allow a greater degree of interpretation as compared to *in vivo* tests.

There are many *in vitro* testing systems currently in use, however the three main types recommended in the ISO standards are the (i) agar diffusion test, (ii) direct contact test, and (iii) extract dilution test. Thus depending on the type of materials and the site of implantation there are many tests which can be used. Knowledge about the intended site of implantation hence the types of cells that will be predominantly in direct contact with the materials will allow better design of the *in vitro* test methods. *In vitro* cellular response to materials has become a useful tool in developing new materials. It is a quick way of testing if their newly synthesised material is capable of potential development towards clinical application.

Xu *et al.* used Polyurethane and Hyaluronic acid to develop copolymers which allowed control of hydrophilicity and surface energy [17]. This was able to decrease red blood cell adhesion leading to potential anti-thromobgenic properties which could be used in the vascular tissue replacement.

BIOMATERIALS CURRENTLY USED IN MEDICINE AND DENTISTRY

There are many disciplines in medicine and dentistry which use biomaterials. There is a degree of overlap and the list in not finite. The types of materials available to the scientists and surgeons are always expanding which ultimately gives a greater degree of control towards developing the best device for a specific application. Materials such as coral, silk, tofu and bamboo are all candidate materials that are either undergoing testing or have been considered as tissue replacement scaffolds [18-24].

Recently, new nano materials such as Carbon Nanotubes (CNTs) show potential as scaffold components in tissue engineered constructs [25-28]. CNTs have certain materials characteristics which have been shown to improve the mechanical and conductive properties of polymeric biomaterials. The behaviour of cells on materials can be altered by a variety of methods. To what degree interaction with cells is desired will determine the type of structure that is formulated on medical devices. It has to be taken into the equation that the surface is a rapidly changing niche and to predict and then control the type of behaviour require further understanding of the biological mechanisms involved.

Table 1: The different disciplines and applications in medicine utilising biomaterials.

Discipline	Application
Orthopaedics	Knee and hip replacements Fracture fixations Bone defect repair/bone graft Intervertebral disc replacement
Dental	Dental implant Restorative dental materials *e.g.*, crowns and fillings
Soft tissue	Enhance would healing Skin replacement Nerve regeneration guides Ligament and tendon construct
Cardiovascular	Heart valves Percutaneous Transluminal Coronary Angioplasty (PTCA) balloons Pacemakers Catheters Grafts Stents
Organs	Cochlear replacement Intraocular lens
Regenerative	Enhancement of the wound healing process Replacement of tissues

TISSUE ENGINEERING

Regardless of the type of tissue being regenerated the basic requirements of any tissue engineered device remain the same. They require the cells of the tissue to be replaced and a scaffold which will provide a surface for cells to attach. The cells themselves can be extracted from the patient from the mature differentiated tissue or be differentiated from a stem cell source. The type of tissue formed will be dependent on the rate of degradation of the scaffold. A controlled rate of degradation is needed. This can be achieved using different degradable polymers as block polymers or copolymers which vary in molecular weight. Table **2** gives some common example of some materials currently being used as possible tissue engineering scaffolds. These are either degradable or non-degradable and of synthetic origin or natural origin.

Table 2: Examples of natural and synthetic materials currently used as cell carriers or devices in tissue engineered applications

Natural materials	Synthetic materials
Collagen	Polylactic acid (PGA)
Fibrin	Polyglycolic acid (PLA)
Fibronectin	Polycaprolactone (PCL)
Gelatin	Poly(propylene fumarates) (PFF)
Hyaluronate	Polyanhydrides
Glycosaminoglycans	Poly(acrylic acid) PAA
Chitin	Poly(ethylene glycol) PEG
Chitosan	Ceramics including Hydroxyapatite and Calcium phosphates
Dextran	Bioactive glasses and phosphate based glasses
Agarose	
Alignates	
Starch	
Polyhydroxyalkanoates	
Coral	
Demineralised bone	

Modern day materials are being developed to mimic natural tissue structures. The concept of is based on 'learning from nature' and using nanotechnology to develop a mechanically stable, smart nano-composite materials that resemble specific tissues. This involves developing materials with specific properties including composition, surface chemistry and architecture (*e.g.* electrospun shape-memory polymers (SMPs), whose design is based upon mimicking the complexity of the natural materials in the body. Electrospun nanoscaffolds for example can be functionalized by biomimetic peptides and/or nanometric fillers in order to obtain 3D matrices with high surface area and optimal structural and mechanical characteristics. The rational design criteria of the current day testing of materials has to take numerous factors into consideration for example, whether, the material/scaffold incorporates selected signalling molecules targeted for specific actions such as lineage differentiation of stem cells in order to create a viable, bio-functional scaffold.

The majority of biological studies to test TE systems are based on a specific cell type response; hence, it would be advantageous to develop co-culture systems of different cell types to mimic the natural physiological environment. More sophisticated testing using bioreactors now allows exploration of the temporal and spatial effects on cell behaviour in a 3D tissue environment and the effect of mechanical and biological cueing on cell proliferation and migration. The impact of these parameters on differentiation (gene expression) and tissue maturation for tissues will allow the realization of creating viable tissues [29-32]. Cartilage tissue for example is one of the hardest tissues to repair and replace. The model scaffold must allow the placement and maintenance of viable chondrocytes within the matrix. This then poses the problem of diffusion as most matrices which allow encapsulation will be of gel form. This limits the size of the construct and leads us to consider vascularizing the scaffold. This can be initiated *in vitro* prior to implantation and can be achieved by co-culturing angiogenic cells with cells of the host tissue that needs to be repaired. Many co-cultures of human cells have been developed, since the *in vivo* situation involves many cell types. The three dimensional scaffold is also very important in the test system as these mimic the principles of the extracellular matrix (ECM) [33-36].

The use of stem cells is one of the most promising areas of research for the treatment of disease and tissue regeneration. Stem cells have the capacity to differentiate into multiple lineages by responding to both biological (growth factors, cytokines) and mechanical cues (tension, compression, sheer force) [37-42], Stem cells incorporated into degradable gels can be differentiated towards different lineages. However the biggest problem of this approach is that due to diffusion problems with nutrients cell death begins to occur in the centre of the scaffolds.

In tissue engineering biodegradable materials are favoured and these are usually of natural origin. Biodegradable materials allow the slow replacement of the scaffold with host cell infiltration and extracellular matrix formation. One example is in the use of both Hydroxyapatite, (HA) and Biphasic Calcium Phosphate (BCP). HA has become an important calcium phosphate based bioceramic which has traditionally been used as a filler or replacement material [43, 44]. (BCP) is made up of a mixture of HA and (β-tricalcium phosphate, (βTCP) [45]. The properties of these materials make them ideal as candidates for tissue regeneration rather than just replacement materials. They can be loaded with growth factors to encourage faster and better integration with the host tissue and the eventual renewal by host cells or cells seeded within the scaffolds [46-49].

CHANGING TRENDS IN BIOCOMPATIBILITY TESTING

Current test methods are becoming more specific and address various aspects of biofunctionality. Human cell lines are the preferred cells for testing devices for human contact and there is a move away from animal cells for this purpose. It is now possible to obtain a variety of cell types commercially. These are primary human cells and so it is no longer necessary to use a mouse sarcoma cell line to test a material that is being developed for use in for example dermal implantation in a human. From the ECACC and other suppliers one can purchase a human primary fibroblast cell line if access to primary tissue in not possible.

It is still a very valid point that although *in vitro* tests are very good at addressing some of the aspects towards the overall biocompatibility of a material; they still remain a poor indicator of the overall biocompatibility when the device is implanted. *In vitro* tests can never fully replace *in vivo* tests but they do offer a rapid, cost efficient, ethical way of obtaining basic data on materials with regard to cytoxicity saving a lot of animals, time and cost. To take the decision to go straight to an *in vivo* model without the basic knowledge on cell adhesion and cell toxicity should be considered carefully.

In cases such as materials with nano topographies, special considerations have to be given as to their physico-chemical properties may influence cell response. *In vivo* the systemic cellular response to a material following implantation swamps any subtle changes that occur early on. *In vitro* it is easy to see differences in the amount of protein adsorption, cell adhesion, and the degree of cell spread and the rate of cell proliferation. These are very difficult to assess *in vivo* as it is difficult to extract a finite number of cells and measure these. During *in vitro* testing a fixed number of cells can be seeded on the surface of a biomaterial and be compared to a control or a number of other materials over a period of time. The size of the materials are defined and comparable, enough replicates can be achieved to allow for statistical analysis. Expression of different proteins can be compared individually without contamination from other cellular processes. The subtle differences in the surface structure can be "sensed" by cells *in vitro* and differences can be observed clearly [23, 50-57]. There is now a growing body of evidence that suggests that stem cells due to the interaction with their environment are able to differentiate down specific lineages. This will have a significant impact on the design of material surfaces for tissue engineering applications [58].

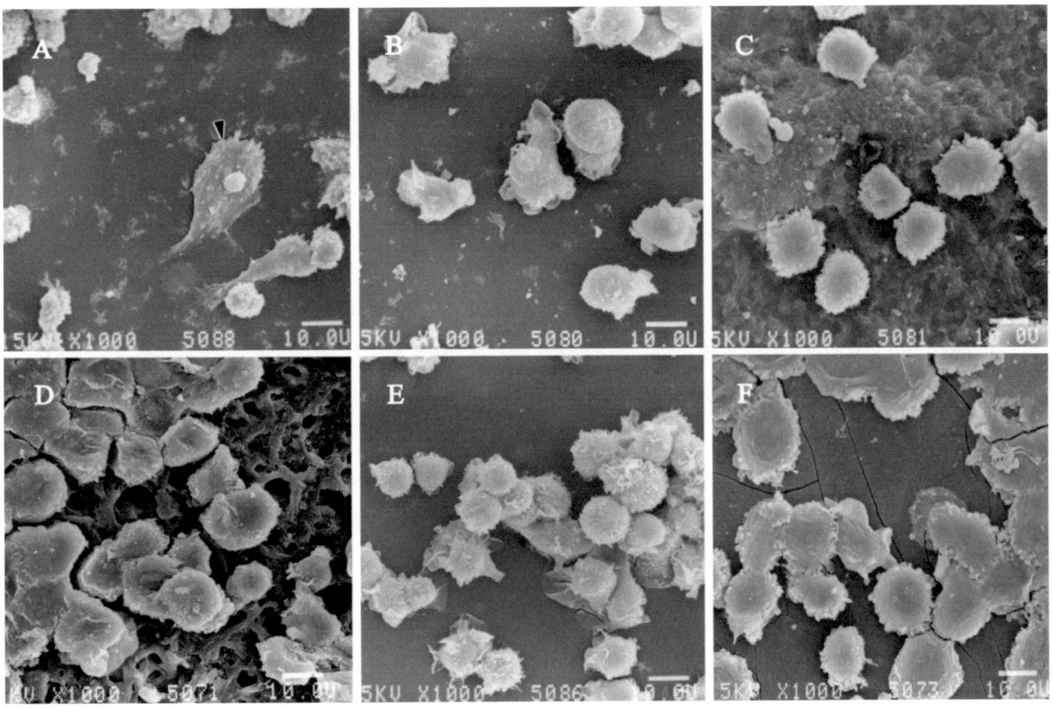

Figure 1: SEM micrographs of THP-1 cells on different degradable polymer surfaces with glass as a non degradable control. A: Glass, B: Gamma irradiated Poly Lactic Acid, C: Poly Hydroxybutyrate, D: Polyhydroxybutyrate-Hydroxyvalerate copolymer, E: Non gamma irradiated Poly Lactic Acid and F: Polycaprolactone. The micrographs show the morphological differences that can occur due to differences in structure of the material surfaces. These cells are monocytes and are able to become activated into macrophages. Fig. **1A**, shows an activated monocyte exhibiting a macrophage like morphology (Arrow head). The cells on Figs. **1B**, **1C** and **1E** show similar morphologies although a ruffling of the cytoplasm can be seen on material E. Cells on **1D** show clumping and cells on **1F** appear to be fusing together and have a very different morphology compared to the other surfaces.

The micrographs in Fig. **1** show that a simple *in vitro* test can yield quite differing results however, interpretation of these initial results requires further testing. Further biochemical tests need to be carried out to determine the nature of this response. When testing a material for biocompatibility its effect with respect to overall cytotoxicity, sensitization, systemic toxicity, irritation is required. Of these only cytotoxicty can be tested *in vitro*. The outcome depends upon the chemical and physical nature of the device and its component materials, the type of tissue that will be in immediate contact with the device and also the duration of this contact for example, the length of implantation. This is a very important consideration in determining biocompatibility as the duration of implantation varies for each type of device. For example, a hip implant will remain *in situ* until failure as opposed to a fracture fixation device, which will be removed as soon as the bone has healed.

MICROARRAYS

Even though microarray technology is in its infancy it is already being used in the field of biomaterials to test cell-biomaterial interactions [59-62]. The basic principal of microarray technology consists of a small coin sized chip, that is made of either glass or other solid although membrane arrays have also been developed which utilise a nylon membrane. These are sometimes called macroarrays. These are sometime called macroarrays, as they have a larger area for DNA spotting. These arrays contain the sequence for any number of genes ranging for a few to several thousand. Following a particular experiment; for example exposure of stem cells to two different formulations of a biomaterial the mRNA from the cells can be extracted and cDNA produced labelled with fluorescent dyes. The labelled cDNA is then hybridized to the DNA on the array. A camera and a microscope capture the signal and the change in intensity of each dot is measured.

A recent review by Hanagata *et al.* provides up to date literature on the use of gene analysis for evaluating biomaterials. It divides the review into two groups; direct and indirect effects of the biomaterials on the cells. Several studies are presented where materials have been exposed to different cell types. They point out that *in vitro* data do not always correlate with *in vivo* data and this is partly due to the use of single cell types in *in vitro* experiments. In future the use of combining techniques and improved technologies and the molecular level will expand the use of DNA microarrays [63, 64].

For example, with the direct contact test it is possible to assess cell adhesion and proliferation but also the gene and protein expression of cells can be determined. In a recent study van der Zande *et al.* [65)], were able to assess the addition of carbon nanotubes (CNTs) and microhydroxyapatite (muHA) incorporation into degradable polymers and test if there was an effect on rat bone marrow cells. They studied cell proliferation and differentiation using the genetic profiles of the cells examined by microarray investigation. The results allowed them to conclude that CNT- and muHA-containing composite materials were not recommended as a bone restorative material. *In vitro* tests using microarray technology allows for a wide screening of the cellular response allowing for modification in the overall biomaterial design at an early stage in the developmental process. Expression of specific cytoskeletal patterns can be used to investigate the effect due to changes in surface properties [65-68].

PROTEOMICS

Proteomics is the study of proteins present in the body and how they interact with each other. A gene can be expressed then translated into mRNA but the resultant protein is what decides the function within the cell. This study of this expression in terms of the amount of protein produced and how it interacts with other proteins is key to understanding how cells behave. [69]. Proteomics used for testing materials can ascertain subtle differences in cell response not normally seen. Xu *et al.* [70)], assessed the levels of proteins using iTRAQ-coupled 2-D LC-MS/MS analysis, They reported the first comparative proteomics profiling of human osteoblast cells cultured on plain HA and CNT reinforced HA. Cytoskeletal proteins, metabolic enzymes, signaling, and cell growth proteins previous associated with cell adhesion and proliferation were found to be differentially expressed on these two surfaces. The level of these proteins was generally higher in cells adhered to HA surface, indicating a higher level of cellular proliferation in these cells.

CONCLUSION

In vitro testing of biomaterials remains at the root of not only determining how cells interact with biomaterials but also in developing the desired response of biomaterials so that, each material can be enhanced by changing its surface structure and chemistry, to achieve the best possible device. This may be a material that doesn't require cell adhesion or one that requires both adhesion and proliferation. Most materials fail due to our lack in understanding of how to control and predict the biology rather than the physics and chemistry of the material. Gene expression and protein expression analysis will change this significantly in the future. Microarray technology is a rapidly developing field and will have a great impact in the field of biomaterials. Its uses in this field are already numerous however, interpretation of the vast amount of data that are generated remains key to fully unlocking the usefulness of microarray technology.

ACKNOWLEDGEMENTS

Declared none.

CONFLICT OF INTEREST

The author(s) confirm that this chapter content has no conflict of interest.

REFERENCES

[1] Williams DF. On the mechanisms of biocompatibility. Biomaterials 2008 Jul; 29(20): 2941-53.
[2] Kirkpatrick CJ, Bittinger F, Wagner M, *et al.* Current trends in biocompatibility testing. Proc Inst Mech Eng H 1998; 212: 75-84.
[3] Kure R, Haustveit G. [Standardization, control and approval of surgical implants in Europe]. Tidsskr Nor Laegeforen. 1994; 114: 591.
[4] Haustveit G, Torheim B, Fystro D, Eidem T, Sandvik M. Toxicity testing of medical device materials tested in human tissue cultures. Biomaterials 1984; 5: 75-80.
[5] Robertson CW. The medical device directive. Health Estate 1998 ; 52: 26-7.
[6] Dobbs H. Current Developments and the Medical Device Directive - A Regulator's Perspective. Drugs 2007; 10: 253-5.
[7] Huang H, Oizumi S, Kojima N, Niino T, Sakai Y. Avidin-biotin binding-based cell seeding and perfusion culture of liver-derived cells in a porous scaffold with a three-dimensional interconnected flow-channel network. Biomaterials 2007; 28: 3815-23.
[8] Figallo E, Flaibani M, Zavan B, Abatangelo G, Elvassore N. Micropatterned biopolymer 3D scaffold for static and dynamic culture of human fibroblasts. Biotechnol Prog 2007; 23: 210-6.
[9] Cho EH, Yang YI, Mun CW, Kim JK. Tissue-engineered semi-microporous segmented polyetherurethane vascular prostheses. J Biomater Sci Polym Ed 2005; 16: 775-90.
[10] Prokop A, Prokop Z, Schaffer D, *et al.* NanoLiterBioReactor: long-term mammalian cell culture at nanofabricated scale. Biomed Microdevices 2004; 6: 325-39.
[11] Sittinger M, Perka C, Schultz O, Haupl T, Burmester GR. Joint cartilage regeneration by tissue engineering. Z Rheumatol 1999; 58: 130-5.
[12] Wilsnack RE. Quantitative cell culture biocompatibility testing of medical devices and correlation to animal tests. Biomater Med Devices Artif Organs 1976; 4: 235-61.

[13] Hench LL, Thompson I. Twenty-first century challenges for biomaterials. J R Soc Interface 2010; 7 Suppl 4: S379-91.

[14] Ahmad M, Jones JR, Hench LL. Fabricating sol-gel glass monoliths with controlled nanoporosity. Biomed Mater 2007; 2: 6-10.

[15] Oh S, Brammer KS, Li YS, *et al.* Stem cell fate dictated solely by altered nanotube dimension. Proc Natl Acad Sci U S A. 2009; 106: 2130-5.

[16] Albert DE. The important role of material and chemical characterisation in device evaluation. Med Device Technol. 2004; 15: 15-8.

[17] Xu F, Li Y, Deng Y, Xiong J. Porous nano-hydroxyapatite/poly(vinyl alcohol) composite hydrogel as artificial cornea fringe: characterization and evaluation *in vitro*. J Biomater Sci Polym Ed 2008; 19: 431-9.

[18] Shahgaldi BF. Coral graft restoration of osteochondral defects. Biomaterials. 1998 ; 19: 205-13.

[19] Tran CT, Gargiulo C, Thao HD, Tuan HM, Filgueira L, Michael Strong D. Culture and differentiation of osteoblasts on coral scaffold from human bone marrow mesenchymal stem cells. Cell Tissue Bank 2011;12:247-61.

[20] Yuan J, Zhang WJ, Liu G, *et al.* Repair of canine mandibular bone defects with bone marrow stromal cells and coral. Tissue Eng Part A 2010; 16: 1385-94.

[21] Thalhammer A, Edgington RJ, Cingolani LA, Schoepfer R, Jackman RB. The use of nanodiamond monolayer coatings to promote the formation of functional neuronal networks. Biomaterials 2010; 31: 2097-104.

[22] Kalbacova M, Rezek B, Baresova V, Wolf-Brandstetter C, Kromka A. Nanoscale topography of nanocrystalline diamonds promotes differentiation of osteoblasts. Acta Biomater 2009; 5: 3076-85.

[23] Yang L, Sheldon BW, Webster TJ. The impact of diamond nanocrystallinity on osteoblast functions. Biomaterials 2009; 30: 3458-65.

[24] Xing Y, Dai L. Nanodiamonds for nanomedicine. Nanomedicine (Lond). 2009; 4:207-18.

[25] Shi X, Hudson JL, Spicer PP, Tour JM, Krishnamoorti R, Mikos AG. Injectable nanocomposites of single-walled carbon nanotubes and biodegradable polymers for bone tissue engineering. Biomacromolecules 2006; 7: 2237-42.

[26] Shi X, Sitharaman B, Pham QP, *et al. In vitro* cytotoxicity of single-walled carbon nanotube/biodegradable polymer nanocomposites. J Biomed Mater Res A 2008; 86: 813-23.

[27] Sitharaman B, Shi X, Tran LA, *et al.* Injectable *et al.* cross-linkable nanocomposites of biodegradable polymers and carbon nanostructures for bone tissue engineering. J Biomater Sci Polym Ed 2007; 18: 655-71.

[28] Sitharaman B, Shi X, Walboomers XF, *et al. In vivo* biocompatibility of ultra-short single-walled carbon nanotube/biodegradable polymer nanocomposites for bone tissue engineering. Bone 2008; 43: 362-70.

[29] Wang Y, Bella E, Lee CS, *et al.* The synergistic effects of 3-D porous silk fibroin matrix scaffold properties and hydrodynamic environment in cartilage tissue regeneration. Biomaterials 2010; 31: 4672-81.

[30] Hagenmuller H, Hitz M, Merkle HP, Meinel L, Muller R. Design and validation of a novel bioreactor principle to combine online micro-computed tomography monitoring and mechanical loading in bone tissue engineering. Rev Sci Instrum 2010; 81: 014303.

[31] Valonen PK, Moutos FT, Kusanagi A, *et al. In vitro* generation of mechanically functional cartilage grafts based on adult human stem cells and 3D-woven poly(epsilon-caprolactone) scaffolds. Biomaterials 2010; 31: 2193-200.

[32] Donegan GC, Hunt JA, Rhodes N. Investigating the importance of flow when utilizing hyaluronan scaffolds for tissue engineering. J Tissue Eng Regen Med 2010 ; 4: 83-95.

[33] Chong MS, Chan J, Choolani M, Lee CN, Teoh SH. Development of cell-selective films for layered co-culturing of vascular progenitor cells. Biomaterials 2009; 30: 2241-51.

[34] Valarmathi MT, Davis JM, Yost MJ, Goodwin RL, Potts JD. A three-dimensional model of vasculogenesis. Biomaterials 2009; 30: 1098-112.

[35] Jiang J, Nicoll SB, Lu HH. Co-culture of osteoblasts and chondrocytes modulates cellular differentiation *in vitro*. Biochem Biophys Res Commun 2005; 338: 762-70.

[36] Unger RE, Halstenberg S, Gunther H, Sartoris A, Brochhausen C, Kirkpatrick CJ. *In vitro* trials with single and co-cultures of osteoblasts and endothelial cells : evaluation of new biomaterials for bone reconstruction and regeneration]. Orthopade 2009; 38):1020-8.

[37] Teo BK, Ankam S, Chan LY, Yim EK. Nanotopography/mechanical induction of stem-cell differentiation. Methods Cell Biol 2010; 98: 241-94.

[38] Salvi JD, Yul Lim J, Donahue HJ. Increased mechanosensitivity of cells cultured on nanotopographies. J Biomech 2010; 43:3058-62.

[39] Potier E, Noailly J, Ito K. Directing bone marrow-derived stromal cell function with mechanics. J Biomech 2010; 43: 807-17.

[40] Nagel T, Kelly DJ. Mechano-regulation of mesenchymal stem cell differentiation and collagen organisation during skeletal tissue repair. Biomech Model Mechanobiol; 9: 359-72.

[41] Huang AH, Farrell MJ, Mauck RL. Mechanics and mechanobiology of mesenchymal stem cell-based engineered cartilage. J Biomech 2010; 4: 128-36.

[42] Boonen KJ, Langelaan ML, Polak RB, van der Schaft DW, Baaijens FP, Post MJ. Effects of a combined mechanical stimulation protocol: Value for skeletal muscle tissue engineering. J Biomech. 2010; 43: 1514-21.

[43] Saad B, Keiser OM, Welti M, Uhlschmid GK, Neuenschwander P, Suter UW. Multiblock copolyesters as biomaterials: *in vitro* biocompatibility testing. J Mater Sci Mater Med 1997; 8: 497-505.

[44] Vallet-Regi M. Revisiting ceramics for medical applications. Dalton Trans 2006; (44): 5211-20.

[45] Daculsi G, Weiss P, Bouler JM, Gauthier O, Millot F, Aguado E. Biphasic calcium phosphate/hydrosoluble polymer composites: a new concept for bone and dental substitution biomaterials. Bone 1999; 25(2 Suppl): 59S-61S.

[46] Lv Q, Nair L, Laurencin CT. Fabrication, characterization, and *in vitro* evaluation of poly(lactic acid glycolic acid)/nano-hydroxyapatite composite microsphere-based scaffolds for bone tissue engineering in rotating bioreactors. J Biomed Mater Res A 2009; 91: 679-91.

[47] Detsch R, Uhl F, Deisinger U, Ziegler G. 3D-Cultivation of bone marrow stromal cells on hydroxyapatite scaffolds fabricated by dispense-plotting and negative mould technique. J Mater Sci Mater Med 2008; 19: 1491-6.

[48] Braccini A, Wendt D, Jaquiery C, *et al.* Three-dimensional perfusion culture of human bone marrow cells and generation of osteoinductive grafts. Stem Cells 2005; 23: 1066-72.

[49] Roeker S, Bohm S, Diederichs S, *et al.* A study on the influence of biocompatible composites with bioactive ligands toward their effect on cell adhesion and growth for the application in bone tissue engineering. J Biomed Mater Res B Appl Biomater 2009; 91: 153-62.

[50] Zhou WY, Guo B, Liu M, Liao R, Rabie AB, Jia D. Poly(vinyl alcohol)/halloysite nanotubes bionanocomposite films: Properties and *in vitro* osteoblasts and fibroblasts response. J Biomed Mater Res A 2010; 93: 1574-87.

[51] Zaveri TD, Dolgova NV, Chu BH, *et al.* Contributions of surface topography and cytotoxicity to the macrophage response to zinc oxide nanorods. Biomaterials 2010; 31: 2999-3007.

[52] Fan J, Chu PK. Group IV. Nanoparticles: Synthesis, Properties, and Biological Applications. Small 2010; 4:2080-98.

[53] Andersson AS, Backhed F, von Euler A, Richter-Dahlfors A, Sutherland D, Kasemo B. Nanoscale features influence epithelial cell morphology and cytokine production. Biomaterials 2003; 24: 3427-36.
[54] Curtis A. Tutorial on the biology of nanotopography. IEEE Trans Nanobioscience 2004; 3: 293-5.
[55] Curtis A, Wilkinson C. New depths in cell behaviour: reactions of cells to nanotopography. Biochem Soc Symp 1999; 65: 15-26.
[56] Curtis A, Wilkinson C. Nantotechniques and approaches in biotechnology. Trends Biotechnol. 2001; 19: 97-101.
[57] Curtis AS, Casey B, Gallagher JO, Pasqui D, Wood MA, Wilkinson CD. Substratum nanotopography and the adhesion of biological cells. Are symmetry or regularity of nanotopography important? Biophys Chem 2001; 94: 275-83.
[58] Chen CS. Using microenvironment to engineer stem cell function. Conf Proc IEEE Eng Med Biol Soc 2004; 7: 4964.
[59] Neuss S, Apel C, Buttler P, *et al.* Assessment of stem cell/biomaterial combinations for stem cell-based tissue engineering. Biomaterials 2008; 29: 302-13.
[60] Anderson DG, Putnam D, Lavik EB, Mahmood TA, Langer R. Biomaterial microarrays: rapid, microscale screening of polymer-cell interaction. Biomaterials 2005; 26: 4892-7.
[61] Klapperich CM, Bertozzi CR. Global gene expression of cells attached to a tissue engineering scaffold. Biomaterials 2004; 25: 5631-41.
[62] Allen LT, Fox EJ, Blute I, *et al.* Interaction of soft condensed materials with living cells: phenotype/transcriptome correlations for the hydrophobic effect. Proc Natl Acad Sci U S A. 2003; 100: 6331-6.
[63] Hanagata N, Takemura T, Minowa T. Global gene expression analysis for evaluation and design of biomaterials. Sci Technol Adv Mater. 2010; doi:10.1088/1468-6996/11/1/013001
[64] Gupta K, Kim DH, Ellison D, *et al.* Lab-on-a-chip devices as an emerging platform for stem cell biology. Lab Chip 2010; 10: 2019-31.
[65] Dalby MJ, Gadegaard N, Curtis AS, Oreffo RO. Nanotopographical control of human osteoprogenitor differentiation. Curr Stem Cell Res Ther 2007; 2: 129-38.
[66] Dalby MJ, Yarwood SJ, Johnstone HJ, Affrossman S, Riehle MO. Fibroblast signaling events in response to nanotopography: a gene array study. IEEE Trans Nanobioscience 2002; 1: 12-7.
[67] Dalby MJ, Riehle MO, Sutherland DS, Agheli H, Curtis AS. Use of nanotopography to study mechanotransduction in fibroblasts--methods and perspectives. Eur J Cell Biol 2004; 83: 159-69.
[68] Allen LT, Tosetto M, Miller IS, *et al.* Surface-induced changes in protein adsorption and implications for cellular phenotypic responses to surface interaction. Biomaterials. 2006; 27: 3096-108.
[69] Gallagher WM, Lynch I, Allen LT, *et al.* Molecular basis of cell-biomaterial interaction: insights gained from transcriptomic and proteomic studies. Biomaterials. 2006; 27: 5871-82.
[70] Xu J, Khor KA, Sui J, Zhang J, Tan TL, Chen WN. Comparative proteomics profile of osteoblasts cultured on dissimilar hydroxyapatite biomaterials: An iTRAQ-coupled 2-D LC-MS/MS analysis. Proteomics 2008; 8: 4249-58.

CHAPTER 8

Biological Characterization of the Materials: *In Vivo* Tests

Roberto Giardino[1,2,*], Milena Fini[1], Nicolò Nicoli Aldini[1] and Anna Paola Parrilli[1]

[1]Preclinical and Surgical Studies Laboratory – Rizzoli Orthopaedic Institute, Codivilla Putti Research Institute, via di Barbiano 1/10 - 40136 Bologna, Italy and [2]University of Bologna, Medical School, Bologna, Italy

Abstract: The *in vivo* evaluation of a material to be used in clinics is a long trial and requires a well planned sequence of steps. *In vivo* tests are complementary to those performed *in vitro* that provide necessary and useful results to be added to those found in the *in vivo* testing, and strongly reduce the number of animals used. Animal studies require first of all the compliance to the ethical and legal rules on animal experimentation. The procedures are standardized at an international level. In particular the rules of ISO 10993 are the guidelines for these investigations. In the procedure of validation of a material, the choice of more than one animal species for the performing of the tests, namely a small/medium and a large size animal should be appropriate. Considering the research on biomaterials, the use of small animals can be acceptable in the early stage of testing, but for the evaluation of prototypes biofunctionality large animals should better approximate the human environment. According to the European and International rules, the experiments must be performed in authorized laboratories, and surgical interventions in well equipped operatory rooms and in general anesthesia. Clinical monitoring and X ray investigations allow the scheduled follow-up on living animals. After the sacrifice histology and histomorphometry provide useful information about the interactions between the material and the tissues, together with biomechanical test. More recently computerized microtomography, a non destructive procedure, is added to the methods of study on laboratory animals allowing high resolution images and the visualization of the internal structures of the objects.

Keywords: Preclinical studies, *in vivo* models, biomaterials, microtomography, histomorphometry.

INTRODUCTION

Appropriate preclinical investigations are a necessary step in the long trip towards

***Address correspondence to Roberto Giardino:** Preclinical and Surgical Studies Laboratory – Rizzoli Orthopaedic Institute, Codivilla Putti, Research Institute, *via* di Barbiano 1/10, 40136 Bologna, Italy; Tel: +39 51 6366786; Fax: +39 51 6366580; E-mail: roberto.giardino@ior.it the validation of a material intended for the use in patients. A preclinical study can be defined as a laboratory or animal study that's done to provide with sufficient evidence that a new treatment is safe and effective enough to be administered in humans [1].

Nowadays no one can think to conduct animal research without a deep knowledge not only of the scientific and technical background, but also of the laws and ethical requirements related to the use of laboratory animals. For this reason the first part of this chapter will be an introduction to this important topic.

LEGISLATION AND ETHICS OF ANIMAL EXPERIMENTATION

The use of laboratory animals for biomedical research is subject to precise rules, to ensure the appropriate use of the animals and the humane treatment of them [2]. The United States approved in 1966 the *Animal Welfare Act*, and more recently the *Public Health Service Policy on Human Care and Use of Laboratory Animals* and the *NIH Guide for the care and Use of Laboratory Animals*. All these documents contains guidelines about the environment, housing and management as well as veterinary care of the animals.

The European Union promulgated in 1986 the *Directive 86/609 – Protection of Experimental Animals* indicating to the member states guidelines to be applied in the regulation of animal research. According to this directive, all the European countries adapted their rules and in particular the Italian Government approved the Law by decree n°116 of January 27, 1992 about the protection of the animals used for scientific purpose [3]. Strategies must be adopted to minimize the number of used animals and to avoid pain, suffering and stress (Art. 2). In particular animal tests cannot be performed before appropriate preliminary *in vitro* tests have been carried out with favourable results. Animal experiments must be carried out in authorized laboratories, with appropriate facilities for animal welfare (Art.3.2). Cage size should be appropriate for small, medium and large animals. Surgical procedures must be performed in general anaesthesia and in well equipped operatory rooms, under the control and responsibility of well trained scientists (Art. 5, 6, 7). Recently the European Union approved a new directive for the protection of the animals used for scientific purposes to whom the state members must adapt their regulations.

The part 2 of ISO 10993 "Animal Welfare Requirements" [4] about the biological evaluation of medical devices indicates the prerequisites to be accomplished when animal tests should be performed. The test are justified only when the expected results are not otherwise available and essential for the material characterization, when no other scientifically validated method not involving animals is available, and when strategies to minimize pain, suffering, distress and lasting harm have been carefully considered and for innovative materials, after the execution of *in vitro* tests.

ISO 10993-2 emphasize also the humane prerequisites of any *in vivo* study as the animal welfare, the anaesthesia and the procedures of euthanasia. The research protocol must describe in detail the scientific aims of the research, the features about the material to be investigated, the rationale for the use of animals, specifying that alternative ways to animal experiments are no possible, and a careful and complete study documentation.

The research protocol must also be approved by an Ethical Committee and submitted to the public authority, according to the regulations of each country.

SELECTION OF THE ANIMAL MODEL

According to Einhorn, "the appropriate use of animal models begins with careful considerations of the question asked." [5]. Planning a research project is based on the contributions of many skills, and the reliability of the results is the direct effect of the proper development of the experimental design. *In vitro* and *in vivo* tests must be considered complementary steps and properly planned [6]. Indeed *in vitro* testing does not exclude *in vivo* testing; *in vitro* methods provide necessary and useful results to be added to those found in the *in vivo* testing, and strongly reduce the number of animals used (Table **1**). Only by adopting both *in vitro* and *in vivo* methods it is possible to reduce limits and increase benefits of each single approach [7]. In particular, the benefits expected from the *in vitro* methods are the high sensitivity, the possibility to create standard experimental conditions, the use of human cells and tissues, the fast and inexpensive screening; they are also easy to perform, manage, and read, and allows a reduction of animal use and suffering. However they can give only partial information on biofunctionality, there is a lack of systemic factors of cell and tissue responses, and there is a difficulty in long term studies.

On the contrary animal studies allows long term investigations, permit the evaluation of biofunctionality, create a biological environment in which the presence of systemic factors interacts on cells and tissue responses and are effective for the final prototype evaluation. They presents otherwise ethical drawbacks and a legal authorization is required; appropriate facilities are also necessary, as well as complex professional expertise. *In vivo* study involves prolonged times and often high costs. It is necessary also should be aware of the differences between humans and animals.

It should be appropriate, in the procedure of validation of a material, the choice of more than one animal species for the performing of the tests, namely a small/medium and a large size animal. Considering the research on bone biomaterials, the use of small animals can are acceptable in the early stage of testing, but the healing characteristics of the large animals should better approximate those of humans during the last stages of research. Indeed the rate of bone growth is inversely related to the position of a specie on a phylogenetic scale. According to these assumptions, in our laboratory mouse is used for toxicity tests and studies about osteoinductive properties of materials; rat is considered for models of osteoporosis, bone healing, peripheral nerve repair, biomaterial implantation in soft and hard tissues, microsurgery; guinea pig for sensitization tests, models of osteoarthritis, studies on biophysical stimulation; rabbit for biomaterial implantation in soft and hard tissues, osteoarthritis, bone defect and fracture healing, cartilage repair, biophysical stimulation, infections, guided bone regeneration; pig and mini-pig for biomaterial implantation, vascular surgery, haemorragic shock, hemoperfusion, sheep for biomaterial implantation, ligament and tendon reconstruction, spinal surgery, bone defect and fracture healing, banked bone and derivatives, cartilage repair, biophysical stimulation, vascular surgery.

Even if the Italian laws admits in particular cases and after a ministerial derogation and authorization the use of dogs, cats and non human primates for the experiments, in our experience never was the use of such species required and reliable results were always obtained only using the above listed animals.

In conclusion, the choice of an animal model must take into account assessment on ethics, availability of the animal, housing requirements, easy of handling, costs, susceptibility to the disease and must be supported by an appropriate background and literature data [8-10].

IN VIVO CHARACTERIZATION OF MATERIALS

The procedures to evaluate the biocompatibility of a material are standardized at an international level. In particular in the European Community the rules of ISO 10993 are the guideline for each investigation in this field [11]. These procedures encompass tests for genotoxicity, carcinogenicity and reproductive toxicity [12], tests for interactions with blood [13], tests for cytotoxicity (*in vitro*), test for irritation and sensitization [14], tests for systemic toxicity [15] and tests for local effect after implantation [16].

The ISO 10993 refers to the fundamental principles leading the biological evaluation of medical devices. In this perspective consideration should be given to the properties of the material and to the prospective applications, taking into account the knowledge provided by scientific literature and previous clinical applications, as well as the type, duration and conditions of the exposure in the human body, the physical and chemical properties of the product and the potential toxicity. With respect to the contact with the body, the devices can be divided into three groups: surface devices, external communicating devices and implant devices. With respect to the duration of this contact the considered time periods are: limited contact (< 24 hours), prolonged contact (24 hours – 30 days) and permanent contact (> 30 days). A material or device will be submitted only to those tests who are related to the expected contact with the body environment.

Biocompatibility tests can be performed using extracts of the material. The samples are obtained incubating the device or material in the extraction medium, with well definite ratios between surface area and extraction medium volume. The extraction must be performed in containers clean, chemically inert and closed. Polar and non polar solvents can be used for the extraction. Polar solvents are water, physiological saline and culture media without serum. Non polar solvents are vegetable oils. The extraction time will be suitable to maximize the amount of the extracted material.

Sterilization of the Sample: The samples should be properly sterilized. The most common sterilization procedures are: Steam and dry heat, Ethylene oxide and others chemical agents, Gamma Irradiation, Electron (E-beam) radiation. Each of these methods presents advantages and disadvantages depending on the type of the material, and consequently the choice of the most suitable procedure of sterilization is of great importance. Mechanical properties of a material could be significantly affected by sterilization methods. For instance sterilization of ultra-high molecular weight polyethylene replacement by gamma irradiation in air causes chain scission and start a long term oxidative process that degrades the chemical and mechanical properties of the polymer. Ethylene Oxide causes less microstructural damage than gamma radiation to ultra high molecular weight polyethilene resulting in higher toughness and percent elongation. Gamma irradiation and thermal annealing increase creep resistance of hydroxiapatite- reinforced polyethilene composites. The effects of manufacturing adds to those of sterilization procedures. For instance the chitosan, obtained from chitin, and other polysaccharide derivates that has many potential fields of application as a biomaterial, with Gamma irradiation showed a modification of the mechanical properties, with a reduction of mechanical strength due to the scission of the proteic chains; the main problem of Ethylene Oxide treatment was related to the residual of the gas that remains present in the material. Steam autoclaving and gutaraldehyde treatment appears to be adequate methods of sterilization. The influence of sterilization on de surface properties of the material should be also considered. This was observed after ethylene oxide treatment. Some residuals of Ethylene Oxide remains in the polymeric matrix, and can diffuse to the surface with adverse biological consequences, expecially when contacting blood. In conclusion, the choice of a method of sterilization involves: susceptibility to heat water vapor; thermolability; absorption and desorption of volatile gases and radiation sensitivity. and the method may be selected according to the material features and properties [17].

Irritation and Sensitization Test: The irritation reaction is the primary and direct effect of a substance on the skin while the sensitization reaction is an immune-mediated contact dermatitis. *Irritation test.* The potential of the material to be tested to produce dermal irritation can be evaluated in rabbits applying on the shaved back both the sides of the spine the test material with a semi-occlusive dressing. For acute tests (single exposure) the appearance of the site of application must be checked at 1, 24, 48 and 72 hours following the removal of the patches. For repeated exposures the appearance of application site will be recorded 1 hour after the removal of the patches and, after the last exposure, at 1, 24, 48 and 72 hours. The score for the evaluation of the results is based on two clinical parameters, the erythema and eschar formation and the oedema formation. Another test is the intracutaneous reactivity, that is the assessment of the potential of the material to produce irritation following intradermal injection of extracts. *Sensitization tests* include the Magnusson-Kligman maximization test and the Buehler closed-patch test. Only one of these methods is usually required. The preferred method is the maximization test. This test requires intradermal injections; therefore, if the material is not obtained in an injectable form, other methods can be used, as the closed patch test. The sensitizing capability of substances is defined as the minimum concentration required to induce a predetermined level of sensitization.

Systemic Toxicity: Systemic toxicity is related to the possible adverse effects that can derive from the absorption, distribution or the metabolism of products that originate from a material or device, involving parts of the body or organs with which they are not in direct contact.

The tests are scheduled to determine systemic adverse effects that can be observed after administration of a single dose or multiple doses of a test sample per day during a period of 24 hours (acute toxicity), or by repeated exposures for a prolonged time (subacute, subchronic and chronic systemic toxicity).

***Pirogenicity Tests*:** Pyrogens are substances that cause a febrile reaction. They can be present in a device and, therefore, leachates of materials can cause a febrile responses (material-mediated pyrogenicity). The assessment of the material-mediated pyrogenicity can be included in the biocompatibility tests. ISO 10993-11 recommends the evaluation of the pyrogenicity potential of extractable substances derived from material leaching and includes the pyrogen test in the category of systemic toxicity testing.

Genotoxicity, Carcinogenicity, Reproductive and Development Toxicity: Genotoxicity (mutagenicity), carcinogeniticy and reproductive and development toxicity are considered in the same ISO 10993-3 standard.

Genotoxicity: The possible mutagenic activity of materials can be considered and investigated, according to the ISO 10993-3 for all the devices with a prolonged (>30 days) body contact. The test can be performed both *in vitro* and *in vivo*. The simplest and sensitive *in vitro* assays are those involving bacteria and cultured mammalian cells. Only if scientifically indicated and if *in vitro* test results indicate potential genotoxicity the tests *in vivo* are performed (search of chromosomal aberration in rodents bone marrow or peripheral blood cells).

Carcinogenicity: These tests explore the carcinogenetic potential of materials to either a single or multiple exposures over a period of the total life-span of the test animal. These tests can requires a prolonged period of observation, and can be designed to evaluate in a single experimental study both carcinogenicity and chronic toxicity. They should be considered for a device that will have permanent contact (longer than 30 days) with tissues.

Reproductive and Development Toxicity: tests are aimed to evaluate the potential effects of the material/device on reproductive function, embryonic development and prenatal and early postnatal development. These tests should be performed only when the device has a potential impact on the reproductive ability of the subject, so they are mostly indicated for intrauterine devices and resorbable or leachable materials and devices.

Hemocompatibility: The effects of the interaction of materials with blood can be both at cellular and plasmatic levels. They include the mechanisms of coagulation, thrombosis, embolism, inflammation, the activation of complement, and fibrinolytic systems. The cellular and humoral pathways are closely connected and reciprocally influenced. The hemocompatibility tests are grouped in five categories: thrombosis, coagulation, platelets, hematology, immunology. Standardization requires an evaluation of hemocompatibility for any medical device that has direct or indirect contact with circulating blood. Hemocompatibility tests can be performed both *in vitro* and *in vivo*. *In vitro* studies allows a first screening, but relevant informations are collected *in vivo* after the implant of the device (*i.e.* vasculard grafts, heart valves) into the vascular system for an assigned period. For the purpose of this chapter we will be particularly taken into account those investigations directed to demonstrate the aspects of the organism and tissue response to the materials after implants [18].

TESTS FOR LOCAL EFFECTS AFTER IMPLANTATION

These test are described in the part 6 of the ISO 10993. The tests requires that the sample of the material to be tested is implanted into an appropriate anatomical site or tissue and that the implant is not subjected to mechanical or functional loading; the local effects are evaluated in comparison with the effects caused by materials with already well assessed and clinically accepted properties (controls). It must be kept in mind that the placement of a material or device in the tissues produces a host response that is dynamic and consequently changes with time [19].

Anesthesia and Postoperative Care: General anaesthesia must be performed with different procedures, depending on the animal species, the type and length of the surgical intervention, and the possible degree of suffering. The subcutaneous or intramuscular injection of anaesthetics can be adequate in only in small animals. In the other cases anaesthetical premedication with subcutaneous or intramuscular administration of drugs is followed the induction of anaesthesia by intravenous injection, and by the maintenance with the administration of gas by mask or by orotracheal intubation. Monitoring of ECG and respiratory parameters during the anaesthesia is also advisable and recommended [20].

Postoperative care includes the veterinary control of the animal with the clinical monitoring of vital paramethers, fluid administration and drugs as analgesics or antibiotics if required, periodic medication of the surgical wound. Care should be exercised in handling of the animal, *e.g.* when the implants are inserted in the back, it is advisable not manipulate the skin of that region, to avoid the displacement of samples [21].

Surgical techniques used in animal experiments are superimposable to those used in humans. Consequently the same iatrogenic potential consequences can be expected.

An appropriate surgical technique is a mandatory prerequisite for a correct *in vivo* evaluation of any product. Care must be taken to avoid unnecessary tissue damage, such as excessive pressure, bleeding and dead spaces. An inaccurate technique can produce the migration or mobility of the implant. Samples must be designed with appropriate size and shape according to the site of implant. According to the ISO 10093-6 for implantation in subcutaneous tissue, specimens made of sheet material shall be of 10-12 mm of diameter and 0, 3-1 mm in thickness; bulk materials must have 1, 5 mm of diameter and 5 mm of length and shall have rounded edges. The implants are placed in the dorsal subcutaneous tissue of adult mice, rats, guinea pig or rabbits. For muscle implantations, *e.g.* paravertebral muscle of rabbits, specimens of 1 to 3 mm of width and about 10 mm of diameter with rounded edges are used.

At the scheduled experimental times, the animal sacrifice is performed after anaesthetical premedication and with humane methods. Tissues are explanted and observed for macroscopic and histological investigations. Several macroscopic and microscopic parameters are considered, including the presence of infection, fibrous tissue, degeneration, presence of phagocytic cells, necrosis, fatty infiltration, and foreign debris.

The scheduled time for short term implants is not more than 12 weeks. For long term implants the scheduled time is longer than 12 weeks. In the evaluation of the local effects of an implant it must be considered also the effects of the surgical procedure, and the repair of the surgical wound; that conditions affects the early periods after implantation.

Short term tests (until 12 weeks) can be performed in mice, rats, guinea-pigs and rabbits with implants in subcutaneous tissue and muscle [22]. For long term studies in subcutaneous tissue, muscle and bone, rats, guinea-pigs, rabbits, sheep, goats and pigs are the animal species more utilized. Subcutaneous implants are placed in pockets created in the tissue of the back [23-25] through a midline incision in at least three animals and in a sufficient site number to place 10 samples for each material and experimental time. For muscle implants, in the paravertebral muscles of rabbits are the preferred sites. The use of a minimum of three rabbits for each scheduled time is recommended, to provide a sufficient number of test and control implants (at least eight implant specimens for each implantation period).

The consequences of the implant of a material in living tissue, must be considered at both macroscopic and microscopic level. Macroscopically signs of infection in the site of the implant will be detected as the evidence of a fibrous reaction surrounding the material or necrotic tissues. For the histological assessment the following paramethers are considered: a) the extent of fibrosis, fibrous capsule and inflammation surrounding the implant; b) degeneration - changes in tissue morphology; c) presence, number and distribution of the inflammatory cells in relation with the distance from the material/tissue interface; d) presence of necrosis; e) material debris, fatty infiltration, granuloma; f) for porous materials, quality and quantity of tissue ingrowth; g) for bone implants, study of the bone implant interface.

BONE IMPLANTS

Bone implants are performed when the material is proposed mainly for orthopaedic or dentistry applications [26, 27]. Skeletal tissue presents particular features: it is composed of many tissues, osseous, cartilagineous, mesenchimal, vascular and hemopoietic. Due to the differences in anatomy, vascularization, mechanical forces, healing capacity the implant site greatly affects both osteointegration and the rate of biodegradation of some biomaterials. The osteointegration tends to be grater in cortical than trabecular or medullar region. On the contrary biodegradation rate is greater in medullar and trabecular sites than in cortical bone. From a biomechanical point of view, forces acting on the implants placed in the tibial or femoral diaphysis are mainly shear forces, while force acting on the implants placed in femoral condyles or tibial plateaus are a combination of shear and compression forces. We must also take account of the remodelling of bone. Humans remodel their skeleton to a greater extent than most other mammals. Remodeling activity increases with age: in the adult about 25% of cancellous bone and 3% of cortical bone are remodelled annually. The bone remodelling around an implant is a long lasting process [28].

Finally, is relevant the influence on scheletric apparatus of systemic and local factors, as sex hormones, thyroid hormone parathyroid hormone, calcitonin, glucocorticoids, vitamin D, and insulin, and of many diseases, the most frequent of them is certainly the osteoporosis. All these factors must be taken into account when a material is implanted in the bone.

The site of bone implants can be the diaphisis (cortical bone), the epiphysis (cancellous bone) or both [29]. The samples are usually designed as screw or of cylindrical shape. There are however other possible shapes for the samples. For cement composites, discs of the material are implanted subcutaneously and in cranial defects [30]; similarly disc shaped samples were employed to evaluate the vestibular tissue reaction to some materials for dentures [31].

The implant size is established according to the animal species selected for the tests and of the skeletal segment in which the implant is placed. Cylindrical implants in rabbits may have a diameter of 2 mm and 6 mm in length; in sheep and goats the size may be 3 mm in diameter and 12 in length. Screw type implants in rabbits, sheep, goats and pigs vary from 2 mm to 4 mm. of size. Equivalent sites of implant will be used for test and control specimens using the contralateral bone segments. In rabbits are placed a maximum of six implants (three tests and three controls). In other species (sheep, goat, pig) a maximum of 12 implant (6 test and 6 controls). Femur and tibia are the long bone segments usually selected for the implants [32]. The preparation of holes in which the implants are placed will be performed using a low drilling speed with profuse saline irrigation. Cylindrical specimens are inserted by simple pressure. When screw shaped specimens are inserted in the bone the use of a device for the recording of the insertion torque is recommended. This paramether will be compared with the extraction torque recorded at the sheduled sacrifice time. The difference between the measured extraction and insertion values is an useful index of the mechanical strength of the interaction between the bone and the material [33].

A the time of the implant in the bone, as well as in the following experimental times, X-ray is a very useful mean to evaluate the correct placement of the sample, and during the follow-up to detect signs of bone deposition or resorption around the implant, infection, fractures.

In the case of bone implants, the contact between the tissue and the material at the interface is matter of particular interest. Regarding this aspect, is defined as osteointegration the direct structural and functional connection between living bone and the surface of the implant [34].

In the evaluation of the osteointegration histological studies are relevant [35]. It is important to emphasize that the accuracy in the harvesting of the samples of the skeletal segment in which the implant is inserted is a prerequisite for the achievement of satisfactory histology slides. An adequate amount of bone surrounding the material must be removed to investigate the quality of tissue around the implant. Soft tissues can be removed carefully, when unnecessary for the evaluations, avoiding the damage of the site of implant.

Usually the specimens containing implants must be processed for the histology according to the procedure used for the undecalcified bone. Briefly, after fixation in 4% buffered paraformaldehyde and subsequent dehydratation in graded series of alcohols, they are resin embedded in methyl-metacrilate. To prepare sections of appropriate thickness, the most useful procedure is by the use of a cutting-grinding system [36]. This system is composed by a cutting device based on a diamond-saw and a rolling clamp on which the sample is firmly hold by means of jaws. The slices, even a certain thickness, are glued on a plexiglas slide, and are then thinned until the required thickness using a grinding system with a series of abrasive paper with increasing grit. Longitudinal sections are usually prepared for screw implants; for cylindrical implants longitudinal or tranversal sections can be considered [37].

To evaluate osteointegration, the histological observation of samples is directed to measure some parameters in the region of interest. The histomorphometry measures should be obtained with image analysis systems. These systems are composed by a microscope connected by a camera with a personal computer provided by an appropriate software to perform interactive and automatic measurements. The most commonly parameters considered to evaluate osteointegration are the Affinity Index (the length of bone directly opposed to the implant/the total length of bone implant interface x 100%) the Bone ingrowth (the area of the bone tissue grown into the screw thread/area between the screw thread) [38, 39], and the Bone Mirror Area that indicates the quality of bone in the region closely contiguous to the implant [40]. The presence and amount of fibrous tissue between bone and the implant, if present, will be also evaluated. When the implant is placed in the cancellous bone, the paramethers related to the quality of the tissue are evaluates according to Parfitt [41, 42]: among these paramethers the most significant are the Trabecular Bone Volume (BV/TV, %); the trabecular thickness (Tb.Th, μm), the trabecular number (Tb.N,/mm) and the trabecular separation (Tb.Sp, μm^2) that were calculated from the BV/TV and the perimeter of trabeculae using Parfitt formulae.

The rate of bone deposition and mineralization are evaluated by the so called dynamic histomorphometry. For this purpose are used fluorochromes (*i.e.* tetracycline, alizarin red, calcein blue, xylenol orange) to label the new bone deposition. Fluorochromes administered to the animal by injecting at scheduled times before the sacrifice. The fluorescence evidenced on histological sections at microscopic observation with appropriate filters marks the new deposition of bone (Fig. **1**) [43].

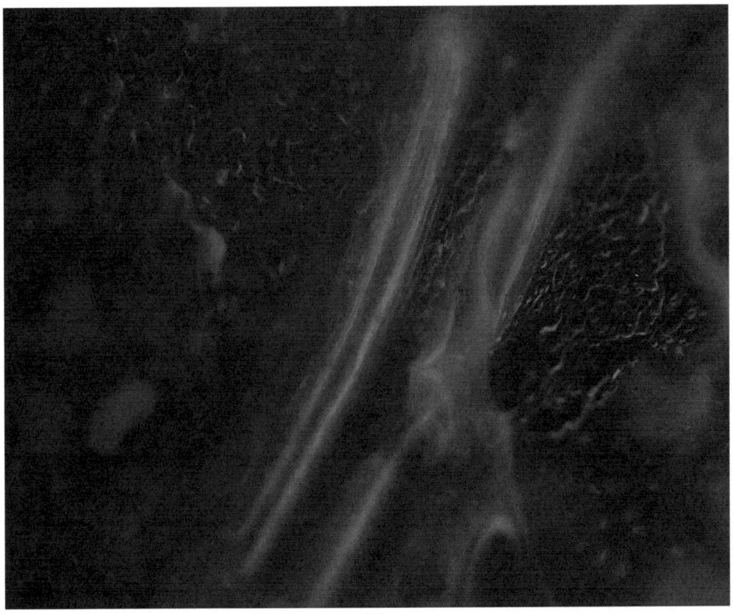

Figure 1: Histological appearance of bone labelled with tetracyclines. The fluorescence bands marks the deposition of new bone.

Mechanical tests are also useful to evaluate the integration of the material with the bone and the bone quality. Implants in bone can be submitted to pull out and push out tests, the whole bone to test of three point bending, torsion, compression [44]. Microhardness investigations, performed on the cutting surface of samples embedded in resin as for histology, allows the assay of the quality of bone surrounding the implant [45].

Besides these investigations, more recently it is possible to perform computerized microtomography on the explanted samples; this is a non destructive technique and the time required to capture the images is short; also, the same sample can therefore be used afterwards for the histology and histomorphometry. More in detail here is some deepening about this technique.

COMPUTERIZED MICROTOMOGRAPHY

Computerized microtomography (micro-CT) is a non destructive technique that represents an evolution of the computerized tomography normally used in medicine. It can reach high resolutions (micrometer range) and allows the visualization of the internal structures of the objects [46].

The conventional microscopy (optical or electron) allows to visualize bidimensional images of the sample surface analyzing its thin sections. Theoretically, a three-dimensional model of the object structure can be interpolated from the bidimensional information of very thin sections of an object visualized by a microscope. Considering that the same structure of the object could have been altered by the technique used for sample preparation, this exacting and intricate method is not further even very reliable. Besides, the distance among the sections is usually too much coarse to avoid losses of 3D information. Currently one of the demands warned more strongly in the productive cycle and in Research is the development of non destructive methodologies for the analysis of the microstructure of different types of material.

Micro-CT with μm level of resolution is set, among the non destructive techniques, as one of the most advanced, due also to the technological advancement in continuous development and to the elevated power of calculation of computers.

Micro-CT is based on the interaction of X rays with matter and derives directly from the radiographic technique in which the complete three-dimensional internal structure of an object is visualized in a bidimensional projection. This happens because the intensity of the non absorbed radiation, after the sample crossing, is recorded on the plan of a detector set at the opposite side of the X ray source. The principal disadvantage of the radiography is overlapping different internal structures in only one bidimensional projection mixing, in this way, the information concerning the depth.

The tomography uses the radiographic images of the object gotten at different angles. In the Fig. **2** is shown, as example, an image of a microtomographic scan. Afterwards, using reconstruction algorithms based on the technique of Filter Back Projection (FBP), it is possible to create virtual sections of the internal structure of the object. With a number of consecutive different reconstructed sections it is possible a three-dimensional visualization of the object.

The obtained microtomographic consecutive sections are a series of images with different grey levels. These grey levels of every image of sections give a map of the distribution of the absorption coefficients of the X-rays related to the analyzed sample. Such coefficients depend on the material density, from the atomic number of the elements that constitute it and from the incidental energy used. In the years, the development of these kinds of equipments, from the first tomographic scanner realized by Godfrey N. Hounsfield in 1972 [47] onwards, allowed to reach a micrometric resolution and the diffusion of the CT, first utilized only in Medicine, also in scientific and industrial fields [48].

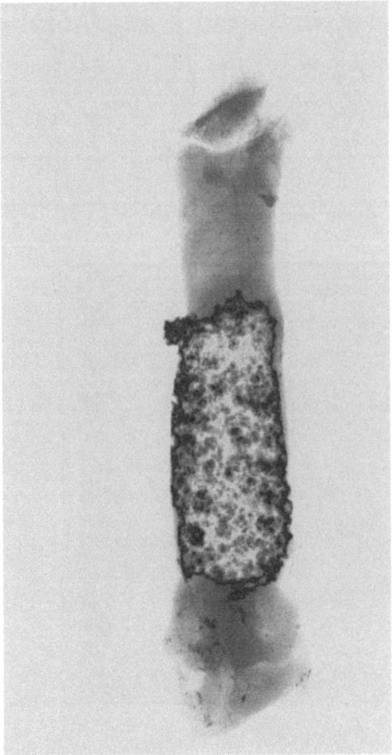

Figure 2: Microtomographic scan of a ceramic device implanted in a tendon.

In fact, the areas of application include biomedical Research (Fig. **3a**), science of biomaterials (Fig. **3b**), development and production of drugs, dental Research, electronic components Research, geology, zoology, botany (Fig. **3c**), construction materials Research, paper manufacture and many others up to paleontology.

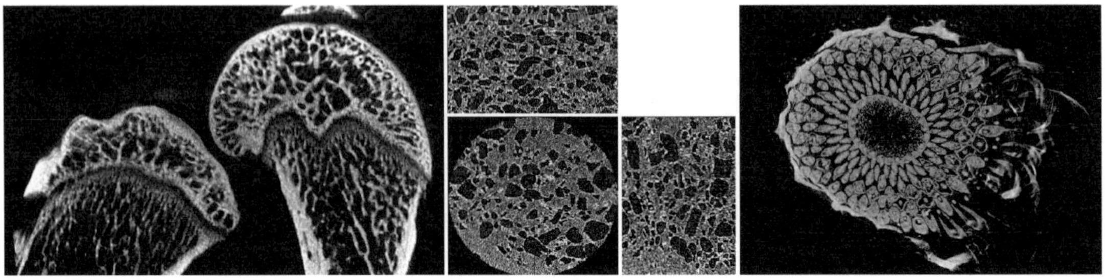

Figure 3: Microtomographic images with different resolutions of (a) rat joint, (b) polymeric biomaterial and (c) daisy flower.

In the only biomedical and dental Research microtomography can be used for numerous evaluation typologies. Biomaterials analysis is carried out before the implant to evaluate porosity or mechanical characteristics; analysis of metallic devices, used frequently both in orthopedics and in dentistry, in which it is very important the study of bone growth through regeneration and osteointegration; analysis of polymeric or ceramic devices (Fig. **4**) that are usually studied for regeneration in defects and for bone losses; or the study of bone morphology, extremely important because a lot of pathologies alter bone microarchitecture (osteoporosis and other pathologies or therapies that influence the bone tissue).

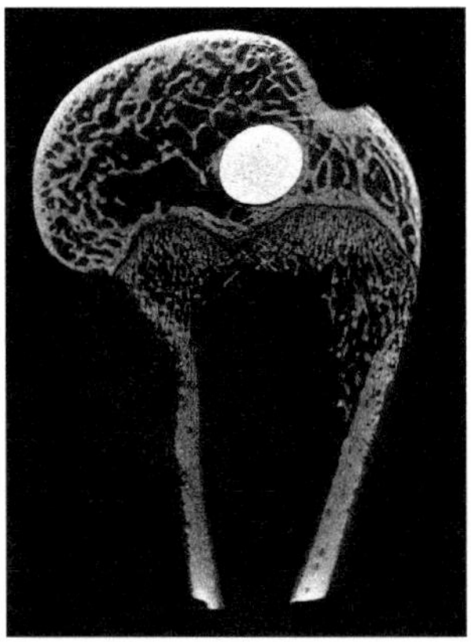

Figure 4: Microtomographic section of ceramic biomaterials implanted in a femoral condyle of a rabbit.

Besides, after the implantation of a biomaterial it is important not only the quantitative analysis of bone growth, but also the quality of the regenerated bone.

From the microtomographic sections it is possible to create virtual 3D models (Fig. **5**) of the analized object that allow realistic visualizations in space. This kind of visualization gives a huge help to fully understand the structures that are normally studied and observed in bidimensional images typical of histology or of classical radiology. The models can be used for creating films of the analyzed object in movement or as a base for prototypes in tissue engineering.

Therefore, micro-CT results to be, especially for the preclinical evaluations of implant materials destined to the musculoskeletal system, an important and valid non destructive method. This technique results to be also very useful to characterize the devices both in pre-implant phase and in the explant phase, in order to appraise possible deformations and/or degradations of the samples.

Figure 5: Three-dimensional model of obturated dental root.

BIOFUNCTIONALITY EVALUATION

An important part in the process of validation of a material is the evaluation of biofunctionality [49]. Biofunctionality is called as the ability of a device to perform its functions after the implantation in the living body and to maintain them during the time. Due to the aggressive biologic environment the behaviour of a final product implanted *in vivo* can differ from that observed after the laboratory tests are only performed. Laboratory simulations can reproduce the mechanical stress to which the device is submitted (i.e joint prosthesis, heart valves), but no simulation test is able to provide the effects of a long stay in the body [50]. After the laboratory tests the *in vivo* implant of the device is therefore required to complete the preclinical informations. Biofunctionality must be evaluated in animal models suitable to predict the clinical application of the final device. Large animals as the pig and the sheep or goats are considered better for this purpose [51].

BIOLOGIC CHARACTERIZATION IN PATHOLOGICAL MODELS

The use of healthy animal models for the biological characterization of the materials is a well assessed procedure, but is not sufficient for a complete preclinical evaluation. In fact, the clinical condition in which the final product will be used is often characterized by a morbid condition that may affect the outcome. This is particularly evident when the material is intended for the use in the skeletal system. For this reason appropriate experimental models were developed to obtain pathological conditions of clinical relevance. One of the most frequent of such disease is osteoporosis, that is related to age, post-menopausal status, or is secondary to systemic diseases and drug treatments [52-54]. Osteopenic animal models are today widely used in the preclinical studies of biomaterials. Bilateral surgical ovariectomy is the better procedure to obtain the pattern of post-menopausal osteoporosis. The female rat and sheep are the species most suitable for this purpose. Bone implants of the testing materials were placed in bone 16 weeks after ovariectomy in rats [55]. In the sheep the development of an osteoporotic state is obtained 12 months after ovariectomy, and is more evident after 24 months [56]. The sheep model is suitable for studies.on long bones and spine (Fig. **6a, 6b**), were the decrease of trabecular and cortical bone mass was observed in the vertebral bodies [57, 58]. The studies on small size animals like the rat are acceptable only in the first phase of a research, because the differences between rat bone quality as compared to humans. Large animal models, as sheep, must be adopted in the further developments, for a better approach to the clinical situation. The use of pathological bone-derived cell culture for *in vitro* tests [59, 60], in association with pathological animal models allows the evaluation of the different response of bone to candidate orthopaedics materials by comparing healthy and pathological conditions.

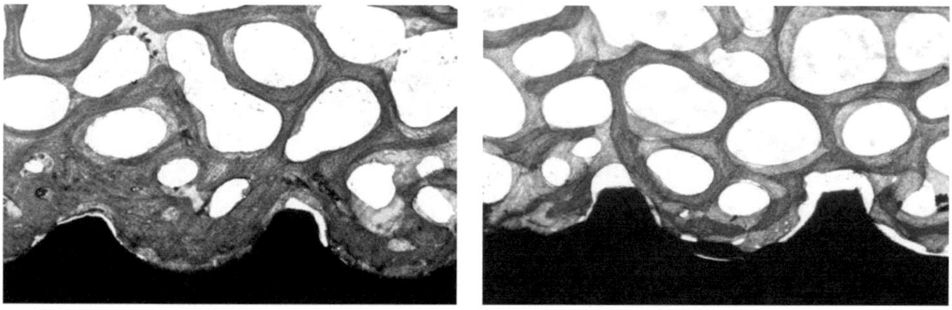

(a) (b)

Figure 6: a - Histological appearance of vertebral pedicular implant of a stainless steel screw in sheep lumbar spine in healthy bone (upper) and osteoporotic bone (down). The better *bone to implant contact* in the healthy bone is evident (Fast Green, 2, 5x).

ACKNOWLEDGEMENTS

Declared none.

CONFLICT OF INTEREST

The author(s) confirm that this chapter content has no conflict of interest.

REFERENCES

[1] Mc Elvany K.D. FDA requirements for Preclinical Studies. In: Woodbury-Harris KM, Coull BM (eds) Clinical Trials in the Neurosciences. Front Neurol Neurosci Karger, Basel, vpl. 25: 46-9

[2] Smith A.C., Fasse R.T. Swindle M.M. Ethics and regulations for the care and use of laboratory animals. In: An Y.H., Friedman R.J. editors. Animal models in orthopaedic research, CRC Press, 1999

[3] Repubblica Italiana. Gazzetta Ufficiale, January 27, 1992

[4] ISO 10993 Biological evaluation of medical devices. Part 2 Animal welfare requirements, 2006.

[5] Einhorn TA. Clinically applied models of bone regeneration in tissue engineering research. Clin Orthop Relat Res 1999; (367 Suppl):S59-67.

[6] Torricelli P, Fini M, Giavaresi G, Giardino R. *In vitro* models to test orthopedic biomaterials in view of their clinical application in osteoporotic bone. Int J Artif Organs 2004 Aug;27(8):658-63.

[7] Muschler GF, Raut VP, Patterson TE, Wenke JC, Hollinger JO. The design and use of animal models for translational research in bone tissue engineering and regenerative medicine. Tissue Eng Part B Rev 2010 Feb;16(1):123-45.

[8] Pearce AI, Richards RG, Milz S, Schneider E, Pearce SG. Animal models for implant biomaterial research in bone: a review. Eur Cell Mater 2007 2;13:1-10.

[9] O'Loughlin PF, Morr S, Bogunovic L, Kim AD, Park B, Lane JM. Selection and development of preclinical models in fracture-healing research. Bone Joint Surg Am 2008 Feb; 90 (Suppl 1):79-84.

[10] Pape HC, Hildebrand F, Krettek C, Green J, Giannoudis PV. Experimental Background - Review of animal studies. Injury 2006 Oct; 37 (Suppl 4):S25-38.

[11] ISO 10993 Biological evaluation of medical devices. Part 1 Evaluation and testing, 2004

[12] ISO 10993 Biological evaluation of medical devices. Part 3 Tests for genotoxicity, carcinogenicity and reproductive toxicity, 2003

[13] ISO 10993 Biological evaluation of medical devices. Part 4 Selection of tests for interactions with blood, 2002

[14] ISO 10993 Biological evaluation of medical devices. Part 10 Tests for irritation and delayed type hypersensivity, 2002

[15] ISO 10993 Biological evaluation of medical devices. Part 11 Tests for systemic toxicity, 2006

[16] ISO 10993 Biological evaluation of medical devices. Part 6 Tests for local effects after implantation, 2007

[17] Giardino R., Nicoli Aldini N., Infection and Sterilization. In: R. Barbucci editor. Integrated Biomaterial Science, Kluwer Academic/Plenum Publisher, 2002

[18] Giardino R. Fini M., Nicoli Aldini N., Parrilli A. Testing the *in vivo* biocompatibility of biocomposites. In: L Ambrosio editor. Biomedical Composites. Woodhead Publishing Limited, 2009

[19] Kasemo B, Lausmaa J. Material-tissue interfaces: the role of surface properties and processes. Environ Health Perspect 1994 Oct;102 (Suppl 5):41-5.

[20] Flecknell P.A. Laboratory Animal Anaesthesia. Third Edition. Elsevier inc. Academic Press. 2009

[21] Jansen J.A. Animal Models for studying soft tissue biocompatibility of biomaterials. In: An Y.H., Friedman R.J. editors. Animal models in orthopaedic research, CRC Press, 1999

[22] Silva RA, Assed S, Nelson-Filho P, Silva LA, Consolaro A. Subcutaneous tissue response of isogenic mice to calcium hydroxide-based pastes with chlorhexidine. Braz Dent J 2009;20(2):99-106.

[23] J Vaisman B, Motiei M, Nyska A, Domb AJ. Biocompatibility and safety evaluation of a ricinoleic acid-based poly(ester-anhydride) copolymer after implantation in rats. J Biomed Mater Res A 2010 Feb;92(2):419-31.

[24] Pulkkinen M, Malin M, Böhm J, Tarvainen T, Wirth T, Seppälä J, Järvinen K. *In vivo* implantation of 2, 2'-bis(oxazoline)-linked poly-epsilon-caprolactone: proof for enzyme sensitive surface erosion and biocompatibility. Eur J Pharm Sci 2009 Feb 15;36(2-3):310-9.

[25] Link DP, van den Dolder J, van den Beucken JJ, Cuijpers VM, Wolke JG, Mikos AG, Jansen JA.Evaluation of the biocompatibility of calcium phosphate cement/PLGA microparticle composites. J Biomed Mater Res A 2008 Dec 1;87(3):760-9

[26] Di Bella C, Nicoli Aldini N, Lucarelli E, Dozza B, Frisoni T, Martini L, Fini M, Donati D. Osteogenic Protein-1 Associated with Mesenchymal Stem Cells Promote Bone Allograft Integration. Tissue Eng Part A 2010 Jun 9.

[27] Pieri F, Lucarelli E, Corinaldesi G, Aldini NN, Fini M, Parrilli A, Dozza B, Donati D, Marchetti C. Dose-dependent effect of adipose-derived adult stem cells on vertical bone regeneration in rabbit calvarium. Biomaterials 2010 May;31(13):3527-35.

[28] Fini M, Carpi A, Borsari V, Tschon M, Nicolini A, Sartori M, Mechanick J, Giardino R. Bone remodeling, humoral networks and smart biomaterial technology for osteoporosis. Front Biosci (Schol Ed) 2010 Jan 1;2:468-82.

[29] Giavaresi G, Ambrosio L, Battiston GA, Casellato U, Gerbasi R, Fini M, Nicoli Aldini N, Martini L, Rimondini L, Giardino R. Histomorphometric, ultrastructural and microhardness evaluation of the osseointegration of a nanostructured titanium oxide coating by metal-organic chemical vapour deposition: an *in vivo* study. Biomaterials 2004;25:5583-91.

[30] Ruhé PQ, Hedberg EL, Padron NT, Spauwen PH, Jansen JA, Mikos AG. Biocompatibility and degradation of poly(DL-lactic-co-glycolic acid)/calcium phosphate cement composites. J Biomed Mater Res A 2005;74:533-44.

[31] Ebadian B, Razavi M, Soleimanpour S, Mosharraf R. Evaluation of tissue reaction to some denture-base materials: an animal study. J Contemp Dent Pract 2008;9:67-74.

[32] Giavaresi G, Fini M, Chiesa R, Giordano C, Sandrini E, Bianchi AE, Ceribelli P, Giardino R. A novel multiphase anodic spark deposition coating for the improvement of orthopedic implant osseointegration: an experimental study in cortical bone of sheep. J Biomed Mater Res A 2008;85:1022-31.

[33] Rocca M, Fini M, Giavaresi G, Nicoli Aldini N, Giardino R. Osteointegration of hydroxyapatite-coated and uncoated titanium screws in long-term ovariectomized sheep. Biomaterials 2002;23:1017-23.

[34] Mavrogenis AF, Dimitriou R, Parvizi J, Babis GC. Biology of implant osseointegration. J Musculoskelet Neuronal Interact 2009;9:61-71.

[35] Wennerberg A, Albrektsson T. Effects of titanium surface topography on bone integration: a systematic review. Clin Oral Implants Res 2009 ;20 (Suppl 4):172-84.

[36] Donath K., Brenner G. A method for the study of undecalcified bone and teeth with attached soft tissue J Oral Pathol 1982; 11: 318

[37] Rebollal J, Soares Gde A, Vidigal GM Jr. Histomorphometric analysis of hydroxyapatite-coated implants in rabbit cortical bone using longitudinal and transverse histologic sections. Implant Dent 2010;19:137-44.

[38] Ramires PA, Wennerberg A, Johansson CB, Cosentino F, Tundo S, Milella E. Biological behavior of sol-gel coated dental implants. J Mater Sci Mater Med 2003;14:539-45.

[39] Giavaresi G, Chiesa R, Fini M, Sandrini E. Effect of a multiphasic anodic spark deposition coating on the improvement of implant osseointegration in the osteopenic trabecular bone of sheep. Int J Oral Maxillofac Implants 2008;23:659-68.

[40] Tavares MG, de Oliveira PT, Nanci A, Hawthorne AC, Rosa AL, Xavier SP. Treatment of a commercial, machined surface titanium implant with H2SO4/H2O2 enhances contact osteogenesis. Clin Oral Implants Res 2007;18:452-8.

[41] Parfitt AM. Bone histomorphometry: proposed system for standardization of nomenclature, symbols, and units. Calcif Tissue Int 1988 ;42:284-6.

[42] Parfitt AM, Drezner MK, Glorieux FH, Kanis JA, Malluche H, Meunier PJ, Ott SM, Recker RR. Bone histomorphometry: standardization of nomenclature, symbols and units. J Bone Miner Res 1987; 2:595-610

[43] Jee WS. The past, present, and future of bone morphometry: its contribution to an improved understanding of bone biology. J Bone Miner Metab 2005;23 Suppl:1-10. Review.

[44] Borsari V, Fini M, Giavaresi G, Tschon M, Chiesa R, Chiusoli L, Salito A, Rimondini L, Giardino R. Comparative *in vivo* evaluation of porous and dense duplex titanium and hydroxyapatite coating with high roughnesses in different implantation environments. J Biomed Mater Res A 2009;89:550-60.

[45] Bacchelli B, Giavaresi G, Franchi M, Martini D, De Pasquale V, Trirè A, Fini M, Giardino R, Ruggeri A. Influence of a zirconia sandblasting treated surface on peri-implant bone healing: An experimental study in sheep. Acta Biomater 2009;5:2246-57.

[46] ASTM E1441-97. Standard guide for computed tomography (CT) imaging. Technical report, American Society for Testing and Materials, 1997.

[47] G.N. Hounsfield. Computerized transverse axial scanning (tomography): Part1. description of system. Brit J Radiol 1973; 46:1016-1022.

[48] A. Sasov and D. Van Dyck. Desktop X-ray microscopy and microtomography. J Micros – Oxford 1998; 191:151-158.

[49] Goyenvalle E, Aguado E, Pilet P, Daculsi G. Biofunctionality of MBCP ceramic granules (TricOs) plus fibrin sealant (Tisseel) *versus* MBCP ceramic granules as a filler of large periprosthetic bone defects: an investigative ovine study. J Mater Sci Mater Med 2010 ;21:1949-58.

[50] Marcacci M, Kon E, Moukhachev V, Lavroukov A, Kutepov S, Quarto R, Mastrogiacomo M, Cancedda R. Stem cells associated with macroporous bioceramics for long bone repair: 6- to 7-year outcome of a pilot clinical study. Tissue Eng. 2007 May;13(5):947-55.

[51] Donati D, Di Bella C, Lucarelli E, Dozza B, Frisoni T, Nicoli Aldini N, Giardino R. OP-1 application in bone allograft integration: preliminary results in sheep experimental surgery. Injury. 2008 Sep;39 (Suppl 2):S65-72.

[52] Dervis E. Oral implications of osteoporosis. Oral Surg Oral Med Oral Pathol Oral Radiol Endod. 2005 Sep;100(3):349-56.

[53] Bornstein MM, Cionca N, Mombelli A. Systemic conditions and treatments as risks for implant therapy. Int J Oral Maxillofac Implants. 2009; 24 Suppl:12-27.

[54] Anakwe RE, Aitken SA, Khan LA. Osteoporotic periprosthetic fractures of the femur in elderly patients: outcome after fixation with the LISS plate. Injury. 2008 Oct;39(10):1191-7.

[55] Fini M, Giavaresi G, Torricelli P, Krajewski A, Ravaglioli A, Belmonte MM, Biagini G, Giardino R. Biocompatibility and osseointegration in osteoporotic bone. J Bone Joint Surg Br. 2001 Jan;83(1):139-43.

[56] Fini M, Giavaresi G, Torricelli P, Borsari V, Giardino R, Nicolini A, Carpi A. Osteoporosis and biomaterial osteointegration. Biomed Pharmacother. 2004 Nov;58(9):487-93.

[57] Fini M, Giavaresi G, Greggi T, Martini L, Nicoli Aldini N, Parisini P, Giardino R. Biological assessment of the bone-screw interface after insertion of uncoated and hydroxyapatite-coated pedicular screws in the osteopenic sheep. J Biomed Mater Res A. 2003 Jul 1; 66(1):176-83.

[58] Nicoli Aldini N, Fini M, Giavaresi G, Giardino R, Greggi T, Parisini P. Pedicular fixation in the osteoporotic spine: a pilot *in vivo* study on long-term ovariectomized sheep. J Orthop Res. 2002 Nov;20 (6):1217-24.

[59] Torricelli P, Fini M, Giavaresi G, Giardino R. *In vitro* models to test orthopedic biomaterials in view of their clinical application in osteoporotic bone. Int J Artif Organs. 2004 Aug;27(8):658-63.

[60] Torricelli P, Fini M, Giavaresi G, Rimondini L, Tschon M, Rimondini R, Carrassi, A, Giardino R. Chronic alcohol abuse and endosseous implants: linkage of *in vitro* osteoblast dysfunction to titanium osseointegration rate. Toxicology. 2008 Jan 14;243(1-2):138-44.

CHAPTER 9

New Concepts Applied to the Development of Biomaterials for Orthopaedic Tissue Regeneration

Anna Tampieri* **and Simone Sprio**

Italian National Research Council, Institute for Science and Technology for Ceramics, ISTEC-CNR, via Granarolo 64 48018 Faenza (Italy)

Abstract: Nowadays long bone healing is still a major socio-economic concern. In fact, although in many fields of bone surgery suitable clinical solutions addressed to the regeneration of bone are achieved, for the healing of load-bearing bones the clinical solutions still consist of bioinert prostheses which only provide a mechanical sustain without restoring the complex biomechanical functionality of bone. As a consequence, the recourse to additional surgery is often unavoidable, especially for the younger patients; additionally the bone surgery in orthopaedics is often very invasive and patients are forced to spend long time in hospital with strong impact on the psychological and physical well-being, as well as on the healthcare costs. In the last decade new regenerative solutions based on the use of bioactive materials and composites and on the set up of innovative synthesis techniques to achieve bone scaffolds endowed with hierarchical organized morphologies are being developed. Such scaffolds exhibit high chemical biomimesis and interesting anisotropic physical and mechanical properties, such as the native bone, so to potentially pave the way for realizing prosthetic devices closer to the extraordinary performance of human tissues addressed to the regeneration of long and load-bearing bones for which no acceptable solutions exist so far.

Keywords: Biomimesis, mechanical properties, loading, bone regeneration.

INTRODUCTION

Bone diseases represent a serious health condition that directly impacts on the quality of life of patients, particularly among the aged. In most cases, the treatment of bone pathologies and defects requires the use of bone grafts, and sometimes in extensive amount [1-5]. Nowadays, more than 2 million bone grafting procedures are happening annually worldwide, with a turn-over of the

***Address correspondence to Anna Tampieri:** Italian National Research Council, Institute for Science and Technology for Ceramics, ISTEC-CNR, *via* Granarolo 64 48018 Faenza (Italy); Tel: +39 0546 699753; Fax: +39 0546 46381; E-mail: anna.tampieri@istec.cnr.it

Lia Rimondini, Claudia Bianchi and Enrica Vernè (Eds)

order of billions of dollars per annum, in order to repair bone defects in orthopaedics, neurosurgery and dentistry, which include the treatment of posttraumatic skeletal complications, such as delayed unions, non-unions, mal-unions. Bone grafting may be also required for the healing of spinal fusions, filling defects following removal of bone tumours and several congenital diseases.

The regeneration of bone tissue is a major concern, compared to other organs and tissues (*e.g.* kidney, heart). In organ grafts, the cells are generally kept alive and the reestablishment of blood supply restores the functions of the whole organ. In bone tissues, however, the blood supply cannot be efficiently restored so that the replacement of bone fragments by foreign bone tissue often leads to poor junction and fast resorption. In consequence, bone scaffolds must provide an efficient chemical, physical and morphological environment for the development of a proper vascular network, which is a key feature to activate an effective bone regeneration.

In addition, bone composition and properties can vary considerably among populations and within a population; thus the remodeling path of a bone implant can be very different from one person to another and from one implant to another; in consequence, the achievement of generic solutions for grafts addressed to bone regeneration is hard to obtain in a worldwide perspective and so, the requests for custom-made solutions are of increasing number and relevance.

Among the bone diseases, those affecting portions of long bone subjected to mechanical loads are the ones which most seriously impact on the quality of life of sufferers. The incidence of such pathologies is particularly relevant among the aged people (osteoporosis), however more recently the number of relatively young patients affected by bone diseases has increased mainly owing to modern life-styles (*e.g.* intense sport activity, obesity, *etc.*); in this case pain and disability also impact on the psychological wellbeing, leading to anxiety, depression, fear for the future, and altered perception of the social role. Such feeling is nowadays shared by the aged people also, because of the increased expectation of an active life and well-being even among the elderly. For this reason, the above-mentioned numbers in terms of socio-economic costs and number of hospitalised people are likely to increase in the next future.

So far, even though in a wide range of applications the current biomaterials are successful in restoring the functionality of bone while being resorbed by the surrounding tissue, the existing solutions in load-bearing applications still consist in prostheses replacing the diseased bone, only capable to function as inert mechanical supports of the biomechanical loads. In the short-term concerns arise due to the low bio-reactivity of such prostheses with existing bone, highlighting clinical implications for the time required for patient rehabilitation [6]. In longer-term studies, biomaterials degrade with time and this may have adverse effects on interfacial apposition to the implant and its mechanical stability [7-9]. Moreover, it is reported that nearly 20% of the hip replacement surgeries performed in past years were to ''revise'' the original implants. For this reason the younger patients often experience partial inability and need multiple surgical interventions in order to replace the bone implant every decade or two.

The very complex composition and morphology of the bone tissue are responsible of its unique ability to continuously adapt to the ever changing mechanical solicitations, so to be simultaneously light, tough and elastic and able to locally regenerate after traumas of limited extension. Thus, with the increasing of the life expectance and well-being, the development of bone substitutes able to restore the whole functionality of the original tissue is of increasing importance; beyond the commercial aspects, the problem has a highly social relevance as the patients can regain a more active and satisfactory life and sooner than in the past.

In response to these pressures and also considering the huge socio-economic impact of long bone diseases, considerable research interest has focused on investigating mechanisms that contribute to implant-prosthesis failure and on developing new biomaterials with extended lifetimes. A new regenerative approach is addressed towards improved implant performance and less invasive surgical procedures also aimed at limiting the recourse to fixation elements, which need to be subsequently removed by further surgery.

The new paradigm for the complete regeneration of long bones involves novel approaches to the selection of biomaterials and prosthesis design to establish nano- and micro-scale features and/or chemistry of the implant, based on bioactive/bio-resorbable phases, mimicking those of living bone tissue, to achieve the early recovery of the bone functionality.

In the last decade a regenerative approach to the healing of bone diseases has led to the progressive replacement of clinical solutions based on the implant of bioinert materials with the development of porous biomimetic phases mimicking the chemical composition of bone, possibly simulating the structure and 3D morphology of the native bone [10]. Such biomaterials exhibit a 3D porous structure conducive to cell attachment and maintenance of cell function, and allow the development of a proper vascular tree able to exchange nutrients and waste materials. However, the scaffold must also exhibit a sufficient mechanical strength to withstand the early biomechanical loads *in vivo*; the construct should be also provided with a rich source of osteoprogenitor cells in combination with selected osteoinductive growth factors, to favour proper cell differentiation and enhance the regenerative processes. When the characteristics of bioactivity, osteoinductivity and osteoconductivity are merged together, the implant is rapidly integrated into the growing bone tissue (osteointegration) and a direct structural and functional connection between ordered living bone and the implant surface is established [11].

BONE REGENERATION IN CONTACT WITH SCAFFOLDS

The stages of bone tissue ingrowth imply different events undertaken by cells: i) adhesion and activation on the scaffold surface; ii) proliferation and migration; iii) differentiation; iv) tissue and vascular remodeling. All these stages are strongly affected by the intrinsic scaffold features in term of bulk chemistry, surface chemistry, geometry and morphological features, biomechanics, osmolarity, etc…

The elective base materials to induce the regeneration of long bones are represented by biomimetic nanostructured hydroxyapatites, since they represent the inorganic part of native bone [12-14]. These apatite phases have the base composition of stoichiometric HA enriched by substituting ions like carbonate, magnesium, silicon, which in specific amounts into the HA lattice enhance and promote the bone regeneration and remodeling.

The scaffold must exhibit suitable chemical and biochemical cues to properly drive cell activity: such signals can be derived by the dissolution products of the scaffold itself, or by appropriate nanometric structures (*e.g.* dendrimers) decorating the scaffold surface. In particular the chemistry of the scaffold surface must allow the adhesion of proteins which is then followed by cell attachment and proliferation; the osmolarity induced by the scaffold chemistry must evolve with time to allow a proper cell activity directed to the regeneration of a mature bone and to avoid any inflammatory effects.

Beyond the chemical aspects, it is now well known that generally the cell adhesion, migration, proliferation and specific differentiation are driven by scaffold functioning as space holders for cell growth, providing mechanical support for the new forming tissue and for the remodeling of a vascular tree. Hence, the biological performances of the prosthesis are strongly related to its 3D morphological architecture, according to the tensegrity model [15], which is being widely recognized as the driving concept for the regeneration of organs and tissues. Such concept is based on the fact that cells are flexible building blocks of tissues, and are physically interconnected to form an integrated matrix continuously subjected to various forms of mechanical stress. Such network has a mechano-transductional function in the response to the mechanical stress, which regulates the organization of the structural components of the tissue and also stimulates specific gene activation and transcription factors [16]. It was found that the absence of a correct establishment of a mechano-signaling network has detrimental consequences on the response of cells to mechanical stress, leading to dystrophy of the affected tissue.

In particular, concerning the bone tissue, its form and function are strictly related to the bone internal architecture, which is in turn determined in response to the stress applied [17]. In fact, bone structure is suited to withstand the forces acting upon them and to arrange its internal and external architecture on the microscopic scale, thus rearranging the structure of trabecules and osteons, and on the macroscopic scale, in terms of shape and dimensions, upon changes in external loading conditions. Moreover, the applied strain determines the extents of bone deposition and bone resorption processes, which in healthy bone are in perfect dynamical equilibrium; conversely, a net unbalance and consequent bone diseases like osteopenia and osteoporosis occur when a non-equilibrium strain state is established. The balance between bone deposition and bone resorption occurs through a feedback loop and determines the net bone remodeling.

The detection of strain change by bone tissue and the consequent adaptation occur on a local basis through a mechano-transduction mechanism, which is supposed to be based on the flow of the bone interstitial fluid in the lacunar-canalicular porosity induced by mechanical loading [18-20]. This mechanism is supposed to be associated with osteocytes: the matrix that surrounds them is non-calcified, thus able to sense the mechanical stimuli, and forms a three-dimensional network connected to the bone lining cells and in close proximity to the marrow stromal cells and periosteum [21]. The signal derived by mechanotransduction is finally modulated by genetic, hormonal and metabolic factors, generating the remodeling potential which regulates the recruitment rate and the activity of osteoblasts and osteoclasts, stimulating bone formation and bone resorption.

In the development of bone scaffolds, the reproduction of the mechanisms at the basis of the tensegrity model is inherent to the establishment of hierarchically organized morphologies [22]. Such structures exhibit the best compromise among the arrangement of the structural elements and the bio-mechanical properties and can reproduce the complex biomechanical behaviour of native bone. Generally, such architectures are not achievable by the current manufacturing methodologies especially when dealing with mineralized phases, since their details range in the sub-micron scale. Considering that the continuous application of opportune and complex stress fields is not to be set aside in order to recover the functionality of diseased load-bearing bones, novel paradigms and new concepts must be set up in the development of bone scaffolds for orthopaedic applications.

CHEMICAL BIOMIMESIS

The new concepts at the basis of the complete regeneration of long bones involves novel approaches to the selection of biomaterials and prosthesis design to achieve the early recovery of the bone functionality, through the use of bioactive/bio-resorbable able to activate proper cell differentiation and bone regeneration. The elective materials should be based on nanostructured biomimetic hydroxyapatite, *i.e.* a calcium-deficient, nearly-amorphous apatite phase presenting partial substitutions with ions relevant for the bone regeneration. The crystal structure of hydroxyapatite allows the incorporation of several foreign ions present in the physiological environment, replacing calcium (mono- and divalent ions like Na, K, Mg, Sr, Fe, ...), phosphate (B-carbonation, hydrogen-carbonate, silicate, ...) and hydroxyl (A-carbonation) in their crystal sites.

A stoichiometric HA, where Ca/P molar ratio is 1.67, exhibits nearly no solubility in physiological environment, thus no capability to be resorbed and to participate to the remodeling/regeneration process; the presence of calcium vacancies destabilizes the crystal structure, leading to an increase of the solubility as soon as Ca/P ratio tends to the limit value of 1.5 [23]. Moreover, the presence of doping ions provokes a further destabilization of HA lattice, so that crystal growth is reduced and the resulting phase is extremely nanostructured (grain size ~ 10-20 nm) and reactive in physiological media. The features of nanostructured substituted HA, like crystallinity and the solubility properties at physiological conditions are directly influenced by the type and amount of foreign ions [24].

The amount of CO_3 ions in the mineral phase of bone varies from ~2–8 wt% and ~1 wt% in Mg up to ppm concentrations for Na, Si, Sr. Although these levels of substitutions are small, it is well understood that these elements are associated with the properties of biological apatites and play a major role in the biochemistry of bone, enamel and dentin [25-27]. Magnesium is associated with the calcification process and determines the features of the bone matrix so that its deficiency adversely affects all stages of skeletal metabolism, causing cessation of bone growth, decrease of osteoblastic and osteoclastic activities, osteopenia and bone fragility [28]. Silicon is a trace elements essential in the processes of bone formation and calcification, enhancing osteoblast cell activity and promoting bone apposition and bone remodeling [29, 30]. In contact with the physiological media, silicon coordinates with hydroxyl ions forming $Si(OH)_4$ functional groups, which is the event initiating the nucleation and deposition of a layer of biomimetic HA [31]. Silicon also influences the cartilage synthesis and the integrity of the extracellular matrix, as well as the biomineralization process. It has dose-dependent effects on the differentiation, proliferation and collagen synthesis of osteoblasts, with a direct influence on the bone remodeling process and osteoclast development and resorption activities [32].

The achievement of a chemical biomimesis implies relevant surface modifications, which make the behaviour of inorganic HA *in vivo* extremely dynamic and "vital", while influencing the cell activity during all the stages of the remodeling path. In fact, substituted apatites are characterized by a very poor crystallinity and HA nanocrystals are surrounded by a hydrated layer arranged in a non-apatitic environment where ions are continuously exchanged with the growing apatite lattice during its maturation. The high surface reactivity of such poorly crystalline HA reflects in a high bioavailability of ions relevant for the biological processes at the apatite surface, which can be exchanged with the physiological environment while the maturation of HA proceeds transforming the apatite phase in a progressively more stable calcium phosphate phase [33]. Additionally, the surface state of the biomimetic HA enhances the ability to bind proteins such as albumin and growth factors, acting as promoters of the biomineralization. Thus, the changes in the HA surface structure and charge due to the ion substitution provide chemical cues increasing the ability of synthetic HA to be involved in the natural bone remodeling process.

B-type carbonation is abundant in young individuals, where the mineral phase of bone is poorly crystallized and the maturation layer is thick and very active, and enhances the apatite solubility at physiological pH, yielding a surface polar charge which enhances the affinity of osteoblastic cells [34, 35]. Magnesium is continuously and reversibly exchanged with the amorphous layer surrounding HA particles, where it is able to coordinate water molecules in a higher extent compared to calcium [36]. The increased surface hydrophilicity allows a easier attachment of proteinic molecules which in turn favour the adhesion of progenitor cells and the consequent activation of the remodeling processes [37].

The increased reactivity of biomimetic nano-structured HA phases with the existing bone drives to a higher stability of the implanted region due to the formation of a strong mechanical bond between the implant and the surrounding regions in the body. In orthopaedics bioactive/bioresorbable prostheses based on biomimetic HA can completely integrate in the existing bone and eliminate all drawbacks related to prosthesis mobilisation and detachment, prosthesis replacement and insufficient limb mobility due to the only partial functionality of bioinert prostheses. This approach overcomes the current paradigm based on bioinert prosthesis and invasive surgery and promotes the development of new solutions able to reduce the hospitalisation time and the patient pain, also drastically cutting the healthcare costs.

Sintering processes are needed to obtain 3D consolidated scaffold to be employed as prosthetic devices in orthopaedics; in this case, the increased solubility and

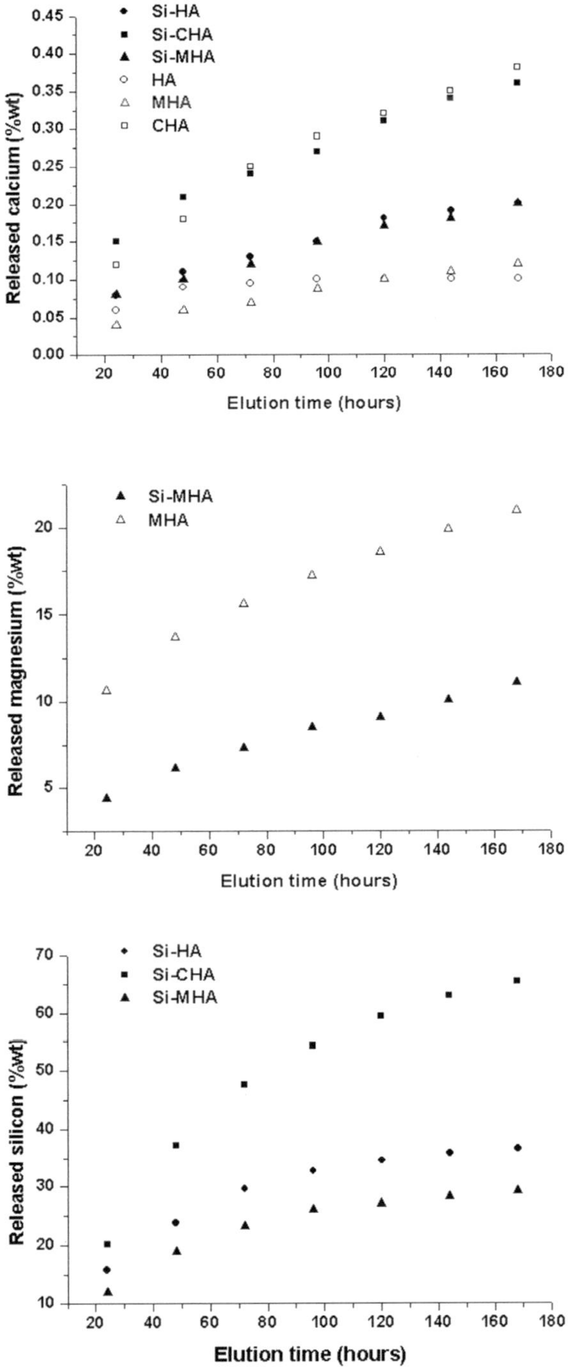

Figure 1: Ion release by multi-substituted HA powders.

reactivity of biomimetic HA phases make them more unstable upon thermal treatments, inducing the formation of undesired secondary phases [38]. In consequence, the applicability of sintering processes to obtain 3D scaffolds is reduced and alternative methods are required, such as solid state reactions at high temperatures ([39], see also paragraph: Composite scaffolds).

Biomimetic hydroxyapatites can be synthesized by wet methods, employing calcium hydroxide, ortophosphoric acid and suitable soluble salts containing the ions of interest [24] or by using SBF as reacting media to introduce small amounts of monovalent ions [40]. Biomimetic HAs are able to progressively release ions relevant for the regeneration/remodeling processes, with a release kinetic depending on the HA crystallinity, type and amount of foreign ions [24]. In fact the simultaneous presence of magnesium and silicon mutually stabilise the two ions upon release in physiological environment; in any case the solubility of single-doped HA (*i.e.* Mg, Si,) is further accelerated in presence of carbonate ions [24], with a potential enhancement of the remodeling processes (Fig. **1**).

The increase in solubility of non-stoichiometric HA is also highlighted by electroacoustic measurements: electrical conductivity is higher in substituted apatites at physiological pH (Fig. **2**, left), revealing an increased ion dissolution and bioavailability; moreover the isoelectric point moves towards more basic pH compared to stoichiometric HA (Fig. **2**, right).

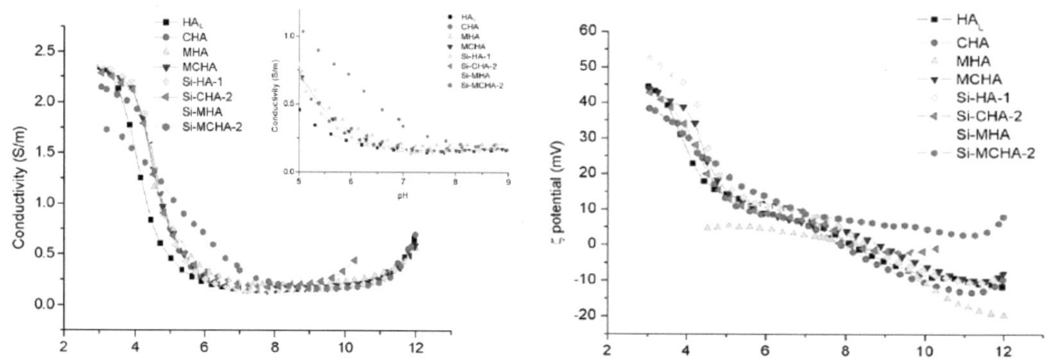

Figure 2: Electrical conductivity (left) and ζ potential of multi-substituted HA powders (right).

Silicate and carbonate ions in B-position compete in the occupancy of phosphate site [24] and stabilise the HA structure upon sintering, preventing phase decomposition, contrary to what occurs in presence of single ion doping [38]. In contact with human osteoblast cells the multi-doping resulted more effective in providing higher biocompatibility and osteogenesis stimulation compared with carbonate-free materials [41]. The improvement of the bioactivity and bio-reabsorbability of multi-doped HA-based materials has been also assessed *in vitro* and *in vivo* [37].

The release of strontium, replacing calcium in the HA lattice [42], increases bone formation and reduces bone resorption, leading to a gain in bone mass and increase of trabeculae thickness, thus improving the mechanical properties of bone [43]. Besides, carbonate ions replacing phosphate can be simultaneously introduced in Sr-doped HA: it has been shown that porous scaffolds made of Sr-doped carbonated HA display an increased dissolution extent, while maintaining compressive strength similar to that of cancellous bone [44].

Nanostructured bio-molecules can be attached to the scaffold surface to provide additional functionalization and triggers able to control the evolution of the osmolarity in proximity of the scaffold and to mediate the *in vivo* response of the scaffold towards specific biological processes like selective cell differentiation in osteoblasts.

MORPHOLOGICAL AND GEOMETRICAL BIOMIMESIS

Beyond the chemical and biochemical aspects, from a morphologic perspective, a bone scaffold should exhibit a 3D highly porous structure with an interconnected pore network for cell growth and flow transport of nutrients and metabolic waste [45, 46]. The 3D structural organization of bone scaffolds for orthopaedics is a key parameter for the activation of the regeneration processes in long bones. So far, although extensive efforts have been put into the development of porous scaffolds for bone regeneration, with encouraging results, all porous materials have a common limitation: the obtained porosity is often uniformly distributed and scaffolds have isotropic mechanical properties, contrary to those of bone. Moreover, high porous materials, as required for bone replacement, exhibit an inherent lack of strength which limits the current application only to non-load bearing applications. Although for such applications the existing solutions can be adequate for bone regeneration, prostheses for the replacement of long bone portions must activate the remodeling activity at cell level so that a scaffold presenting a morphological biomimesis is needed.

In the last decades several techniques have been set up and developed to achieve porous scaffolds with anisotropic porosity (Fig. **3**), in order to better simulate the biomechanical performance of bone. An interesting technique recently developed for the development of bone scaffolds is based on the freeze-casting of aqueous suspensions of HA [47, 48]: different morphologies, cellular and lamellar, can be obtained, by varying the freezing temperature (Fig. **4**). Lamellar HA structures are obtained due to the proceeding of the cold front from bottom to the top of the suspensions, which progressively freezes the water along a columnar direction, forming channel like elongated porosity. The pore size of both the cellular and lamellar portions are large enough to allow the growth of new bone inside the scaffold and the columnar pores are well interconnected by micro-channels suitable for the exchange of nutrients and waste materials.

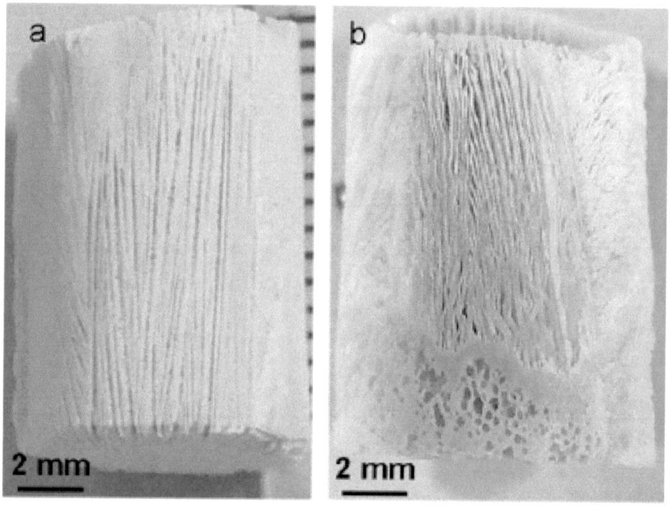

Figure 3: Different morphology of HA scaffolds obtained by freeze-drying at different conditions.

The use of ice as porogen makes this process relatively simple and "clean", avoiding any chemical porogen or artificial template, which have to be eliminated in a successive step by burning out or leaching, complicating the production and increasing the risk of failure. Further, the brittleness of porous ceramic materials and their tendency to form debris can be reduced by adding suitable natural polymers, like gelatine, which also offer a conducive way to the proliferation of cells inside pores, due to its bio-resorbability. In this case the resorption kinetics can also be tailored by employing suitable cross-linking agents.

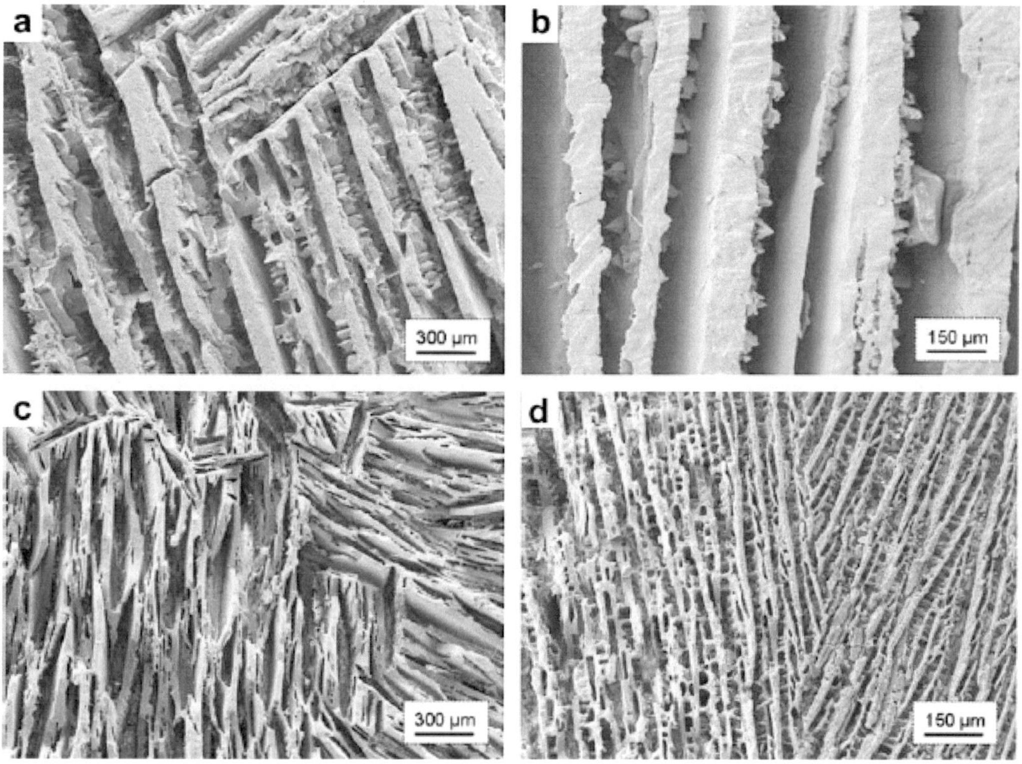

Figure 4: Effect of freezing temperature on the transverse and longitudinal morphology of lamellar HA scaffold obtained by freeze-drying.

COMPOSITE SCAFFOLDS

In load-bearing applications, the implant must withstand the early biomechanical loads *in vivo* while driving the formation and spatial organization of new bone tissue, and being in turn progressively resorbed [49-51]. Thus, the intrinsic mechanical strength of the scaffold material is a key parameter for further development of porous scaffolds, coupled with an adequate bioactivity to ensure the early adhesion and proliferation of cells and an even partial resorbability to drive the bone regeneration towards the 3D scaffold morphology and the establishment of new well-organized tissue.

Although calcium phosphates, in particular hydroxyapatite and tricalcium phosphate (TCP), represent the elective class of materials for bone replacement, as they closely mimic the mineral part of the bone tissue, applications for long bone substitution are hindered by their insufficient strength. Dense HA is characterized by a fracture toughness slightly below the lower limit of cortical bone (see Table **I**); Conversely, Young modulus is typically higher [31] and the flexural strength is limited to ~80 MPa [52-54], thus not useful for load-bearing applications, involving long bones like femur.

Nowadays the elective clinical solution is still represented by the bioinert implants, based on alumina, titanium, etc..., provided with a bioactive coating to improve the cell adhesion and osteointegration. Although such solutions allow the complete replacement of the diseased bone and the recovery of the sustain function, this positive effect suffers in the long term of a number of drawbacks, which often force the patient to a second surgery for the replacement of the prosthesis. In the short term the low bio-reactivity of the scaffold with the existing bone leads to increased time required for patient rehabilitation [6]. In the long term the scarce integration of a poorly bioactive implant with the surrounding bone often has adverse effects on interfacial apposition of bone to the implant and its mechanical stability [7-9]. Moreover there is an insufficient load transfer from a material with a too high elastic modulus to the adjacent remodeling bone, which may result in localized disease or death of bone due to stress shielding, bone resorption and eventual loosening and detachment of the prosthetic device [55]. In younger patients, the concerns are even more relevant, since the still active process of bone growth is not compatible with the absence of any activity by the implanted scaffold. In this case the patient can also lose the capacity to walk properly and several repeated invasive surgery are needed.

In the perspective of long bone regeneration, the leading idea in the development of composite scaffolds for orthopaedics is the intimate joining of bioactive/bioresorbable calcium phosphate phases with a bioinert/bioactive hard ceramic functioning as reinforcement. A series of ceramic composites was investigated so far, *e.g.* bioglasses [56, 57], HA + ZrO_2 [58, 59], HA + calcium silicates [39], HA + TiO_2 [60], HA + Al_2O_3 and other composites with highly resistant bioactive ceramics [61, 62].

The main concerns arising when developing such composites refer to phase decomposition, due to the chemical interactions between HA and the reinforcing phases at high temperatures; in these cases, HA undergoes uncontrolled phase decomposition and transformation in β- and α-TCP and, although in many cases the mechanical strength of the composite is higher than that of HA, the formation of secondary phases alters the degree of bioactivity and bioreabsorbability of HA, as they are characterized by higher solubility. Moreover, the formation of α-TCP yields a volume increase in the ceramic body, compared with HA, and it can induce micro-cracks in the final scaffold, impairing mechanical properties.

The consolidation of ceramic materials endowed with high mechanical strength usually occurs at higher temperatures compared to HA, thus to achieve a complete sintering of the material the temperature must be raised well above 1200-1300 °C, usually up to 1500 °C. At these temperatures the extent of HA decomposition is huge, with formation of phases stable at high temperatures like α-TCP and tetracalcium phosphate (TeCP), and in some cases an incongruent melting of HA can be observed. Another inconvenient arising with ceramic composites is that in most of the cases the higher thermal stability of the reinforcing phases is also coupled with a higher chemical stability, so that the reinforcement behaves as a bioinert body, which in some cases can also induce the formation of a HA layer on its surface [63], but it cannot be reabsorbed by osteoclasts neither in the long term. The main limitations to the development of an ideal 3D bone scaffold are so due to the difficulty to obtain a material exhibiting bioactivity, bioreabsorbability and mechanical resistance in the meantime.

In biomimetic composites, the reinforcing phase should exhibit capacity to be integrated in the surrounding bone tissue and mechanical properties, in particular elastic modulus and fracture strength, should not exceed those of bone. This requirement limits the spectrum of materials which can be used as reinforcing phase so to minimise the above described drawbacks; in particular materials endowed with high density, high toughness and high elastic modulus should be avoided.

Although glass-based ceramics are characterized by high biological activity, their brittleness limits its mechanical strength, so that crystalline materials are more adequate for load-bearing applications. In this perspective, calcium silicates are materials widely employed in the development of concretes (*i.e.* Ca_2SiO_4, Ca_3SiO_5, $CaSiO_3$, …) which also exhibit the capacity of inducing the nucleation of a HA layer at physiological conditions and also mechanical strength higher than HA, thus promising as reinforcing phases in bioactive composites for bone scaffolding. Wollastonite phase (a polymorph of the $CaSiO_3$ family) is a phase commonly found in partially crystallised bioglasses, thus already employed as a biomaterial and exhibit very interesting mechanical strengths, compared to the native bone [64]. Generally, biomaterials containing $CaO–SiO_2$ appear to have excellent bioactivity and were found to bond to living bone and soft tissue, through the development of a biologic HA layer on their surface [65-67]. Another study illustrates the analogies between the hydration behaviour of calcium silicates and calcium phosphates [68], when forming the calcium silicate hydrated ($xCaO \cdot SiO_2 \cdot nH_2O$, $0.833<x<\sim1.7$) and HA phase respectively. The analogies in the hydration behaviour are significant, as the very early modifications of the implant surface, when *in vivo* it is in contact with the physiological environment, are determinant for the development of the HA layer, and the adhesion of proteins and other growing factors. Since composite ceramics require consolidation *et al.* thermal treatment at high temperature, chemical reactions at the interface should be minimised in order to reduce the extent of HA phase decomposition. Recently it has been shown that sintered HA/Ca_2SiO_4 composites can be obtained by Fast Hot-Pressing technique in vacuum, avoiding any phase decomposition and achieving improved flexure strength [39]. Fast sintering of HA/silicate composites in controlled environment can be a promising route to achieve reinforced materials for long bone substitutions. Moreover, the sintering of HA-based composites can provide additional positive features to the apatite phase. In fact, multi-substituted biomimetic HA phases exhibit a reduced structural stability, leading to a higher extent of ion release at physiological conditions but also an increased tendency to the formation of undesired secondary phases [38]. During sintering at high temperatures, surface or bulk ion diffusion can be activated and the reinforcing phase can exchange ions with the HA lattice, as also demonstrated in case of composites of stoichiometric HA with Ca_2SiO_4 where silicate ions substituted phosphate during hot pressing [39], potentially improving its bio-reabsorbability when implanted *in vivo*. Following this concept, silicate phases containing ions relevant for the regeneration processes and with mechanical properties close to cortical bone [31, 64, 69, 70] (see Table **1**) can be used and potentially exchange ions with stoichiometric HA, exploiting its reduced stability and increased capability to accept foreign ions at high temperatures.

Table 1: Mechanical properties of different bioactive calcium silicates in comparison with cortical bone.

	Fracture Toughness K_{IC} (MPa m½)	**Young Modulus E(GPa)**	**Bending Strength (MPa)**
Ca_2SiO_4	~2	~40	~200
$CaSiO_3$	~2	~40-50	~300
$Ca_7MgSi_4O_{16}$	~1.6	~40-50	~150
HA	~1	~80-110	~80-100
Cortical bone	~ 2 - 12	7-30	50-150

Titanium oxides (anatase and rutile), such as silicates, are light and high-resistance bioactive materials, able to increase the kinetic of apatite nucleation on their surface and to form a tight chemical bond with the bone tissue [71, 72]. In most application, titania has been employed in composite with HA to coat load-bearing prostheses made of titanium, in order to increase the surface bioactivity and improve osteointegration. Notwithstanding, ceramic prosthetic devices based on HA matrix reinforced with TiO_2 are scarcely reported in literature [60] but they can potentially found application in bone surgery where improved mechanical features are required. Bulk titania in fact exhibits a Young modulus about 200 GPa, a compression strength of ~ 700 MPa and a fracture toughness > 3 $MPa \cdot m^{1/2}$. Powder mixtures of stoichiometric HA and nanostructured titania can be sintered to obtain consolidated ceramics exhibiting high density and controlled phase composition: it was recently found that commercial titania powder, made of a mixture of anatase and rutile polymorphs, induce the formation of β-TCP and the perovskite phase ($CaTiO_3$) when sintered in contact with HA, following the reaction:

$$Ca_{10}(PO_4)_6(OH)_2 + TiO_2 \rightarrow 3Ca_3(PO_4)_2 + CaTiO_3 + H_2O \uparrow$$

When excess titania is added, TiO_2 phase in form of rutile is also present in the final composite, acting as reinforcing phase. In such composites β-TCP represent the bioactive/bioresorbable phase and TiO_2 is the reinforcement, while $CaTiO_3$ can act as a promoter of the apatite growth [72]. The mechanical properties of HA/TiO_2 composites resulted improved compared to those of sintered HA bodies characterized by similar density. In particular, the flexural strength of HA was about 40 MPa while HA/TiO_2 composite (80: 20 vol%) exhibited a value of about 130 MPa.

On the basis of these preliminary results it can be concluded that the joining of the bioactive features of calcium phosphates and the mechanical strength of titania-based materials can represent an innovative solution for the development of composites for load-bearing bone scaffolds which can be successfully applied in orthopaedic application.

MORPHOSYNTHESIS OF HIERARCHICALLY ORGANIZED BONE SCAFFOLDS

The establishment of an interconnected pore network is fundamental for the development of a vascular tree able to allow a proper blood circulation throughout the whole scaffold. It has been demonstrated that the fluid transport system in living bodies is subjected to principles of energy and mass minimization, coupled with high efficiency in fluid transport and exchange of oxygen [73-75]. The hierarchically organized morphologies typical of many native tissues are structured in function of optimized biomechanics and for the achievement of an efficient vascular tree.

This concept, based on the tensegrity model, is supposed to be at the basis of the morphology and functioning of all living tissues and organs. Bone tissue is a hybrid composite with a very complex morphology, anisotropic and hierarchically organized at different size levels, from nanometer to millimetre: this feature allows to distribute the mechanical stresses along a huge number of small trabeculae, which can be more easily replaced after their rupture. Although highly porous anisotropic scaffolds have already been developed, it is impossible to build such complex morphologies by conventional bottom-up approach with a complete control of the microstructural features at all size levels, especially when dealing with mineralized materials.

A recently proposed solution is based on inspiration by biologic structures already existing in Nature. Such structures have been optimized in million years of evolution so to allow the store and transfer of information at molecular level which are hardly achievable in artificially designed structures. Hence, the inspiration by natural processes and structures can be a very promising way to develop smart devices not achievable with the conventional manufacturing techniques.

Such a new approach in the development of bone scaffold exploits the native structures of vegetable species as organic templates to be transformed into inorganic bioactive devices [76, 77]. Wood structures in particular are very promising in this view since they exhibit anisotropic structures and different pore size distribution, very similar to those of bones. In particular, rattan (*Calamus Rotang*) is endowed with a channel-like porosity whose size ranges in the hundredth of microns, simulating the Haversian channels; these channels in turn are interconnected by a network of channels whose size is in the range of tenth of microns, thus the structure results adequate for cell colonisation and for the development of an adequate vascular network. Several methods were set up to transform wood structures into inorganic materials [78-81] for various applications; in most cases these so called biomorphic materials are simple oxides or nitrides, obtained through sol-gel infiltration of native woods by suitable organo-metallic precursors and sintering. A similar approach was recently employed to prepare HA scaffolds derived by transformation of rattan wood [77]. Biomorphic scaffolds made of carbonated HA derived by rattan and other woods were also obtained by multi-step transformation process [76] (Fig. **5**), in which each reaction was carried out among a gaseous phase and a solid template derived by native wood, so that the transformation occurred at a molecular scale, resulting in a very good reproduction of the starting template, minimising the occurrence of defects and cracks (Fig. **6**).

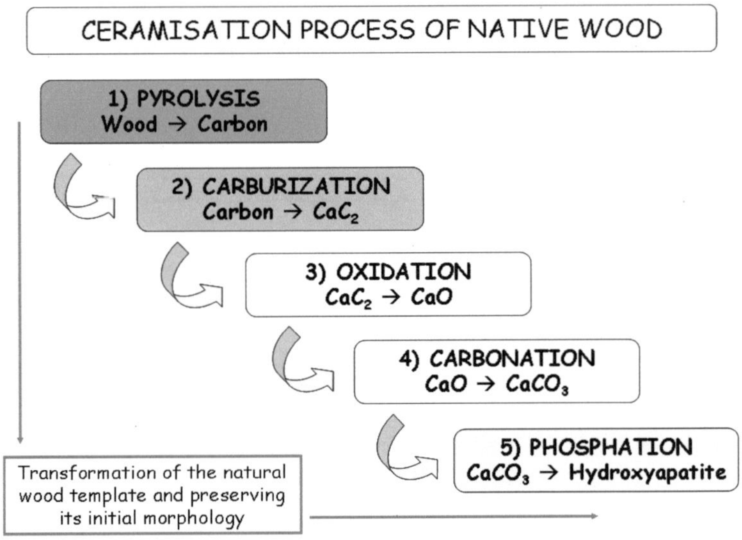

Figure 5: Multi-step process to transform native wood into biomorphic HA.

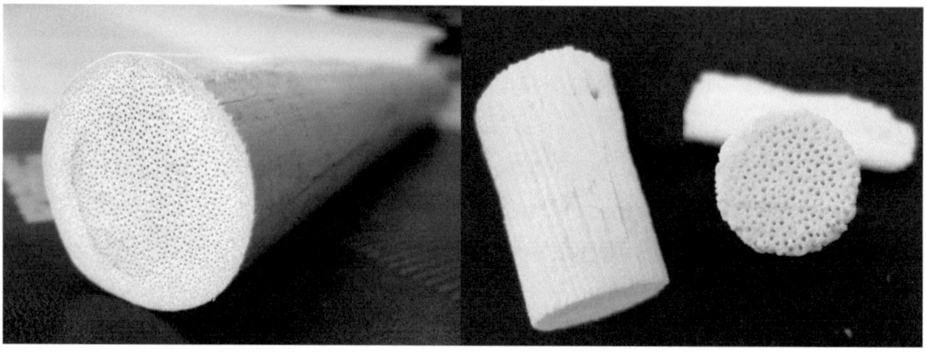

Figure 6: Native rattan wood (left) and the biomorphic HA scaffolds reproducing the starting wood morphology (right). Following the different chemical reactions at the basis of the transformation process are indicated:

$$2C + Ca \rightarrow CaC_2$$

$$CaC_2 + H_2O \rightarrow CaO + C_2H_2$$

$$CaO + CO_2 \rightarrow CaCO_3$$

$$10CaCO_3 + 6(NH_4)_2H(PO_4) + 2H_2O \rightarrow Ca_{10}(PO_4)_6(OH)_2 + 6(NH_4)_2CO_3 + 4H_2CO_3$$

Each step of the transformation process is mediated by the diffusion of the reacting gaseous phase into the solid biomorphic template. In the conversion of carbon into CaC_2 the formation of micropores due to the volume increase facilitate the penetration of the gaseous calcium; it is also evidenced that the reaction initiates inside the pores, then extending throughout all the carbon template (Fig. **7**).

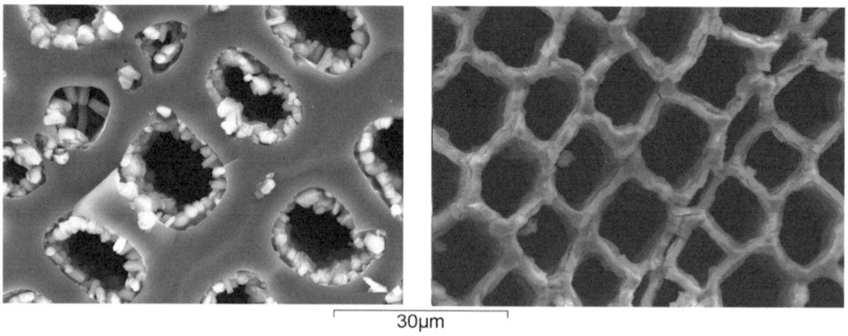

Figure 7: Nucleation of CaC_2 phase inside the pore of pinewood-derived carbon template.

The oxidation step, carried out in room atmosphere, takes place through the formation of a layer of CaO onto the CaC_2 template by heterogeneous gas-solid reaction, following by volume diffusion: it was detected that higher temperatures of oxidation induce a higher density of the superficial CaO layers, making harder the subsequent O_2 penetration inside the CaC_2 template. The carbonation step can be carried out in furnace or in autoclave under high CO_2 pressure. Also in this case a first layer of $CaCO_3$ forms at the template surface by heterogeneous CO_2-CaO reaction: the high temperatures employed in furnace are effective in providing a higher energy for $CaCO_3$ formation, nevertheless, room pressure is insufficient to achieve complete gas penetration into the surface layers of $CaCO_3$ formed on the surface of the CaO template; in autoclave the application of higher pressures employed enhances the diffusion process but due to the lower temperatures the kinetic of $CaCO_3$ formation results reduced, leading to long times of reaction for the achievement of a fully carbonated biomorphic template. The phosphation process can be carried out in autoclave or at room temperature. The reaction is found to occur by dissolution-reprecipitation mechanism, allowing the maintenance of the original native structure (Fig. **8**), which yield a poorly crystalline carbonated hydroxyapatite. The reaction kinetic is enhanced by the reaction temperature and by a higher availability of phosphate ions; in consequence of the faster reaction kinetic, the mechanical strength resulted negatively affected, so that a compromise must be found in order to optimize the mechanical strength and the conversion time. Sintering processes can be applied, in CO_2 atmosphere, to consolidate the biomorphic construct without losing the carbonate ions present in the HA lattice; however such processes can partly decompose the apatite phase into β-TCP, resulting in an increased solubility *in vivo* and increased mechanical strength. Preliminary tests revealed a strong anisotropic behaviour of biomorphic scaffold, similar as that of native bone: rattan-derived HA exhibited a compression strength of 4-5 MPa along the direction perpendicular to the channels orientation, and 1-2 MPa when measured along the transversal direction.

The hierarchical organization and chemical composition of biomorphic bone scaffolds can reproduce the pore size distribution and biomechanical performances of the native wood, thus a suitable environment capable to induce the regeneration of long bones. Preliminary *in vivo* tests were carried out on rabbits: porous cylinders made of rattan-derived HA were implanted into the trabecular bone of the distal femur. After one month. it resulted that biomorphic HA scaffolds exhibited no cytotoxicity and the porous scaffolds were completely osteointegrated in the existing bone. These devices can be a breakthrough in the development of bioactive/bioresorbable bone scaffolds for load-bearing applications, which are still based on bioinert devices and thus not representing an acceptable clinical solution in the long term.

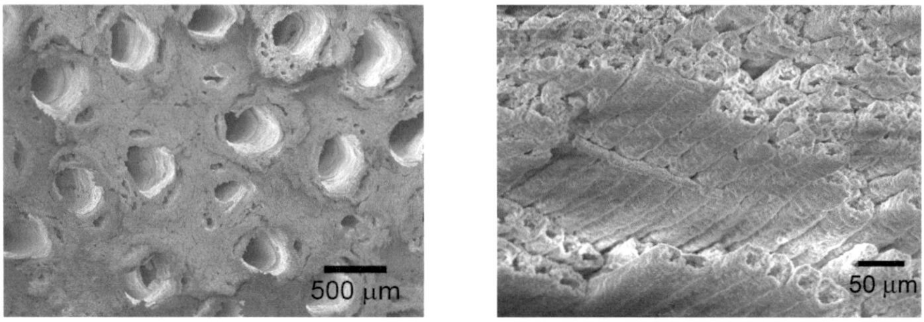

Figure 8: Biomorphic HA derived by rattan (left) and pinewood (right).

BIO-HYBRID COMPOSITES FOR BONE AND OSTEO-CHONDRAL REGENERATION

The role of bioceramics during reconstructive bone surgery is fundamentally osteoconductive; nevertheless, the morphological features of the bone scaffold are crucial to induce cell proliferation extended to the inner part of the scaffold itself. This is one of the major limitations to the development of advanced bone scaffold, since the reproduction of the chemistry and the very complex morphology of bone is rather difficult and it is not achievable through the current manufacturing technologies. The characteristics of the commonly available artificial bone tissue are drastically different from those of natural ones; this is mainly due to the absence of the self-organizing interaction between the apatite crystals and the collagenic components, leading to peculiar ultrastructural organization [82]. Bone is the typical example of an "organic matrix-mediated" mineralization process [83], in which the formed mineral phase is deeply influenced at a physico-chemical level by the matrix itself. During the mineralization process of bone tissues, collagen is first synthesized, extruded from the cell, and then self-assembled in the extra-cellular space before mineralization begins. Several proteinic matrix can be prepared, differing in chemical composition and molecular architecture with the aim to design organic/HA composites. However, a biologically inspired approach in the synthesis of collagen/HA composite is the only procedure capable to reproduce the same structure of the native tissue at all size level, where the apatite phase is directly nucleated on the collagen fibrils during their self-assembling, exploiting the ability of the collagen negatively charged carboxylate groups to bind the calcium ions of HA [84]. Through this approach, it is possible to synthesize 3D bio-hybrid composites in which the information are exchanged among the organic and inorganic phase at molecular level, such as the native tissues.

Type I collagen fibres are commonly employed; before use, collagen is purified by telopeptides and glycosylated regions are removed. The mineralization process is performed using H_3PO_4 solution, mixed with collagen gel (1% wt) and dropped in a basic suspension containing $Ca(OH)_2$ and other salts containing ions relevant for the bone regeneration (Mg, Si, ...) to obtain mineralization with biomimetic HA [32]. This specific procedure implies the dispersion of collagen into the acid solution and the drop-wise addition of this mixture into the basic dispersion so that to maintain basic condition (pH ~ 8) for almost the whole process. The maintenance of these conditions is crucial to allow the fibril formation before the precipitation of calcium phosphate, so that the collagen fibrils act as templates for the mineralization. Upon decrease of pH below ~ 8, amorphous HA forms onto the fibrils before their assembling into fibres. As the pH approaches a neutral value, two processes enter in competition, involving the same binding chemical groups on the surface of the fibres: the organization of collagen fibres into a three-dimensional network and the proceeding HA crystallization [84-86]. Finally, a biocomposite is obtained made of fibrillated collagen embedding nanocrystals of doped HA (see Fig. **9**); the water in excess is further eliminated by a freeze-drying process. The final porosity of the spongy mineralized composites is strongly dependent on the freezing temperature, heating/cooling ramp and the content of water in the starting gel. Cross-linking agents can also be employed (BDDGE, Genipin, ...) to further adjust the porosity and the mechanical strength of the construct, as well as to retard its bio-degradation. X-ray micro-CT with a local resolution of 10 μm can be successfully employed to evaluate the total porosity of HA/Coll composites (usually ranging between 80–85%), and the pore wall thickness (usually 15–20 μm).

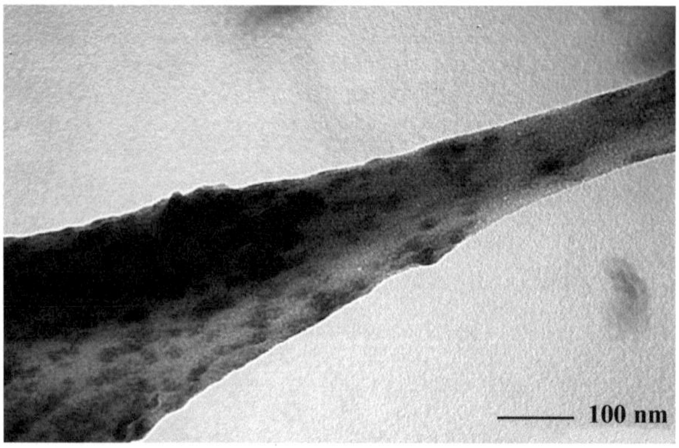

Figure 9: Biologically inspired nucleation of nanosized HA onto collagen fibres.

Since the collagen biomineralization here described reproduces the biological processes, all the mechanisms active during the natural biomineralization process occur: i) chemical: the foreign ions are spontaneously incorporated into the neo-forming apatite lattice; ii) structural: the apatite lattice results poorly ordered [87], due to the interactions of the mineral phase with the organic matrix in which it is embedded (see Fig. **10**); moreover, the HA nanocrystals grow inside collagen fibres with their **c** axis preferentially oriented parallel to the direction of the long axis of collagen, as occurs in the native tissues (see insert in Fig. **10**); iii) morphological: the HA crystals nucleated on the organic fibre exhibit a rod-like structure (Fig. **11**) similar as the nanocrystal of the newly deposited bone mineral.

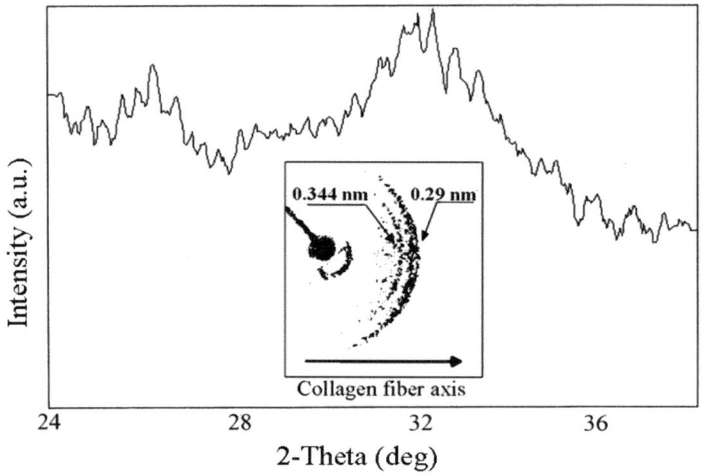

Figure 10: X-ray diffraction of bio-hybrid collagen/HA composite.

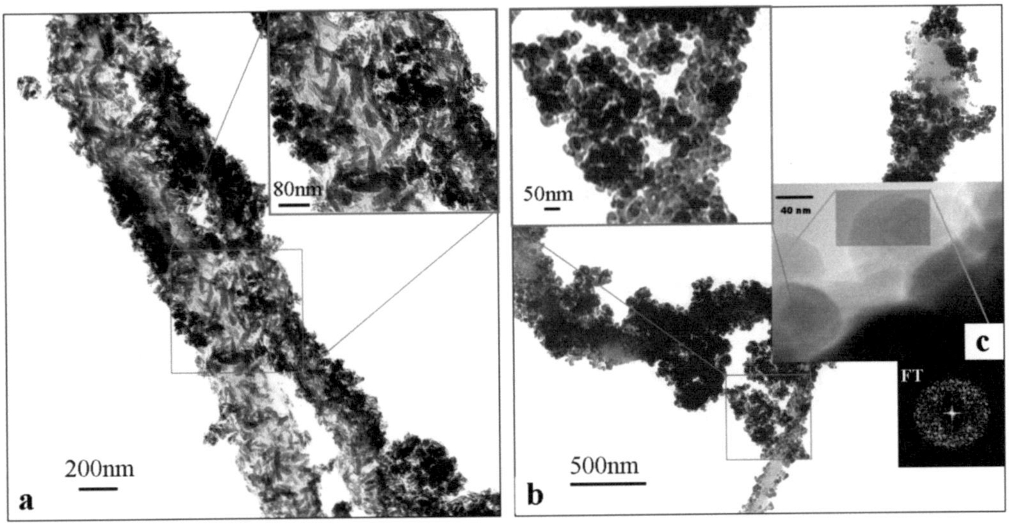

Figure 11: Morphology of HA nanoparticles nucleated onto collagen fibers in bio-hybrid Collagen/HA composite.

Bio-hybrid composites with a high level of mineralization (70:30 wt%) exhibit a pseudo-plastic behaviour, which simulate the behaviour of the native bone; in particular, calculations made on the basis of regression curves gave elastic modulus reproducing the value found for trabecular bone at correspondent values of porosity [32].

The degree of mineralization of the collagen matrix can be adjusted, ranging from a mineral-free 3D structure, reproducing the chemical and structural features of cartilage, to a high mineralized composite which is able to rapidly regenerate bone tissue. It is possible to apply biologically inspired nucleation of HA on self-assembling collagen matrices to develop suitable materials for the repair of osteo-chondral defects [88]. A graded material can be developed by merging differently mineralised HA/collagen composites in form of layers so to simulate the composition of hyaline cartilage, tidemark and subchondral bone.

The cartilage-like layer can be prepared by adding pure collagen with hyaluronic acid to create bridges between the collagen fibres and improve the hydrophilicity of the system. The porosity of the cartilaginous layer is represented by pores more isotropic than the mineralized composite and the average pore diameter is in the range of 100–150 μm.

Microscopic investigation of the graded composite by ESEM (Fig. **12**) displays the structure of the three-layered composite, where a columnar-like structure converging towards the external surface where it forms horizontal flat ribbons (see insert in Fig. **12**), resembling the morphology of the *lamina splendens*. Such morphology is induced by the propagation of the planar ice front during freeze drying. The increased structural disorder of the mineralized layer is due to the higher density and more scarce flexibility of the bone-like layer, compared to the non-mineralized collagen.

Figure 12: Morphology of the three-layered osteo-chondral substitute.

The different features of such a graded composite provide physico-chemical and morphological cues able to support the selective cell differentiation in osteoblasts and chondrocytes, so to properly regenerate the different anatomical compartments of the osteo-chondral district. In preliminary biologic assessment, chondrocyte-seeded scaffolds were prepared and cultured for 2 weeks; after this time cells in the cartilaginous layer displayed a chondrocytic morphology, whereas cells in the subchondral bone layer remained fibroblastic and did not accumulate histologically detectable amounts of glycosaminoglycans. It is possible to conclude that the cartilaginous layer of the composite scaffold is permissive to human articular chondrocyte differentiation and cartilaginous matrix deposition, whereas only a fibrous tissue could develop in the subchondral bone layer. Conversely, bone tissue formation is supported within the subchondral layer of the composite, but not in the cartilaginous region [32].

In vivo tests were performed on large animals (horses, *n* = 2); in the knee articulation, a defect type 4 was created and repaired with the three-layered osteo-chondral substitute. It resulted that the natural formation of tissue occurred after 3 months with abundant regeneration of subchondral bone (Fig. **13**, left) and good integration of new fibrous cartilage and native cartilage (Fig. **13**, right). In a clinical case, the three-layered osteo-chondral substitute was employed in the healing of a focal lesion type IV in the femoral condyle of knee (Fig. **14**, left): after 1 year the regeneration of the osteo-cartilaginous defect is evident (Fig. **14**, right).

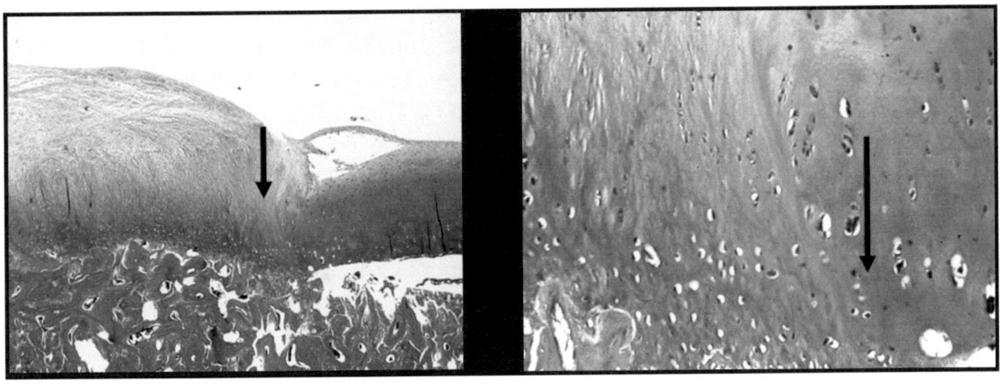

Figure 13: Histology of horse articular cartilage after 6 month from implant of three-layered osteo-chondral substitute.

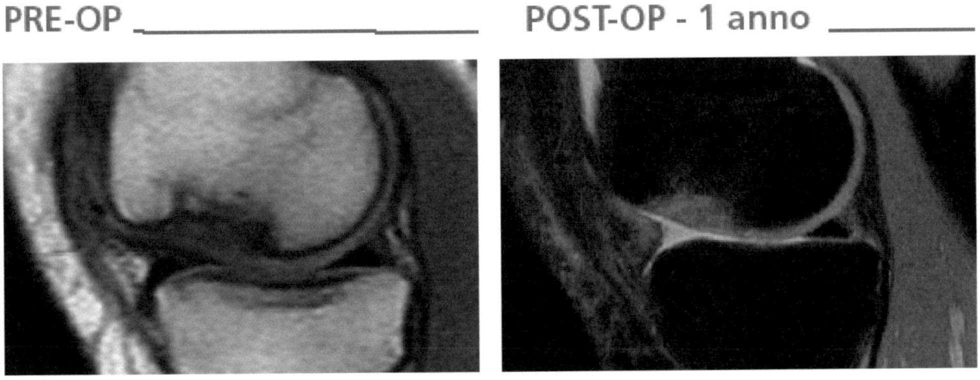

Figure 14: Pre-operatory and 1 year post-operatory histology of the femoral condyle of human knee after implant of three-layered osteo-chondral substitute.

ACKNOWLEDGEMENTS

The authors gratefully acknowledge the European Commission for financial support under the FP6 contract number NMP4-CT-2006-033277 (TEM-PLANT).

CONFLICT OF INTEREST

The author(s) confirm that this chapter content has no conflict of interest.

REFERENCES

[1] Kokubo T, Kim HM, Kawashita M. Novel bioactive materials with different mechanical properties. Biomaterials 2003; 24: 2161–2175.
[2] Giannoudis PV, Dinopoulos H, Tsiridis E. Bone substitutes: An update Injury. Int. J. Care Injured 2005; 36S, S20-S27.
[3] Gross KA, Berndt CC. Biomedical Application of Apatites. In: Kohn MJ, Rakovan J, Hughes JM, Eds. Reviews in mineralogy and geochemistry. The Mineralogical Society of America and The Geochemical Society, 2002: vol. 48, pp. 631-672. ISBN 0-939950-60-X
[4] Hench L. Bioceramics. J Am Ceram Soc 1998; 81 (7): 1705-1728.
[5] Rey C. Calcium phosphates for medical applications. In Amjad Z, Ed. Calcium phosphates in biological and industrial systems. Kluwer Academic Publishers, Dordrecht, The Netherlands, 1998.
[6] Oonishi H. Orthopaedic applications of hydroxyapatite. Biomaterials 1991; 12: 171-178.
[7] Bloebaum RD, Becks D, Loor LD, Savory CG, DuPont JA, Hoffmann AA. Complications with hydroxyapatite particulate separation in total hip arthroplasty. Clin. Orthop. 1994; 298: 19-26.
[8] Bloebaum RD, Dupont JA. Osteolysis from a press-fit hydroxyapatite-coated implant. A case study. J Arthroplasty 1993; 8(2): 195-202.
[9] Baltag I, Watanabe K, Kusakari H, Taguchi N, Miyakawa O, Kobayashi M, Ito N. Long-term changes of hydroxyapatite-coated dental implants. J. Biomed. Mater. Res. (Appl. Biomater.) 2000; 53: 76-85.
[10] Tampieri A, Celotti G, Sprio S, Delcogliano A, Franzese A. Porosity-graded hydroxyapatite ceramics to replace natural bone. Biomaterials 2001; 22: 1365-1370.
[11] LeGeros RZ, Craig RG. Strategies to Affect Bone Remodeling: Osteointegration. J Bone Min Res 1993; 8, Suppl. 2: S583-S596.
[12] Suchanek W, Yoshimura M. Processing and properties of hydroxyapatite-based biomaterials for use as hard tissue replacement implants. J Mater Res 1998; 13: 94-117.
[13] Bigi A, Cojazzi G, Panzavolta S, Ripamonti A, Roveri N, Romanello M, Noris Suarez K, Moro L. Chemical and structural characterization of the mineral phase from cortical and trabecular bone. J Inorg Biochem 1997; 68: 45-51.
[14] Driessens FCM. The mineral in bone, dentin and tooth enamel. Bull Soc Chim Belg 1980; 89: 663.
[15] Ingber DE. Cellular tensegrity: defining new rules of biological design that govern the cytoskeleton. J Cell Sci 1993; 104: 613-627.

[16] Houben F, Ramaekers FCS, Snoeckx LHEH, Broers JLV. Role of nuclear lamina-cytoskeleton interactions in the maintenance of cellular strength. Biochim Biophys Acta 2007; 1773 (5): 675-686.
[17] Wolff J, Hirchwild A. The law of remodeling. Springer-Verlag, Berlin, 1986.
[18] Pavalko FM, Norvell SM, Burr DB, Turner CH, Duncan RL, Bidwell JP. A model for mechanotransduction in bone cells: The load-bearing mechanosomes. J Cell Biochem 2003; 88: 104-112.
[19] Cowin SC. Bone poroelasticity. J Biomech 1999; 32: 217-238.
[20] Sikavitsas VI, Temenoff JS, Mikos AG. Biomaterials and bone mechanotransduction. Biomaterials 2001;22: 2581-2593.
[21] Burger EH, Klein-Nulend J. Mechanotransduction in bone – role of the lacuno-canalicular network. FASEB J 1999; 13S: S101-112.
[22] Liebschner MAK. Biomechanical considerations of animal models used in tissue engineering of bone. Biomaterials 2004; 25: 1697-1714.
[23] Dorozhkin SV. A review on the dissolution models of calcium apatites. Prog Cryst Growth Charact Mater 2002; 44: 45-61.
[24] Sprio S, Tampieri A, Landi E, Sandri M, Martorana S, Celotti G, Logroscino G. Physico-chemical properties and solubility behaviour of multi-substituted hydroxyapatite powders containing silicon. Mater. Sci. Eng. C 2008; 28: 179-187.
[25] LeGeros RZ. Calcium phosphates in oral biology, medicine. New York, NY: Kruger, 1991; pp. 82-107.
[26] Elliot JC. Structure, chemistry of the apatites and other calcium orthophosphates. Amsterdam, The Netherlands: Elsevier, 1994; pp. 80-94.
[27] Smith DK. Calcium phosphate apatites in nature. In: Brown PW, Constanz B, Eds. Hydroxyapatite and related materials. Boca Raton, FL: CRC Press, 1994; pp. 29-44.
[28] Bigi A., Foresti E., Gregorini R, Ripamonti A, Roveri N, Shah JS. The role of magnesium on the structure of biological apatites. Calc Tissue Int 1992; 50: 439-444.
[29] Carlisle EM. Silicon: A possible factor in bone calcification. Science 1970; 167: 179-280.
[30] Porter AE, Patel N, Skepper JN, Best SM, Bonfield W. Effect of sintered silicate-substituted hydroxyapatite on remodeling processes at the bone-implant interface. Biomaterials 2003; 24: 4609-4620.
[31] Hench LL. Bioceramics. J Am Cer Soc 1998; 81 (7): 1705-1728.
[32] Tampieri A, Sprio S, Landi E, Sandri M Developing biocomposites as scaffolds in regenerative medicine. In: L. Ambrosio Ed. Biomedical composites, Woodhead Publishing, Abington Hall, Abington, Cambridge, CB21 6AH, United Kingdom 2009; pp. 547-572.
[33] Cazalbou S, Combes C, Eichert D, Rey C, Glimcher MJ. Poorly crystalline apatites: evolution and maturation *in vitro* and *in vivo* J Bone Miner Metab 2004; 22: 310–317.
[34] Rey C, Renugopalakrishnan V, Collins B, Glimcher M. Fourier transform infrared spectroscopic study of the carbonate ions in bone mineral during aging. Calcif Tissue Int 1991; 49: 251-258.
[35] Rey C, Collins B, Goehl T, Dickson IR, Glimcher M. The carbonate environment in bone mineral: a resolution-enhanced Fourier transform infrared spectroscopy study. Calcif Tissue Int 1989; 45: 157-164.
[36] Bertinetti L, Tampieri A, Landi E, Martra G, Coluccia S. Punctual investigation of surface sites of HA and magnesium-HA. J Eu Ceram Soc 2006; 26: 987-991.
[37] Landi E, Logroscino G, Proietti L, Tampieri A, Sandri M, Sprio S. Biomimetic Mg-substituted Hydroxyapatite: from synthesis to *in vivo* behaviour. J Mater Sci: Mater Med 2008; 19: 239-247.

[38] Tampieri A, Sprio S, Landi E, Celotti G, Sandri M. Synthesis and properties of multisubstituted biomimetic apatites containing silicon, 10th Ceramics, Cells and Tissues, Ravaglioli A, Krajewski A, Eds., Faenza: Italy 2006: 366-374.
[39] Sprio S, Tampieri A, Celotti G, Landi E. Development of Hydroxyapatite/calcium silicate composites addressed to the design of load-bearing bone scaffolds. J Mech Behav Biomed Mater 2009; 2: 147-155.
[40] Landi E, Tampieri A, Celotti G, Langenati R, Sandri M, Sprio S. Nucleation of biomimetic apatite in synthetic body fluids:dense and porous scaffold development. Biomaterials 2005; 26: 2835–2845.
[41] Landi E, Uggeri J, Sprio S, Tampieri A, Guizzardi S. Human osteoblast behavior on as-synthesized SiO_4 and B-CO_3 co-substituted apatite. J Biomed Mater Res A 2010; 94: 59-70.
[42] Landi E, Tampieri A, Celotti G, Sprio S, Sandri M, Logroscino G. Sr-subtituted hydroxyapatites for osteoporotic bone replacement. Acta Biomater 2007; 3(6): 961-969.
[43] Marie JP, Ammann P, Boivin G, Rey C. Mechanisms of action and therapeutic potential of strontium in bone. Calcif Tissue Int 2001; 69: 121–129.
[44] Landi E, Sprio S, Sandri M, Celotti G, Tampieri A. Development of Sr and CO_3 co-substituted hydroxyapatites for biomedical applications. Acta Biomater 2008; 4: 656-663.
[45] Hutmacher DW. Scaffolds in tissue engineering bone and cartilage. Biomaterials 2000, 21: 2529-2543.
[46] Porter JR, Ruckh TT, Popat KC. Bone tissue engineering: a review in bone biomimetics and drug delivery strategies. Biotech Prog 2009; 25(6): 1539-1560.
[47] Deville S, Saiz E, Tomsia AP. Freeze casting of hydroxyapatite scaffolds for bone tissue engineering. Biomaterials 2006; 27: 5480–5489.
[48] Landi E, Valentini F, Tampieri A. Porous hydroxyapatite/gelatine scaffolds with ice-designed channel-like porosity for biomedical applications. Acta Biomater 2008; 4: 1620–1626.
[49] Williams D. Benefit and risk in tissue engineering. Mater Today 2004; 7: 24–29.
[50] Mano JF, Sousa RA, Boesel LF, Neves NM, Reis RL. Bioinert, biodegradable and injectable polymeric matrix composites for hard tissue replacement: state of the art and recent developments. Compos Sci Technol 2004; 64: 789–817.
[51] Shin H, Jo S, Mikos AG. Biomimetic materials for tissue engineering. Biomaterials 2003; 24: 4353–4364.
[52] Kothapalli C, Wei M, Shaw MT. Influence of temperature and concentration on the sintering behaviour and mechanical properties of hydroxyapatite. Acta Mater 2004; 52: 5655-5663.
[53] Juang HY, Hon MH. Effect of calcinations on sintering of hydroxyapatite. Biomaterials 1996; 17: 2059-2064.
[54] Kobayashi S, Kawai W, Wakayama S. The effect of pressure during sintering on the strength and the fracture toughness of hydroxyapatite ceramics. J Mater Sci: Mater Med 2006; 17: 1089-1093.
[55] Davis JR. Hand Book of Materials for Medical Devices, 1st ed., ASM International, 2003.
[56] Chen OZ, Thompson ID, Boccaccini AR. 45S5 Bioglass R-derived glass.ceramic scaffolds for bone tissue engineering. Biomaterials 2006; 27 (11): 2414-2425.
[57] Hench LL. Bioceramics: From concept to clinic. J Am Ceram Soc 1991; 74(7): 1487-1510.
[58] Abdelrazek KK, Won KS, Yong KH. Consolidation and mechanical properties of nanostructured hydroxyapatite.(ZrO_2 + 3 mol% Y_2O_3) bioceramics by high- frequency induction heat sintering. Mater. Sci. Eng. A 2007; 456 (1-2): 368-372.

[59] Sung YM, Shin YK, Ryu JJ. Preparation of hydroxyapatite/zirconia bioceramic nanocomposites for orthopaedic and dental prosthesis applications. Nanotechnology 2007; 18(6): 065602-065607.

[60] Que W, Khor KA, Xu JL, Yu LG. Effects of Titania Content and Sintering Temperature on Structural, Mechanical and Bioactive Behaviors of Titania Reinforced Hydroxyapatite Nanocomposites. Adv Eng Mater 2008; 10(10): B53-B59.

[61] Chiba A, Kimura S, Raghukandan K, Morizono Y. Effect of alumina addition on hydroxyapatite biocomposites fabricated by underwater-shock compaction. Mater. Sci. Eng. A 2003; 350 (1-2) : 179-183.

[62] Choi JW, Kong YM, Kim HE. Reinforcement of hydroxyapatite by addition of Ni_3Al and Al_2O_3. J Am Ceram Soc 1998 ; 81 (7): 1743-1748.

[63] Uchida M, Kim HM, Kokubo T, Miyaji F, Nakamura T. Bonelike Apatite Formation Induced on Zirconia Gel in a Simulated Body Fluid and Its Modified Solutions. J Am Ceram Soc 2001; 84: 2041-2044.

[64] Long LH, Chen LD, Bai SQ, Chang J, Lin KL. Preparation of dense β-$CaSiO_3$ ceramic with high mechanical strength and HAp formation ability in simulated body fluid. J Eu Cer Soc 2006; 26: 1701–1706.

[65] Hench LL, Splinter RJ, Allen WC, Greenlee TK. Bonding mechanisms at the interface of ceramic prosthetic materials. J Biomed Mater Res Symp 1971; 36: 117.

[66] Örjan HA, Kaj HK. On the bioactivity of silicate glass. J Non-Cryst Solids 1991; 129: 145.

[67] De Aza PN, Guitian F, De Aza S. Bioactivity of wollastonite ceramics: *in vivo* evaluation. Scr Metal Mater 1994; 31: 1001.

[68] Brown PW. Hydration behavior of calcium phosphates is analogous to hydration behavior of calcium silicates. Cem Concr Res 1999; 29: 1167–1171.

[69] Gou Z, Chang J, Zhai W. Preparation and characterization of novel bioactive dicalcium silicate ceramics. J Eu Cer Soc 2005; 25: 1507–1514.

[70] Wu C, Chang J, Wang J, Ni S, Zhai W. Preparation and characteristics of a calcium magnesium silicate (bredigite) bioactive ceramic. Biomaterials 2005; 26: 2925-2931.

[71] Ducheyne P, Qiu Q. Bioactive ceramics: the effect of surface reactivity on bone formation and bone cell function. Biomaterials 1999; 20: 2287-2303.

[72] Coreno J, Coreno O. Evaluation of calcium titanate as apatite growth promoter. J Biomed Mater Res A 2005; 75A(2): 478-484.

[73] McCulloh KA, Sperry JS and Adler FR. Water transport in plants obeys Murray's law. Nature 2003; 421 (6926): 939–942.

[74] Williams HR, Trask RS, Weaver PM and Bond, IP. Minimum mass vascular networks in multifunctional materials. J Royal Soc Interf 2008; 5(18): 55–65.

[75] Sherman TF. On connecting large vessels to small. The meaning of Murray's law. J Gen Phys 1981; 78(4): 431–453.

[76] Tampieri A, Sprio S, Ruffini A, Celotti G, Lesci IG, Roveri N. From wood to bone: multi-step process to convert wood hierarchical structures into biomimetic hydroxyapatite scaffolds for bone tissue engineering. J Mater Chem 2009, 19 (28): 4973-4980.

[77] Eichenseer C, Will J, Rampf M, Wend S, Greil P. Biomorphous porous hydroxyapatite-ceramics from rattan (Calamus Rotang). J Mater Sci: Mater Med 2010; 21(1): 131-137.

[78] Rambo CR, Sieber H. Novel synthetic route to biomorphic Al_2O_3 ceramics. Adv Mater 2005; 17(8): 1088-1091.

[79] Rambo CR, Cao J, Rusina O, Sieber H. Manufacturing of biomorphic (Si, Ti, Zr)-carbide ceramics by sol-gel processing. Carbon 2005; 43: 1174-1183.
[80] Greil P, Lifka T, Kaindl A. Biomorphic Cellular Silicon Carbide Ceramics from Wood: I. Processing and Microstructure. J Eu Ceram Soc 1998; 18: 1961-1973.
[81] Parfen'eva LS, Orlova TS, Kartenko NF, Sharenkova NV, Smirnov BI, Smirnov IA, Misiorek H, Jezowski A, Varela-Feria FM, Martinez-Fernandez J, De Arellano-Lopez AR. Thermal conductivity of the SiC/Si biomorphic composite, a new cellular ecoceramic. Phys. Solid St 2005; 47(7): 1216–1220.
[82] Addadi L, Weiner S. Design strategies in mineralized biological materials. J Mater Chem 1996; 7: 689–702.
[83] Lowestam HA, Weiner S. On biomineralization. New York: Oxford University Press; 1989.
[84] Tampieri A, Celotti G, Landi E, Sandri M, Roveri N, Falini G. Biologically inspired synthesis of bone-like composite: Self-assembled collagen fibers/hydroxyapatite nanocrystals. J Biomed Mater Res A 2003; 67: 618–625.
[85] Gelinsky M, Welzel PB, Simon P, Bernhardt A, Konig U. Porous three dimensional scaffolds made of mineralized collagen: preparation and properties of a biomimetic nanocomposite material for tissue engineering of bone. Chem Eng J 2007; 137(1): 84–96.
[86] Mann S. Biomineralization: principles and concepts in bioinorganic materials chemistry. Oxford University Press, 2001 ISBN: 0198508824.
[87] Bouyer E, Gitzhofer F, Boulos MI. Morphological study of hydroxyapatite nanocrystal suspension. J Mater Sci: Mater Med 2000; 11: 523–531.
[88] Tampieri A, Sandri M, Landi E, Pressato D, Francioli S, Quarto R, Martin I. Design of graded biomimetic osteochondral composite scaffolds. Biomaterials 2008; 29: 3539–3546.

CHAPTER 10

Metallic Surfaces for Osteointegration

Silvia Spriano* and Sara Ferraris*

Institute of Materials Engineering and Physics, Applied Science and Technology Department – DISAT, Politecnico di Torino, Corso Duca degli Abruzzi 24, 10129, Torino, Italy

Abstract: Metallic materials are widely employed for bone contact applications (dental and orthopaedic ones) because of their good mechanical properties and load bearing ability. Titanium and its alloys are the most diffused biomedical metals due to their biocompatibility, on the other hand bare metals are almost inert and cannot stimulate tissue-integration processes. In order to improve bone integration ability of metallic implants several solutions have been proposed both in the scientific literature and also among the commercial clinical applications. A synthetic review of these techniques is described in the present chapter. Surface modification strategies have been divided in morphological, chemical and biological, considering their main aim: realization of a particular surface topography, introduction of chemical elements/reactive groups or, finally, addition of specific biomolecules.

Keywords: Osteointegration, oral implant, titanium, titanium alloys, orthopaedic implants.

INTRODUCTION

Biomaterials must induce an appropriate response in physiological tissue. The suitability of the response depends on material function. Surfaces intended for bone integration (such as orthopaedic or dental prostheses) have to restore the physiological function of the replaced bone and at the same time to interact with surrounding tissues supporting integration and regeneration.

When a material comes in contact with physiological fluids, interactions play mainly at the interface. In the first nanoseconds water molecules cover material surface and different ions are subsequently incorporated into the formed water layer. Few seconds after proteins reach the hydrated synthetic surface. Only later

***Address correspondence to Silvia Spriano and Sara Ferraris:** Institute of Materials Engineering and Physics, Applied Science and Technology Department – DISAT, Politecnico di Torino, Corso Duca degli Abruzzi 24, 10129, Torino, Italy; Tel: +39 011 0905768; Fax: +39 011 090 4699; E-mails: silvia.spriano@polito.it; sara.ferraris@polito.it

Lia Rimondini, Claudia Bianchi and Enrica Vernè (Eds)

cells arrive and interact mainly with the formed protein layer. Surface characteristics, such as topography and chemistry, affect material ability to adsorb water and proteins and consequently to interact with cells [1]. A biomaterial can send to biological components both inorganic and biological signals depending on the typology of its surface.

A schematic representation of bone-implant interaction is given in Fig. **1**.

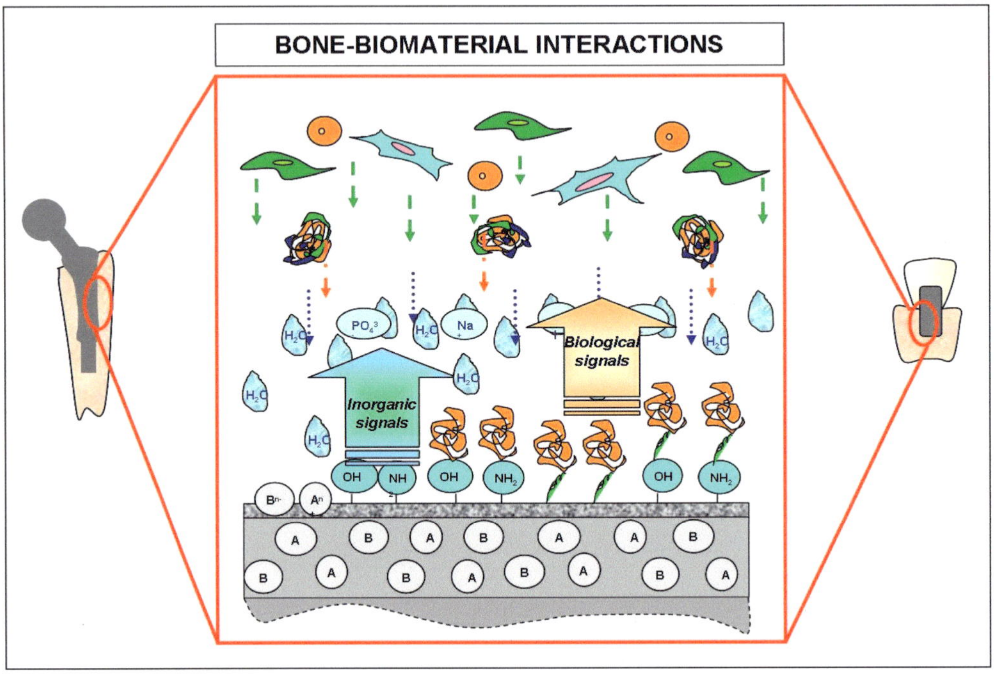

Figure 1: Schematic representation of bone-biomaterial interaction.

As far as metallic biomaterials for bone-integration are concerned, several solutions have been proposed in order to improve and fasten the osseointegration process.

They can be classified in three main groups:

- Morphological modifications – that affect mainly surface roughness and create particular topography patterns able to improve the mechanical anchorage of bone and stimulate cells at the cellular and subcellular levels.

- Compositional modifications – that affect chemical characteristic of the surface simply introducing favourable functional groups (*e.g.* hydroxyls groups) or creating a surface layer that differs in composition from the substrate. It can improve bone formation on implant surface (*e.g.* bioactive coatings).

- Biological functionalizations – that introduce biological functionalities on the inorganic surface in order to stimulate cells for tissue regeneration. It is a versatile tool to tailor surface properties in order to obtain specific functionalities.

The different strategies, described above, are schematically represented in Fig. **2** (a- morphological modifications, b1-b2 chemical modifications and c- biological functionalizations).

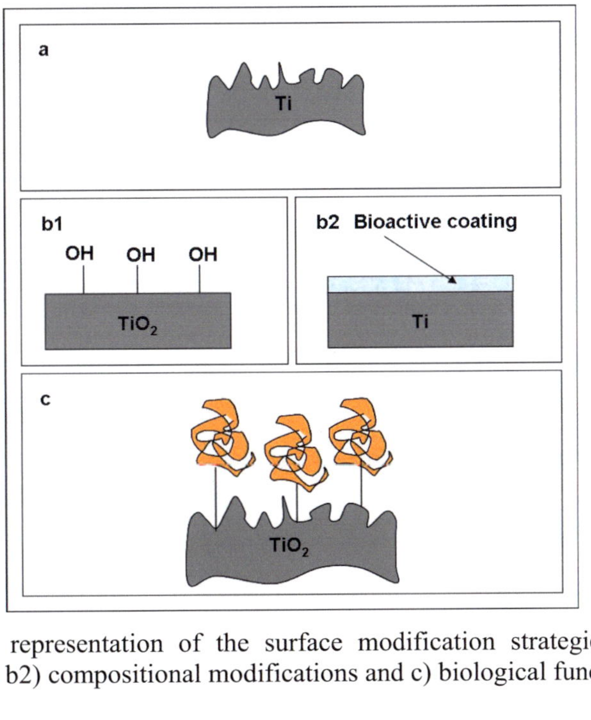

Figure 2: Schematic representation of the surface modification strategies. a) morphological modifications, b1) and b2) compositional modifications and c) biological functionalizations.

MORPHOLOGICAL MODIFICATIONS

It is now widely accepted that surface roughness deeply affect cell shape and behaviour, in particular it has been suggested that cellular gene expression is influenced by roughness [2]. So the morphological changes of the surfaces are essentially aimed at increasing its roughness, in order to facilitate secondary stabilization of the prosthesis. They can be classified as surfaces for bone on-growth, when the new bone grows onto the surface of the implant, or surfaces for bone-in growth if they present macropores (100-400 microns) in which vascular tissue can penetrate. It must be considered that highly porous/rough surfaces lose fatigue resistance. So they are usually applied to the proximal area, which is devoted to osteointegration.

The techniques employed for bone on-growth are mainly blasting and etching.

Blasting is produced by bombarding the Ti surface with hard particles (corundum, zirconia, silica, titania). The typical final roughness is in the range 4-10 micron and the topography presents sharp features. Particle size is the crucial parameter to determine the final roughness, for instance blasting with alumina particles of 25-75 micron results in a R_a of 0.5-1.5 micron, while alumina of 200-600 micron induces a R_a of 2-6 micron. Blasting lead to local plastic deformation and removal of the material, so there is a surface hardening (cold working) lowering the limit of fatigue strength, for instance a R_a in the range 4-6 micron reduces fatigue strength by 20%. There is also a danger of distortion of thin-walled components. Alumina and silica particles are embedded into the Ti surface, so blasted surfaces are chemically treated in order to remove contaminations or blasting with hydroxyapatite or zirconia is used for less dangerous contamination. The resulting surface consists of a thin surface oxide, so the excellent corrosion resistance of titanium is not altered.

Etching processes consist mainly of pickling in a mixture of nitric and hydrofluoric acid. The final roughness is in the range from 0.1 micron to several microns. The topography present facetted surfaces, because the different crystalline directions are etched at different rates and low-indexed planes are preferentially exposed. In biphasic alloys beta phase protrudes, because alpha phase is etched faster. Absorption of hydrogen and its incorporation as hydride below the oxide layer can induce embrittlement. Etching can be also applied after blasting.

Acid etching combined with chemical oxidation could result in a reproducible nanotextured surface. Nanci *et al.* [3-5] explored several combinations of acids and oxidants in order to obtain the optimal conditions able to induce nano-pattering onto both commercially pure titanium and titanium alloy (Ti6Al4V). The most widely investigated solution is a mixture containing equal volumes of concentrated H_2SO_4 and 30% aqueous H_2O_2. It has been observed that nanotextured surfaces are able to enhance intra- and extracellular expression of osteopontin and bone sialoproteins in calvaria derived osteogenic cells. Moreover it seems that modified surfaces are able to selectively promote cells adhesion and proliferation, in fact it has been noted that osteoblast proliferation is promoted while fibroblast one is reduced. Finally it has been observed that nano-porous surfaces created by chemical oxidation accelerate the growth of human umbilical cord stem cells when compared with untreated titanium and glass coverslips. This suggests that nano-porosity also has an effect on cell differentiation.

Recently it has been observed [6] that a micro-nano-hierarchical topography of TiO_2 can positively affect osteoblast behaviour. Sputtering deposition of TiO_2 nanonodules (200 nm) onto microtextured commercially pure titanium (obtained by HF-H_2SO_4 acid etching) is able to enhance osteoblast proliferation and spreading without reducing cellular differentiation induced by microtexture. Moreover fibroblast proliferation and function result reduced onto this complex topography.

Porous surfaces for bone in-growth can be obtained by several techniques presenting different features. The main characteristics affecting their performance are not so related to the pore shape, but mainly to the size of interconnecting pores. Pore sizes between 100 and 400 micron are necessary in order to optimise mineralised bone in-growth. An increasing tendency for the formation of fibrous tissue seems to appear when the pore size is increased beyond 1 mm. The porosity of most implants is usually determined to compromise between maintaining the mechanical strength of the implant, while still providing adequate pore size for tissue in-growth.

Traditional techniques for porous surfaces are sintering (sintered beads, cancellous structured, fiber meshed) and plasma spray. But it is known from clinical practice that they have some disadvantages, mainly due to a lower fatigue resistance. Moreover some substrate microstructural changes are possible, such as precipitation of carbides (Co alloys) or swelling of the grains during sintering. Furthermore there is a weak mechanical link between the substrate and the coating and not infrequently it displaces. In this case particles can become lodged in the articulating parts and they act as a third body. Alternative techniques are investment casting, welded meshes (mesh structured), metallic foams, rapid manufacturing or microfabbrication. Some examples of porous surfaces will be described in the following.

A multilayer of sintered beads of Ti can be applied to a Ti or Co based substrate, it consists of beads of 0.2-1mm in diameter and it presents pores of about 400 micron, with a total porosity of 35% [7]. Alternatively the same surface geometry can be obtained by investment casting [8]. Sintering can be also applied for obtaining a cancellous structure, called CSTi [9]. Pure titanium powder is mixed with a polymer pore-former and the mixture fills a pocket of the implant. *CSTi* coating is sintered onto the substrate at high temperature, resulting in a solid solution at the interface. *CSTi* coating is structurally similar to human cancellous bone and provides interconnected pores for bone ingrowth (pore size 400-500 micron, pore volume 60-77%). The *CSTi* process was originally developed for application on a titanium alloy substrate, but it was later applied onto CoCr-alloy femoral knee components. This bimetal combination was a major advancement that combined the biocompatible properties of *CSTi* on the bone-contacting side of the implant with the excellent wear properties of CoCr for the articulating surface. The last application of sintering to porous surfaces is fibre mesh, where a package of titanium wire is sintered on the implant [10]. Alternatively a woven of titanium wires can be welded onto the surface [11], but in any case both sintering and welding affect the fatigue resistance of the substrate.

Vacuum/air plasma spray of titanium/apatite are widely employed in order to obtain rough or porous surfaces. The advantages of these techniques are that they produce uncontaminated and reproducible surfaces on implants of complex geometry with a high productivity. Vacuum plasma spraying of pure titanium produces coating of a thickness from 200 to 500 micron, pore size from 20 to 200 micron, porosity from 15 to 40% and roughness of 10-20 micron. The substrate can be c.p. Ti, Ti alloys, CoCrMo alloys, steel or structural polymers such as PEEK or UHMWPE. The fatigue resistance can be lowered of about 25-30%, which is less than the effect of sintering. An improvement of the coating adhesion and of the cohesive strength within the coating can be obtained if as first a dense coating of Ti (about 50 micron thick) is sprayed in order to cover the roughness of the substrate, then the titanium porous structure (150-250 micron thick) is sprayed on it. In this case the notches at the surface are not interconnected to the notches of the blasted base material. A last layer of apatite can be also deposited (about 50 micron thick), it will be resorbable or permanent according to its crystallinity. The opportunity to deposit a coating with different composition from the substrate will be discussed in more detail in the next session.Trabecular Metal structure has attracted much interest in orthopaedic [12]. It is achieved through the realization of glassy carbon foam. A tantalum layer is deposited on it by CVD (Chemical Vapor Deposition). The bond with the prosthetic support is made through compression molding, if the media is UHMWPE, or diffusion bonding on metallic substrates.

Recently porous coating were realised also by EBM or laser rapid manufacturing [13]. It must be underlined that these techniques result in good mechanical properties, but data are dispersed because of random porosity and the obtained microstructures are quite unusual.

SURFACE CHEMICAL MODIFICATIONS

Chemical modifications introduce onto materials surface elements and/or compounds able to positively affect the biological response. Inorganic substances will be included in this section, while biological ones will be described in the last part (biological functionalizations).

Bioactive elements and/or functional groups can be introduced onto metallic surfaces both by coating and also by enrichment of the first surface layers.

The deposition of a coating is the easiest strategy to introduce a bioactive layer on metal surface, aimed to improve bone integration.

Hydroxyapatite, carbonate-apatite and calcium phosphates (*e.g.* α- and β-tricalcium phosphate, tetracalcium phosphate, octacalcium phosphate) are the most widespread bioactive ceramics employed, because they are close to the natural mineral phase of bone and several literature works suggest that they are able to guide bone growth on their surface by a process called osteoconduction [14, 15]. These materials are not able to support high loads (such as the ones involved in gait and mastication) because they are too brittle, so they are frequently applied as coatings onto the surface of metallic materials, in order to combine the fundamental mechanical properties of metals with the biological ones of ceramics.

Several solutions have been proposed in order to deposit bioactive ceramic coatings onto metallic substrates. Plasma spray [16] is the most widely applied one, followed by electro-deposition; other techniques have been proposed and studied, such as RF magnetron sputtering, electrospray, pulsed laser deposition, sol-gel deposition and biomimetic deposition, but they do not have significant clinical applications at the moment [15].

Plasma spray hydroxyapatite coatings are commercially available for both dental and orthopaedic applications. This kind of coating has shown good results in orthopaedics, while it was essentially a failure in dentistry. This may be because it has been tested prematurely, when the conditions of the deposition process were not yet optimized (high thickness, low crystallinity, poor adhesion) and on an older generation of dental implants with low primary stability. In addition, the dental implant is percutaneous, while orthopaedic prostheses are submerged, so infections are easier and they cause a rapid resorption of apatite resulting in implant mobilization. Plasma spray deposition of bioactive glass coatings is mainly at the research level, even though it has been demonstrated that their adhesion to titanium is higher than the one of hydroxyapatite and their surface can be easily modified in order to induce particular properties, such as antibacterial one [17, 18].

Electro-deposited ceramic coatings are commercially available (BONIT®, Corin).

They consist of an electrodeposited bioresorbable coating, containing Ca and P, performed at room temperature, employing electrolytes containing Ca^{2+} and PO_4^{3-} ions, on substrates previously covered with porous titanium *via* plasma spray coating (VPS) [19, 20]. In this case the bioactive coating supports only the early phases of bone integration and then gradually dissolves. The average thickness of the coating is at about 15 μm and can preserve the starting roughness of the substrate.

Deposition of nanometric calcium-phosphates crystals, on dual-acid-etched Ti surfaces, by sol gel process, has been considered and commercialized (NANOTITE®, BIOMET 3i) in order to improve osteogenesis around implant increasing without the introduction of a coating [21, 22].

The main drawback of coatings is the introduction of an interface between two different phases that can represent a weak point for the implant and lead to coating detachment.

Surface fluorine enrichment is now widely investigated [23] in order to improve osteoblast adhesion and differentiation. This kind of process does not induce a conventional bioactive behaviour (hydroxyapatite precipitation onto metal surface), but it seems that it can positively affect the cellular response both *in vitro* and *in vivo*. This process is the base of commercial Astra Tec AB Osseospeed™ [22, 24].

Aita *et al.* [25] recently propose ultraviolet functionalization of titanium in order to improve its bone integration ability. It has been shown that UV irradiation (1 - 48h) decreases carbon surface contamination and induces super-hydrophilicity. UV-treated surfaces show enhanced protein adsorption ability, they increase osteoblast adhesion and proliferation and bone formation in *in vivo* models, if compared to traditional machined titanium.

Particular production and storage technologies have recently been developed in order to preserve an active titanium surface, free from contaminants (SLAactive, Straumann®). These materials present a higher Ti and O content if compared to traditional one and a super-hydrophilic behaviour, thanks to the reduced contamination. Good fibronectin and albumin absorption has been observed on these surfaces together with osteoblast differentiation *in vitro* and bone apposition *in vivo* [21, 22].

The formation of a thick and bioactive oxide layer onto titanium surfaces can be obtained by anodic oxidation. Treatments above the breakdown limit (spark anodizing) produce a three dimensional oxide layer (up to 20μm in thickness), that is characterized by increased roughness, numerous open pores and a bioactive behaviour [26, 27]. With this technique, depending on the electrolyte employed in the process, it is possible to tailor the oxide composition and introduce elements favourable for bone integration.

In this field, Ti-Unite ™ (Nobel Biocare), can be enumerated as commercial example of anodic oxidation. It is a highly crystalline and phosphorous enriched titanium oxide and presents a microstructure without sharp features characterized by the presence of uniformly distributed open pores in the low micrometer range [22, 28].

Chemical or thermo-chemical treatments able to surface modify Ti and its alloys without the introduction of a coating and induce bioactive behaviour are widely investigated in the international scientific community, but nowadays they are still at the research level and are not yet commercially available.

Historically it was firstly proposed by Kokubo and co-workers [29, 30]. This treatment foresees sample soaking in alkaline solution (NaOH) for the formation of a sodium titanate layer, subsequently stabilized and crystallized by a thermal treatment. The resulting surface presents a micro-texture and the ability to induce hydroxyapatite precipitation, in simulated body fluid, and improvement of bone contact to metallic implant was evidenced *in vivo*. On the other hand so modified surfaces show a significant decrease in the fatigue resistance [31]. The opportunity to remove sodium from the modified layer, in order to reduce cellular stress without modifying bioactivity is now under investigation [32, 33].

Hydrogen peroxide oxidation of titanium is another widely studied technique for the obtaining of bioactive titanium surfaces. Osaka, Tsuru and co-workers [34, 35] showed that H_2O_2 treatments up to 72 hours can allow the formation of a bioactive TiO_2 layer and that Tantalum salt addition favours low temperature crystallization and enhance bioactive behaviour. It has been observed that a sever control of reaction kinetics can lead to the formation of a porous oxide layer.

Recently Spriano *et al.* developed and patented an innovative surface modification treatment aimed to improve bone integration ability of Ti-based biomaterials [36, 37, 38]. It is based on a first acid etching in hydrofluoric acid, a subsequent controlled oxidation in hydrogen peroxide and a final thermal treatment. The process is applicable to both commercially pure titanium and also to Ti-alloys (Ti6Al4V). It results in the modification of both surface topography and chemistry. Modified metal, in fact, presents a multi-scale topography that includes a micro-roughness (at about 0.2 μm) with a nano-texture (at about 10 nm) superimposed on it. These two features can be coupled with the macro-roughness of conventional blasted implants. So treated titanium is rich in hydroxyls groups, which induce hydroxyapatite precipitation and can favour protein absorption. Moreover a high density of OH groups can be successfully exploited for biological functionalizations, as described later.

BIOLOGICAL FUNCTIONALIZATIONS

Biological functionalization introduces biomolecules (such as proteins, enzymes or growth factors) onto material surface with the aim to stimulate cells for a faster and physiological tissue regeneration.

Active molecules can be coupled to artificial surfaces by simple absorption, inclusion into a resorbable carrier or covalent grafting [39]. Two main categories of biomolecule can be considered for functionalization: proteins from the extracellular matrix (ECM), or their fragments, [40] or growth factors [41]. In the first case the idea is to reproduce on the artificial substrate the physiological environment of cells in order to favour material integration. On the other hand the employment of growth factors (*e.g.* bone morphogenetic proteins) is intended for the stimulation of cell metabolism for the promotion of proliferation, differentiation and mineralization.

In order to individuate the optimal functionalization strategy it must be considered that the grafting procedures have to preserve molecular functionality (avoiding denaturation and not interfering with the active site), the distance between biomolecule and substrate must be appropriate to allow the correct biomolecular activity (it can be adjusted employing proper spacer molecules) and finally the density of active species must be large enough to induce a biological response [14].

Surface functionalization of metals in order to improve their integration properties and to accelerate the tissue regeneration process is nowadays an active research field for the international scientific community. A lot of different techniques have been investigated, a brief selection of them is reviewed below.

Silanization is one of the most widely employed technique for biomolecular grafting to inorganic surfaces. Bagno *et al.* [42] exploit this approach, with glutaraldehyde activation in order to graft two different synthetic peptides (a RGD-containing peptide and one mapped on human vitronectin - HVP) on titanium (sand blasted and oxidized). They observed that functionalized surfaces are able to promote cellular adhesion, in particular RGD one seems to be more effective, but non osteoblast specific; while HVP one is less effective, but it is osteoblast specific.

The same strategy has been employed by Nanci *et al.* [43] for the grafting of alkaline phosphatase and albumin, on cleaned and oxidized commercially pure titanium. They observed that ALP maintains its activity after covalent binding up to 7 days. Albumin shows a stable binding with antigenicity preservation too.

The main drawback of these approaches is glutaraldehyde employment, due to its toxicity. Several authors consider also the possibility to use silanization techniques free from this substance.

In particular a sintered compound of rutile (TiO_2) and perlite (volcanic silicate) with a macrostructure of interconnecting pores has been realized and surface modified with fibronectin *via* silanization (amminopropyltriethoxysilane) and a homobifunctional amino-specific polyethylene glycol linker [44]. It has been shown that the composite scaffold can support osteoblast growth and that cellular spreading can be improved by fibronectin attachment.

Also collagen has been grafted onto titanium, titanium alloy and cobalt to improve initial osteoblast adhesion and implant–tissue integration [45]. Silanization route has been compared to simple absorption. The application of various oxidation techniques to the metallic substrates resulted in surfaces with different hydroxyl group contents, which directly influenced the amount of immobilized silane. It has been observed that surface topography and surface hydroxyl group content influence the biodegradability of a chemically bound collagen layer. Collagen coating significantly enhances adhesion and proliferation of osteoblast-like cells independently from the stability of the immobilized protein layer.

Other authors suggest the employment of surface activation techniques that enable the biomolecular direct grafting, without the introduction of spacer molecules. Hayakawa *et al.* [46, 47] use surface activation with pure tresyl-chloride (without any solvent) in order to graft fibronectin and collagen to titanium. Cellular experiments underline that collagen immobilized titanium is effective for the enhancement of gingival fibroblast initial adhesion and spreading, while fibronectin immobilized one not.

Plasma activation of the surface has also been explored for biological functionalization of metals. Morra *et al.* [48] grafted collagen onto titanium surface *via* plasma deposition of a hydrocarbon thin film followed by acrylic acid grafting and biomolecule coupling. Biocompatibility of modified implants has been verified both *in vitro* and *in vivo*.

The same authors [49] consider the opportunity to covalently graft hyaluronan (HA) to titanium surface, aminated by means of different approaches (polyethileneinmine adsorption and deposition from allylamine plasma), with the aim of reducing unspecific cellular adhesion.

Plasma activation processes have been effectively applied also for the grafting of bone morphogenetic protein 4 (BMP-4) to Ti6Al4V surface [50]. Plasma polymerization of allyl amine has been used to provide functional amino groups on the metallic surface. It has been shown that the amount of protein weakly and strongly bound to the surface can be controlled by the choice of surface treatment and immobilization chemistry. BMP-4 has been grafted to Ti6Al4V with a high density of amino groups using carbodiimide immobilization scheme. This solution provides a surface able to induce significant osteoblastic activity.

The opportunity to couple the typical osteoblast stimulating properties of bone morphogenetic proteins (in particular bone morphogenetic protein 2 - BMP-2) and the antibacterial one of chitosan has been explored [51]. Chitosan has been covalently grafted to metallic surface *et al.* glutaraldehyde cross-linking and BMP-2 has been subsequently adsorbed on the chitosan layer. With this technique BMP-2 are not firmly bonded to the surface, but are slowly released in solution over time. It has been observed that so modified titanium alloy is able to promote the differentiation of bone marrow-derived stem cells into osteoblasts.

It has been suggested that a strong and irreversible immobilization of biomolecules on the surface of artificial materials can diminish molecular biological functionalities. A destructive hydrophobic interaction between the material surface and the biomolecule may be cause of this inactivation. Alternatively dissociation of the biomolecule from the surface may be necessary for signal transduction. In that context, recently Kashiwagi *et al.* [52] created artificial proteins that contained three copies of a Ti-binding motif, and fused them to the N-terminal of BMP-2. The engineered BMP-2 showed reversible binding to Ti surfaces and induced BMP signalling activity.

In order to obtain an antibacterial surface, also the functionalization of Ti with polyelectrolyte multilayers (PEMs) of hyaluronic acid (HA) and chitosan (CH), has been considered [53]. It has been observed that PEMs coated Ti present a significant antibacterial behaviour towards S. Aureus but the adhesion of osteoblasts cell is reduced if compared to pure Ti. A further functionalization of the PEMs-Ti with RGD-peptide *via* 1-ethyl-3-(3-dimethylaminopropyl) carbodiimide and N-hydroxysuccinimide chemistry can significantly improve osteoblast adhesion and function without reducing the antibacterial power. Only immobilizing RGD-peptide on the outermost layer of the PEMs it is possible to reach this effect, in fact its introduction within the PEMs does not expose its active functions to cell inhibiting its effectiveness.

Another technique that could be successfully employed in order to realize thin biomolecular films on implant surfaces without altering molecular characteristics is electrospray. De Jonge *et al.* [54] applied this method to produce an enzyme (alkaline phosphatase) coating onto titanium surface. They demonstrate that electrospray enable the maintenance of ALP activity and increase material bioactivity in cell culture medium. Finally it has been observed [55] that the simple absorption of the alkaline phosphatase enzyme on plasma sprayed titanium implants could improve bone production in rabbit model.

Also Spriano *et al.* worked on the grafting of alkaline phosphatase enzyme (ALP) on titanium surfaces [36-38]. ALP has been successfully anchored to the surface of Ti6Al4V alloy modified to be bioactive, according to the previously described process. Tresyl chloride activation has been employed in order to obtain molecular direct grafting. It has been observed that the enzyme maintain it activity after grafting. Its presence does not inhibit the starting inorganic bioactivity of the modified metal; moreover it enhances and quickens hydroxyapatite precipitation kinetics. Preliminary results of cellular tests suggest a good proliferation and differentiation of osteoblast on modified metal.

This kind of surface modifications constitutes a large part of the research topics for the international scientific community but, at the moment it is mainly at the research level and no commercial solutions can be enumerated in this field. In fact biologically functionalized materials are critical for sterilization and storage processes and can require more complex procedure for certification and commercialization then traditional material, since they contain biological substances.

ACKNOWLEDGEMENTS

Regione Piemonte is kindly acknowledged for its funding to Biosurf project, regarding functionalisation of Titanium alloys and polymers.

CONFLICT OF INTEREST

The author(s) confirm that this chapter content has no conflict of interest.

REFERENCES

[1] Kasemo B. Biological surface science. Surf Sci 2002; 500: 656-677

[2] Brunette DM. Principles of cell behaviour on titanium surfaces and their application to implanted devices. In: Brunette DM, Tengvall P, Textor M, Thomsen P. Titanium in Medicine. Berlin Heidelberg New York, Springler-Verlag, 2001; pp.561-585

[3] Tambasco de Oliveira P, Nanci A. Nanotexturing of titanium-based surfaces upregulates expression of bone sialoprotein and osteopontin by cultured osteogenic cells. Biomaterials 2004; 25: 403-413

[4] Variola F, Yi J-H, Richert L, Wuest JD, Rosei F, Nanci A. Taioloring the surface properties of Ti6Al4V by controlled chemical oxidation. Biomaterials 2008; 29:1285-1298

[5] Vetrone F, Variola F, Tambasco de Oliveira P, Zalal S F, Yi J-H, Sam J, Bombonato-prado KF, Sarkissian A, Perepichka DF, Wuest JD, Rosei F, Nanci A. Nanoscale Oxidative patterning of metallic surfaces to modulate cell activity and fate. Nano Lett 2009; 9: 657-665

[6] Hori N, Iwasa F, Ueno T, Takeuchi K, Tsukimura N, Yamada M, Hattori M, Yamamoto A, Ogawa T. Selective affinity of biomimetic micro-nano-hybrid structured TiO_2 overcomes the biological dilemma of osteoblasts. Dent Mater 2010; 26: 275-287

[7] Amigó V, Salvador M., Romero F, Solves C, Moreno J F. Microstructural evolution of Ti–6Al–4V during the sintering of microspheres of Ti for orthopedic implants. J Mat Proc Tech 2003; 141: 117-122

[8] www.smith-nephew.com
[9] Dall'Oca C, Zambito A, Aldegheri R. 100 furlong C.S.F. cup (Cancellous Screws Fixation) preliminary results analisys. GIOT 2004;30:73-79
[10] Mayman DJ, González Della Valle A, Lambert E, Anderson J, Wright T, Nestor B, Sculco T P, Salvati EA.Late MD. Fiber Metal Shedding of the First and Second-Generation Harris Galante Acetabular Component. A Report of 5 Cases J Arthroplasty 2007; 22: 624-629
[11] Biswal S, Brighton RW. Results of unicompartmental knee arthroplasty with cemented, fixed-bearing prosthesis using minimally invasive surgery J Arthroplasty 2010; 25: 721-727
[12] www.zimmer.com
[13] Costa Santos E, Shiomi M. Osakada K. Laoui T. Rapid manufacturing of metal components by laser forming. Int J Mach Tools Man 2006; 46: 1459-1468
[14] Xiao S-J, Kenausis G, Textor M. Biochemical modification of titanium surfaces. In: Brunette DM, Tengvall P, Textor M, Thomsen P, Titanium in Medicine, Berlin Heidelberg New York, Springler-Verlag, 2001; pp.417-453
[15] De Jonge LT, Leeuwenburgh SCG, Wolke JGC, Jansen JA. Organic-inorganic surface modifications for titanium implant surfaces. Pharm Res 2008; 25: 2357-2369
[16] Rizzi G, Scrivani A, Fini M, Giardino R. Biomedical coatings to improve the tissue-biomaterial interface. Int J Artif Organs 2004; 27: 649-657
[17] Miola M, Ferraris S, Di Nunzio S, Robotti P, Bianchi G, Fucale G, Maina G, Cannas M, Gatti, S, Masse A, Vitale-Brovarone C, Vernè E. Surface silver-doping of biocompatible glasses to induce antibacterial properties. Part II: plasma sprayed glass coatings J Mater Sci. Mater Med 2009; 20: 741-749
[18] Verné E, Ferraris S, Miola M, Fucale G, Maina G, Robotti P, Bianchi G, Martinasso G, Canuto RA, Vitale Brovarone C. Synthesis and characterization of a bioactive and antibacterial glass-ceramic (II): Plasma Spray coatings on metallic substrates. Adv Appl Cer 2008; 107: 1743-6753
[19] Zeggel P. Bioactive calcium phosphate coatings for dental implants. International Magazine of Oral Implantology 2000; 1: 52-57
[20] Szmukler-Moncler S, Perrin D, Ahossi V, Pointaire Evaluation of BONIT®, a fully resorbable CaP coating obtained by electrochemical deposition, after 6 weeks of healing: a pilot study in the pig maxilla. Key Eng Mater 2001; 192-195: 395-398
[21] www.straumann.com
[22] Junker R, Dimakis A, Thoneick M, Jansen JA. Effects of implant surface coatings and composition on bone integration: a systematic review. Clin Oral Impl Res 2009; 20 (Suppl. 4): 185-206
[23] Guo J., Padilla R.J., . Ambrose W, De Kok I.J., Cooper L.F. The effect of hydrofluoric acid treatment of TiO_2 grit blasted titanium implants on adherent osteoblast gene expression *in vitro* and *in vivo* Biomaterials 2007; 28: 5418-5425
[24] www.astratechdental.com
[25] Aita H, Hori N, Takeuchi M, Yamada M, Anpo M, Ogawa T The effect of ultraviolet functionalization of titanium on integration with bone. Biomaterials 2009; 30: 1015-1025
[26] Sandrini E, Chiesa R, Rondelli G, Santin M, Cigada A. A novel biomimetic treatment for a n improved osteointegration of titanium. J Appl Biomat Biomech 2003; 1: 33-42
[27] Giordano C, Sandrini E, Del Curto B, Signirelli E, Rondelli G, Di Silvio L. Titanium for osteointegration: Comparison between a novel biomimetic treatment and commercially exploited surfaces. J Appl Biomat Biomech 2004; 2: 35-44

[28] www.nobelbiocare.com

[29] Nishiguchi S, Nakamura T, Kobayashi M, Kim HM, Miyaji F. Kokubo T. The effect of heat treatment on bone-bonding ability of alkali-treated titanium. Biomaterials. 1999; 20: 491-500

[30] Kokubo T, Matsushita T, Takadama H. Titania-based bioactive materials. J Eur Ceram Soc 2007; 27: 1553-1558

[31] Spriano S, Bronzoni M, Vernè E, Maina G, Bergo V, Windler M. Characterization of surface modified Ti-6Al-7Nb alloy. J Mat Sci: Mat Med. 2005; 16: 301-312

[32] Fawzy AS, Amer MA. An *in vitro* and *in vivo* evaluation of bioactive titanium implants following sodium removal treatment. Dent Mater 2009; 25: 48-57

[33] Pattanayak DK., Kawai T, Matsushita T, Takadama H, Nakamura T, Kokubo T. Effect of HCl concentrations on apatite-forming ability of NaOH-HCl- and heat treatment titanium metal. J Mater Sci Mater Med 2009; 20: 2401-2411

[34] Wang XX, Hayakawa S, Tsuru K, Osaka A. Bioactive titania gel layers formed by chemical treatment of Ti substrate with H_2O_2/HCl solution. Biomaterials 2002; 23: 1353-1357

[35] Kaneko S, Tsuru K, Hayakawa S, Takemoto S, Ohtsuki C, Ozaki T, Inoue H, Osaka A. *In vivo* evaluation of bone-bonding of titanium metal chemically treated with hydrogen peroxide solution containing tantalum chloride. Biomaterials 2001; 22: 875-881

[36] Spriano S, Vernè E, Ferraris S. Multifunctional titanium surfaces for bone integration. EP2214732, 2008 (TO2007A000719, 2007)

[37] Spriano S, Ferraris S, Bianchi CL, Cassinelli C, Torricelli P, Fini M, Rimondini L, Giardino R. Bioactive titanium surfaces, in: Titanium alloys: preparation, properties and applications, New York, Pedro N. Sanchez, Nova Science Publishers, Inc. 2010; pp.269-294

[38] Spriano S, Ferraris S, Bianchi CL, Cassinelli C, Torricelli P, Fini M, Rimondini L, Giardino R. Synergistic effect between inorganic and biological bioactivity of modified titanium surfaces. Mater Sci Res J 2009; 3: 203-222

[39] Puleo DA, Nanci A. Understanding and controlling the bone-implant interface. Biomaterials 1999; 20: 2311-2321

[40] Morra M. Biochemical modification of titanium surfaces: peptides and ECM proteins. Eur Cell Mater 2006; 1: 1-5

[41] Lee S-H, Shin H. Matrices and scaffolds for delivery of bioactive molecules in bone and tissue engineering. Adv Drug Del Rev 2007, 59: 339-359

[42] Bagno A, Piovan A, Dettin M, Chiarion A, Brun P, Gambaretto R, Fontana G, Di Bello C, Palù G, Castagliuolo I. Human hosteoblas-like cell adhesion on titanium substrates covalently functionalized with synthetic peptides. Bone 2007; 40: 693-699

[43] Nanci A, Wuest JD, Peru L, Sharma V, Zalzal S, McKee M.D. Chemical modification of titanium surfaces for covalent attachment of biological molecules. J Biomed Mater Res 1998; 40: 324-335

[44] Rüger MWM, Rago C, Steffens GCM, Hollander DA, Paar O, Maier HR, Jahnen-Dechent W, Bosserhoff AK, Herli HJ. *In vitro* behaviour of a porous TiO_2/perlite composite and its surface modification with fibronectin. Biomaterials 2005; 26: 2813-2826

[45] Müller R, Abke J, Schnell E, Scharnweber D, Kujat R, Englert C, Taheri D, Nerlich M, Angele P. Influence of surface pretreatment of titanium- and cobalt-based biomaterials on covalent immobilization of fibrillar collagen. Biomaterials 2006; 27: 4059–4068

[46] Hayakawa T, Yoshinari M, Nemoto K. Direct attachment of fibronectin to tresyl chloride-activated titanium. J Biomed Mater Res A 2003; 67: 684-688

[47] Hayakawa T, Nagai M, Yoshinari M, Makimura M, Nemoto K. Cell-adhesive protein immobilization using tresyl chloride-activation technique for the enhancement of initial cell attachment. J Oral Tissue Engin 2004; 2: 14-24

[48] Morra M, Cassinelli C, Cascando G, Cahalan P, Cahalan L, Fini M, Giardino R. Surface engineering of titanium by collagen immobilization. Surface characterization and *in vitro* and *in vivo* studies. Biomaterials 2003; 24: 4639-4654

[49] Morra M, Cassinelli C, Carpi A, Giardino R, Fini M Effects of molecular weight and surface functionalization on surface composition and cell adhesion to hyaluronan coated titanium. Biomed Pharmacother 2006; 60: 365-369

[50] Puleo DA, Kissling RA, Sheu MS. A technique to immobilize bioactive proteins, including bone morphogenetic protein-4(BMP-4), on titanium alloy. Biometerials 2002; 23: 2079-2087

[51] Lim TY, Wang W, Shi Z, Pho K, Neoh KG. Human bone marrow-derived mesenchymal stem cells and osteoblast differentiation on titanium with surface-grafted chitosan and immobilized bone morphogenetic protein. J Mater Sci Mater Med 2009; 20: 1-10

[52] Kashiwagi K, Tsuji T, Shiba K. Directional BMP-2 functionalization of titanium surfaces Biomaterials 2009 30: 1166-1175

[53] Chua PH, Neoh KG., Kang ET, Wang W. Surface functionalization of titanium with hyaluronic acid/chitosan polyelectrolyte multilayers and RGD for promoting osteoblast functions and inhibiting bacterial adhesion. Biomaterials 2008; 29: 1412-1421

[54] De Jonge LT, Leeuwenburg SCG, van den Beuken JJJP, Wolke JGC, Jansen JA. Electrosprayed enzyme coatings as bioinspired alternatives to bioceramic coatings for orthopaedic and dental implants. Adv Func Mater 2009; 79: 1-8

[55] Piattelli A, Scarano A, Corigliano M, Piattelli M. Effects of alkaline phosphatase on bone healing around plasma-sprayed titanium implants: a pilot study in rabbits. Biomaterials 1996; 17: 1443-1449

CHAPTER 11

Tailoring of Tissue-Surface Interaction in Blood Contacting Materials

Roman Major and Boguslaw Major*

Institute of Metallurgy and Materials Science Polish Academy of Sciences, 30-059 Krakow, 25 Reymonta St. Poland

Abstract: The cell–material interaction plays an important role in the biomaterial design. Biomaterials, such as diamond-like carbon (DLC), titanium (Ti), and stoichiometric titanium nitride (TiN) as well as titanium carbo-nitrade (Ti(C, N)), seem to be good candidates for future blood-contact applications. These materials were deposited as thin films by the hybrid pulsed laser deposition (PLD) technique to examine the influence of such surfaces on cell behavior. The cell-material reactions were examined in static conditions with *Dictyostelium discoideum* cells and then subjected to a dynamical test to observe the cell detachment kinetics. For a given cell, detachment occurs for critical stress values caused by the applied hydrodynamic pressure above a threshold which depends on cell size and physicochemical properties of the substrate. Tests revealed differences in behavior with respect to the applied coating material. The strongest cell-biomaterial interaction was observed for the carbon-based materials compared to the titanium and titanium nitride. The research activity was performed on fabrication and diagnostics of materials characterized by reduction or erasing of thrombogenicity. Development of surfaces that both favor endothelial cell monolayer reconstruction and prevent platelets aggregation and binding was under examination. Research study was performed on soft polyurethane surfaces (PU) by application of thin inorganic coatings. Experiments have been performed on surface functionalization with application porous materials produced by electrospinning. Research on scaffolds for precursors of tissue analogs and cell migration channels was presented.

Keywords: Film deposition, microstructure, cell adhesion, biocompatibily, tissue analog, migration channels.

1. INTRODUCTION

Over the past 30 years, much effort in the research of biomaterials has been directed toward the development of materials that are inert to human blood and do

***Address correspondence to Boguslaw Major:** Institute of Metallurgy and Materials Science Polish Academy of Sciences, 30-059 Krakow, 25 Reymonta St. Poland; Tel: +48 12 637-42-00 or +48 12 295 28 00; Fax: +48 12 637 21 92 or +48 12 295 28 04; E-mail: nmmajor@imim-pan.krakow.pl

Lia Rimondini, Claudia Bianchi and Enrica Vernè (Eds)

not react with platelets or coagulation factors [1]. A number of often conflicting approaches have been developed. There is still no consensus about the advantageous hydrophilic or hydrophobic behaviour of a surface. This lack of agreement is largely due to both the incomplete understanding of the biological pathways leading to material failure and the designers' inability to fully evaluate the blood material failure and the blood material responses. In the development of the materials with a lower thrombogenicity, researchers have primarily focused their efforts on modifying the surfaces. This approach is reasonable, since it is only the surface chemistry of the material that should dictate its biological responses and the mechanical characteristics of the material are primarily dictated by the bulk chemistry. Despite the successes in reducing the protein and the cellular deposits on some materials, a truly non-thrombogenic surface does not exist.

Experimental and theoretical work has been dedicated to the cell adhesion from three very different perspectives [2-8]: (i) From the biological point of view, the understanding of the molecular mechanisms is of central importance, which means the understanding of when a cell adheres, rolls, or slides on passive or reactive substrates. Many functions performed by living cells depend on these properties. (ii) From the physical-chemical point of view, bio-adhesion involves cells, a solid substrate, and a liquid medium. The relevant properties of a micro organism are the hydrophobic and the charge of the cell surface, cell size, and possession. A lot of experimental and theoretical work has been dedicated to cell adhesion [3]. (iii) From the physical point of view, even the passive (non-reactive) response of the cells to an external force offers new phenomena that find no equivalent in more traditional materials. This is the case in the bio-adhesion phenomena, where the discrete nature of the contact regions, the weak or non-covalent bonding of the cell to the substrate, and the multi-component variety of the cytoplasm membrane or of its extra cellular matrix (ECM) are the origin of contact forces. Measuring the strength of these bonds is a major challenge in cellular biology, since it allows for the identification of the different factors at work in the adhesion phenomena.

The topography of surfaces is an important property for the contact of a component to its surrounding. The tribological behaviour, the decorative appearance and even the interaction to biological matter (cells, bacteria, and tissue) is strongly influenced by the texture in macro-, micro- and nano-scale. Shape deviations, waviness, grooves, *etc.* are some of the macro- and micro-scaled topography features, while nano-scale features include the roughness due to the microstructure and due to the atomic arrangement (lattice). It's quite clear, that mainly the nano-scale features gain more and more interest for functional surfaces in sensors, electronics, medical and implant technology and mechanical engineering.

Nanostructures between atomic and micrometer size (generally between 1-100 nm) can be achieved by a multitude of techniques, developed since the early 1990s. Different approaches are used both in laboratory and in industrial scale to generate nanostructured surfaces [9-14].

Biomaterials such as: titanium (Ti), and stoichiometric titanium nitride (TiN) as well as titanium carbo-nitrade (Ti(C, N)) and carbon basis DLC (diamond like carbon) and PLC (polymer like carbon), seem to be good candidates for future blood-contact applications [1].

These materials were under examination in this work and they were deposited as thin films by the hybrid technique to examine the influence of the deposited material on the cell behaviour [10-14]. Much attention was put to the diagnostics of the deposited coating in order to establish the optimal technological parameters from the physicochemical parameters of the biomaterials [14, 15]. The kinetics energy of the evaporated particles was controlled by an application of a variation of different reactive and non-reactive atmospheres during the deposition [10-13]. The conventional TEM and the high resolution transmission electron microscopy (HREM) was used to reveal the structure dependence on a specific atmosphere in the reactive chamber [15]. The biomaterial examinations were performed in static conditions with fibroblast cells line and then subjected to the dynamical test to observe the cell detachment kinetics [14, 15]. For a given cell, the detachment occurs for the critical stress values, caused by the applied hydrodynamic pressure above a threshold, which depends on the cell size and the physicochemical properties of the substrate, but it is not affected by the de-polymerization of the actin and tubulin cytoskeleton. The tests revealed differences in behaviour, depending on the applied coating material [14, 15]. The strongest cell-biomaterial interaction was observed for the carbon-based materials, as compared to titanium and titanium nitride. Examinations on haemocompatibily were performed by Sanak *et al.* [16]. The analysis was focused on medical materials and aim to detect adverse interaction between artificial surface and blood, which can activate or destruct blood components.

2. PREPARATION OF THE SURFACES AND DIAGNOSTICS

A development of the surface modification with the use of thin film materials was performed by hybrid PLD and mostly by magnetron sputtering in direct current (DC), unbalanced mode. Titanium (medical grade titanium for titanium and titanium nitride coatings) and carbon targets (for Ti(C, N) and DLC) were used to deposit about 10-20 nm thick films on polyurethane substrates at room temperature in argon atmosphere (Ti) or nitrogen-argon atmosphere (TiN). To ensure the homogenous film thickness over the entire coated surfaces, the substrates were rotated during the deposition at the speed of 5.4 cm s^{-1} through the plasma plumes. A detailed description of the deposition arrangement is given elsewhere [10-15].

2.1. Thermal Effect

Deposition process could have the significant influence on the material properties [17]. It is a crucial aspect in the case of biomaterials designing procedure. The temperature and the atmosphere seem to be the basic parameters which determine the proper structure. The temperature effect was analyzed as the Heat Flow in function of the applied temperature (Fig. **1**). The performed investigations involved a small glass transformation on the level of 110 deg. All the visible effects probably come from the substrate. The following materials which were applied as the surface modification exhibited various reactions:

- Ti, Ti(C, N), TiOx, Si+DLC caused an additional endothermic effect in the lower temperature area,

- TiN, DLC, Ti+DLC influenced the additional endothermic effect but the change of the temperature of the glass transformation is visible as well.

2.2. Surface Analysis

The surface analysis was performed using a scanning electron microscopy (E-SEM) and a scanning acoustic microscopy (SAM) [14].

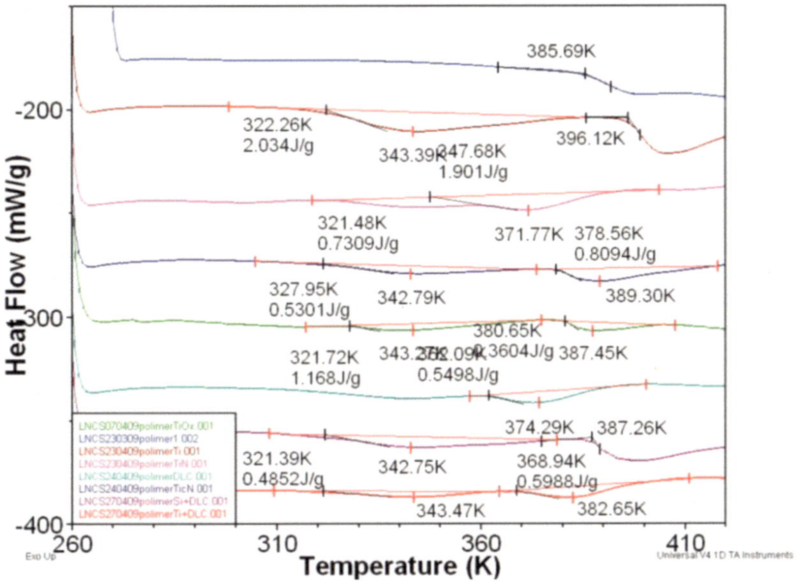

Figure 1: Heat Flow in function of the applied temperature in calorimetric measurements for coatings deposited on PU: (from bottom) Ti+DLC, Si+DLC, Ti(C, N), DLC, TiN, Ti, polymerX, TiOx.

The proper surface structure and the lack of cracks is strongly correlated with the appropriate mechanism of the thin film growth [13]. The coating deposited under the late mechanism of the thin film growth is brittle and exhibits the cracks on its surface [11].

2.2.1. Scanning Electron Microscopy E-SEM

In the E-SEM examinations of Ti, TiN, Ti(C, N) (Figs. **2-3**), it was stated that these films have the proper thickness. The visible scratches come from the substrate. The titanium coating was too thick. It caused numerous cracks, which could influence the platelet activation process. Thus, the thrombus formation could be blurred by the effect coming from the scratches. An optimization of the deposition parameters leads to the limiting of the thickness of the film to the primary mechanism of the deposition [13].

2.2.2. Scanning Acoustic Microscopy SAM

The Scanning Acoustic Microscopy (SAM) is a technique which enables to investigate whether the material has been uniformly or non-uniformly deposited [14,

18-21]. The only problem which was observed was associated with Ti. The other coating materials were deposited properly. In the case of Ti(C, N), a visible effect of the residual stress concentration was seen by the isolines appearance (Fig.**4**).

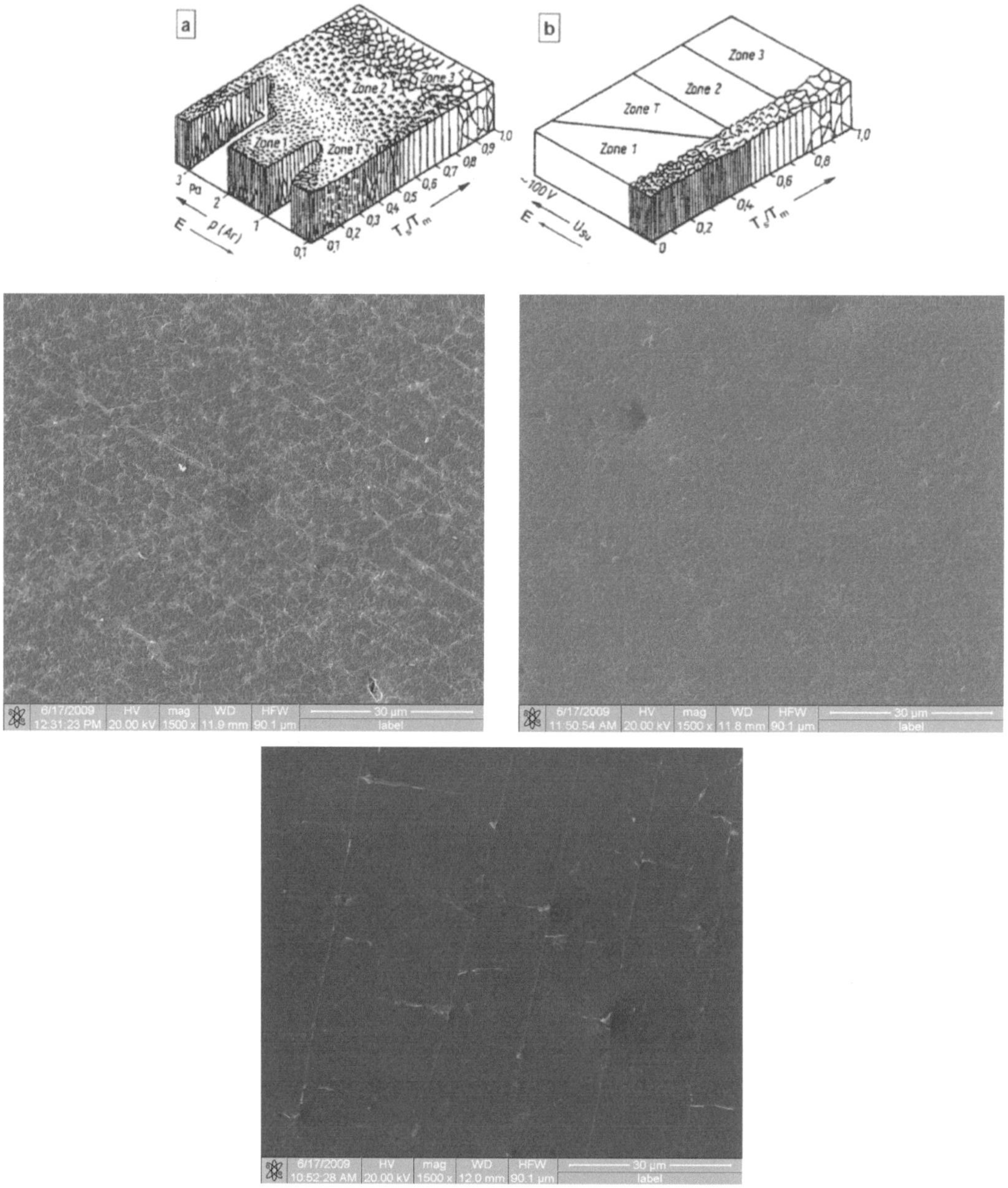

Figure 2: E-SEM (environmental-scanning electron microscopy) examinations of the deposited coatings: Ti, TiN, Ti(C, N) with the Thornton diagram [17].

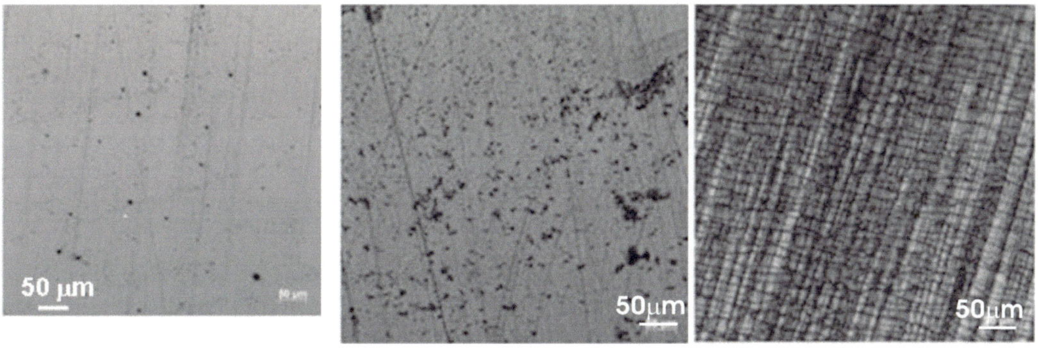

Figure 3: CLSM (confocal laser scanning microscopy) examinations of the deposited coatings with the Thornton diagram [17]; defects in cracking forms can be seen; more cracks in Ti coating; least defects in Ti(C, N), TiN.

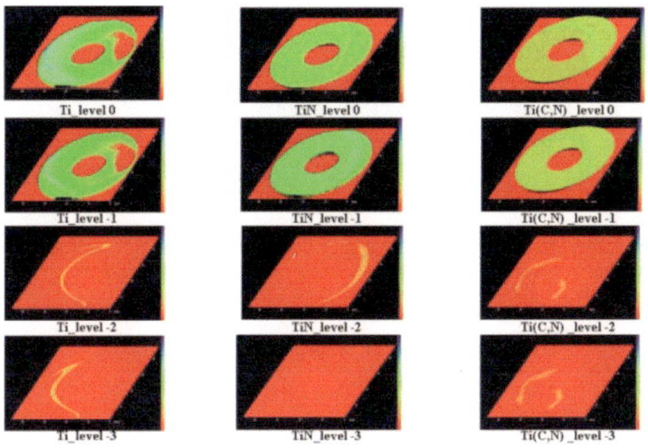

Figure 4: Tomography analysis of Ti, TiN and Ti(C, N) coatings using SAM (scanning acoustic microscopy) on different thickness levels, starting from the surface to bottom of coating at thickness step of about 7 nm; thickness of coating about 20 nm; area of adhesion (light contrast); pictures of 0, 1/3, 2/3 and 1 thickness of coating.

2.2.3. Microstructure Analysis

The intended target of the work was to elaborate materials which would not change the properties of the PU substrate but also effectively separate it from the tissue.

The transmission electron microscopy TEM and the high resolution electron microscopy HREM was used to study the microstructure of the thin foils prepared from the film cross-section, using the focused ion beam (FIB) technique [12, 14].

The examinations focused on the study of the adhesion of the coating to the substrate and thus, the quality of the bio-material.

The microstructure of Ti presents columnar character which proves the late mechanism of the thin film growth (Fig. **5**). It could result in the stiffness increscent of the PU-coating system which automatically would reject such material from the clinical use. Cross-section in the bright field mode of the TiN and Ti(C, N) microstructure, deposited on PU, were similar and exhibit elastic properties of the coating (Figs. **6, 7**). This was possible due to the proper mechanism of the film growth leading to close an amorphous structure.

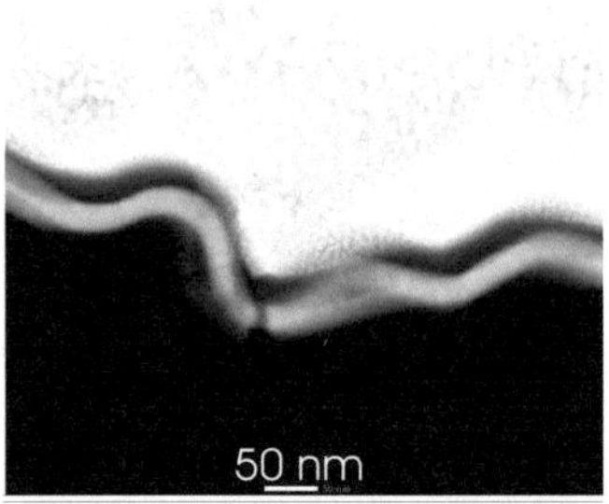

Figure 5: TEM (transmission electron microscopy) microstructure of layer cross-section of Ti deposited on PU.

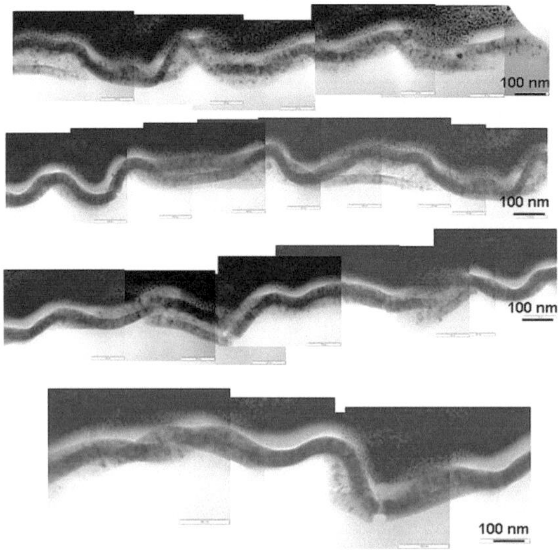

Figure 6: TEM (transmission electron microscopy) bright field of TiN/PU coating, cross-section, sets of pictures along the coating.

The STEM picture (Fig. **7**) revealed the so called Z-contrast (phase contrast). These pictures were obtained with the use of the detector HAADSTEM - *high angle angular dark field STEM.* The more scattered are the electrons, the brighter the picture. The chemical line analysis was carried out by means of the EDS technique (Energy Dispersive Spectroscopy) across the thickness (Fig. **8**).

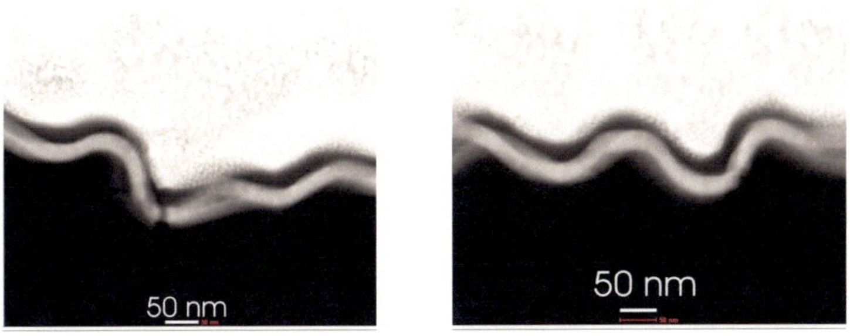

Figure 7: STEM (scanning transmission electron microscope) microstructure of TiN/PU; cross-section.

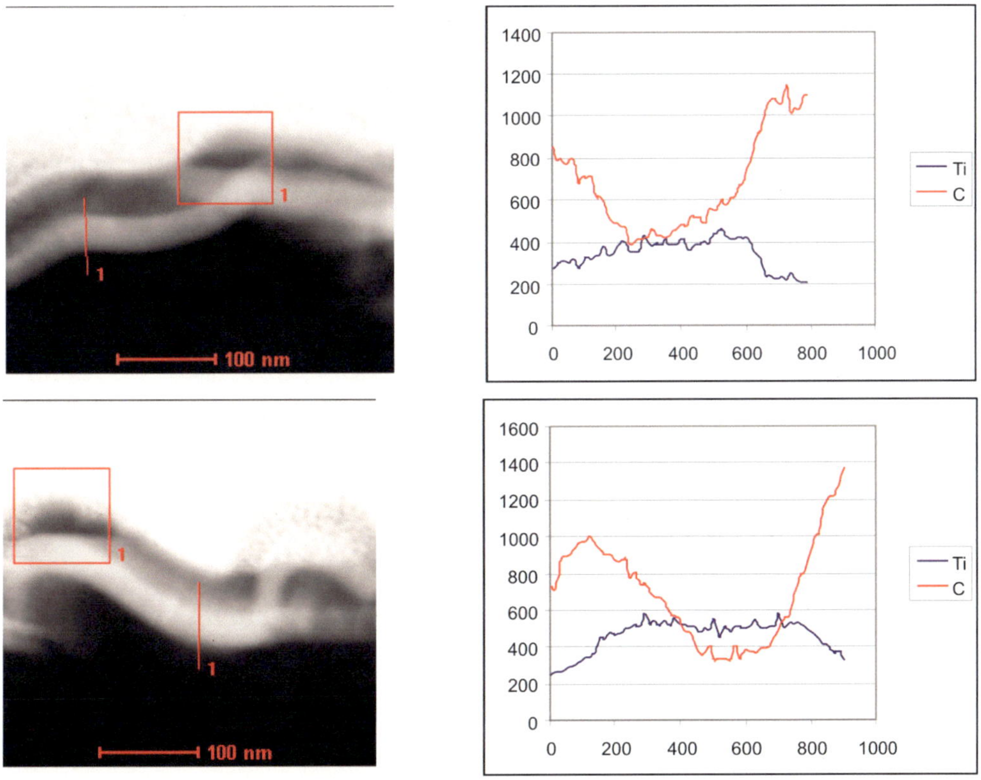

Figure 8: TEM/EDS (TEM/energy dispersive spectrometry) line chemical analysis of the Ti(C, N)/PU coating; two lines of analysis a) and b).

3. Preparation of the Cell Culture

3.1. Cell Culture

For the study of the cell–material interactions, the mouse fibroblast cells line L929 from the ATCC collections were used. The cells were cultured in "Medium 199" supplemented with 10% FCS (Fetal Calf Serum) and with antibiotics streptomycine and penicyline (Sigma) at 37°C with 5% CO_2 in a humidified incubator (Haereus). The media were changed every 2 days. The cells were cultured in 25 cm^2 culture flasks. The media were changed every 2 days. When the L929 cells in the culture had grown till to the 75% of confluence, the cells were ready to use for the cytotoxicity tests [22].

3.2. Seeding Procedure

Before the tests, the samples were subjected to the sterilization process using ethylene oxide. Afterwards, they were ventilated for 2 weeks to remove the rest of ethylene oxide. Before seeding, the cells were harvested from the culture dishes using 0,05% Trypsin/EDTA solutions and re-suspended in the culture medium (Medium 199 supplemented with 10% FCS, 1% L-glutamine) to the concentration of 5×10^{6} cells/ml.

The tested materials were cut into squares and placed in 24 wall culture dishes. To ensure a complete contact between the samples and the wall, the samples were pressed with plastic rings. Then the pellet of the re-suspended cells were plated on the material surface with the density of the cells of around 10 000 cells/well. After 30 min. of incubations, the culture media were completed to a total volume of about 3ml/well. An empty well with fibroblast cells was used for control purposes. The materials with the cells and the control were cultured at 37°C with 5% CO_2 in a complete medium (Medium 199 supplemented with 10% FCS, 1% L- glutamine) for a total of 24h.

3.3. Fluorescent Microscopy

In order to observe the changes in the cell viability and morphology after the cell material interactions (attachment), we applied a fluorescent microscopy. The viability of the cells was analysed with the use of the Live/Dead dye Fluoresceine Diacetate (FDA) and the Propidium Iodide (PI). The viable cells stained green with FDA and the non-viable cells stained red with PI.

4. Cell-Material Interaction

4.1. Cell-Material Interaction in Hydrodynamic Conditions (Radial Detachment Test)

4.1.1. Bio Kinetics - Cell Detachment

Dictyostelium discoideum was used as a model organism to study cell detachment kinetics [2-8]. In the case of *D. discoideum* a unicellular eukaryote also able to complete a pluricellular development cycle, membrane adhesiveness is tightly related to lifestyle. In the vegetative phase, the small (8-μm diameter) ameba feeds upon bacteria and yeast by phagocytosis. Plasma membrane adhesion is therefore directly related to the phagocytic properties of the cell. In the development phase, cell–cell adhesion is more important, and specific contact proteins are expressed [2]. Axenic strains showing an enhanced fluid-phase endocytosis have been obtained and are able to grow in suspension in nutritive medium [3]. The sequencing of *D. discoideum* 34-Mbp genome and Expressed Sequenced Tags from various developmental stages are currently in a completion phase [4]. The available molecular tools make *D. discoideum* a good model organism to study phagocytosis and cellular motility using null or overexpression mutants [3]. Several mutant studies suggested the existence of three kinds of adhesion protein. A glucose-selective lectin, mediating attachment of bacteria to the cell, was revealed by the sensitivity of mutants to 50 mM glucose. A hydrophilic receptor called Phg1, consisting of a 9-helix transmembrane protein was recently identified. As for now, the molecular nature of its ligand is unknown, but it is present in nutritive media, because Phg1 null mutants are adhesion defective in this medium. The third class of the adhesion molecule is sensitive to surface hydrophobicity, as exemplified by the HV32 mutant, which does not bind to polystyrene in the presence of glucose.

For a given cell, detachment occurs for values of the applied hydrodynamic pressure above the threshold. In this purpose, a radial flow detachment essay was used. Cells were re-suspended in the Sörensen buffer and spread at a density of 300 cells/mm^2 on the substrate surface and subsequently submitted to the shear flow under the radial geometry. The temperature of the liquid was kept constant at 21°C. A radial hydrodynamic flow was generated between the stainless disc and sample on which cells adhered (Fig. **9**) [5-8]. The stainless steel disk with a hole in the center was placed above the sample with the 100 μm distance. The action of the shear flow depends on the distance between the disc and the sample and the level of the upper tank influence on the shear stress between the cells and the material surface. The layout of the investigation is presented in Fig. **9**.

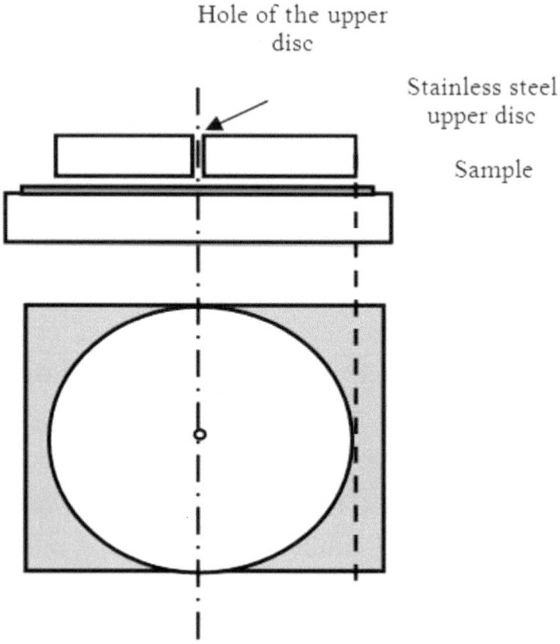

Figure 9: Cell detachment assay equipment layout for radial shear flow test.

After the test, samples were given under the fluorescence microscopy observation. In use of Image-ProPlus software, the cells were counted and on basis of the received values critical shear stress was calculated in use of the following formula (1):

$$\sigma_{50\%} = \frac{3D\eta}{\pi r_{50\%} e^2} \quad \textbf{(1)}$$

$r_{50\%}$ is the distance to the center where 50 % of cells are detached

D - rate during the test, η - buffer viscosity, e – space between disk and plate

In order to receive stagnation point, the considered time of the test changed from 5 to 10 minutes. Fig.**10** presents hydrodynamic stress as a function of detached cells for titanium substrate in 5 and 10 minutes of the test durable. Time increase caused that more cells detached. It was observed that shear stress should be applied for longer times to determine detachment kinetics. Titanium nitride coating was exposing for the three deferent test durable, 5, 10 minutes. The longer the time is the more cells are detached. Comparing the results between the substrate and the coating, 20% more cells remained on the uncoated substrate (Fig.**11**).

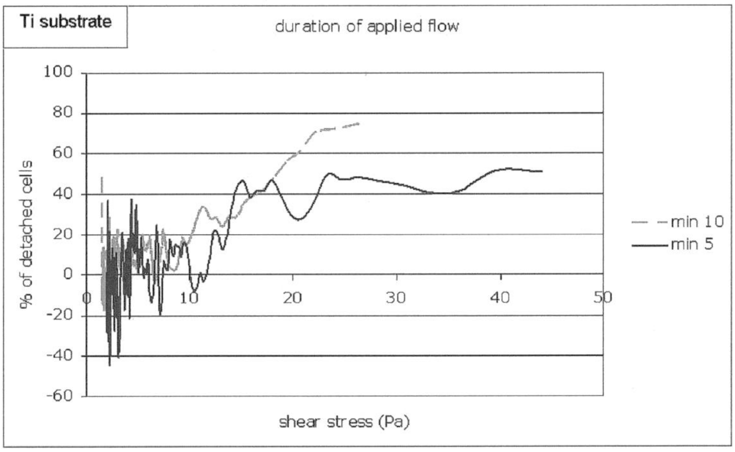

Figure 10: Cell detachment kinetics for Ti substrate in two exposition times.

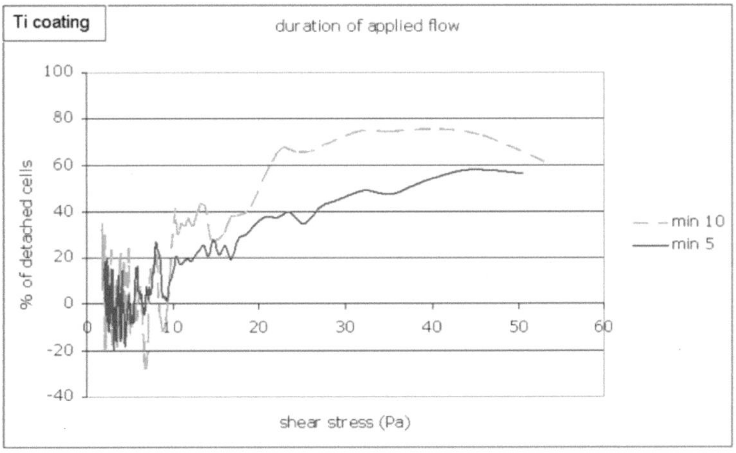

Figure 11: Cell detachment kinetics for Ti coating in two exposition times.

In conclusion to this result, shear stress should be applied for longer times to determine detachment kinetics. Fig.**12** presents detachment kinetics for titanium carbonitride with low carbon content. It was observed no clear difference between 5 and 10 min, thus steady state is almost reached. The results associated with titanium carbonitride with high carbon content are presented in Fig.**13**. Shear stress should be applied for longer times to determine detachment kinetics.

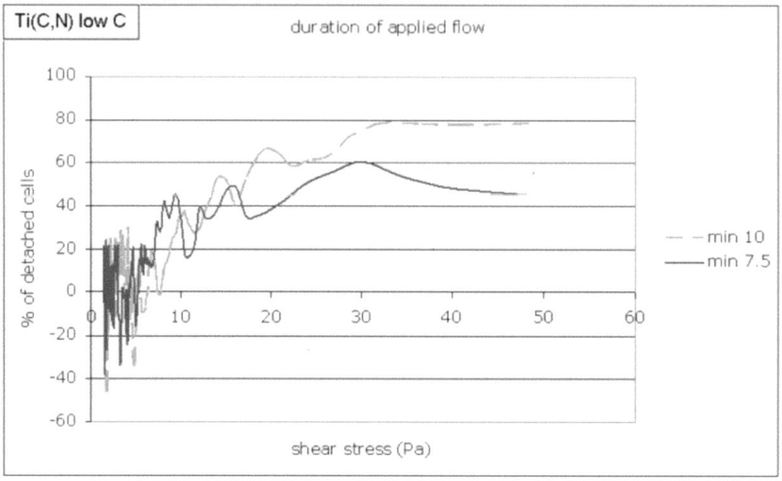

Figure 12: Cell detachment kinetics for Ti(C, N) with low carbon content coating in two exposition times

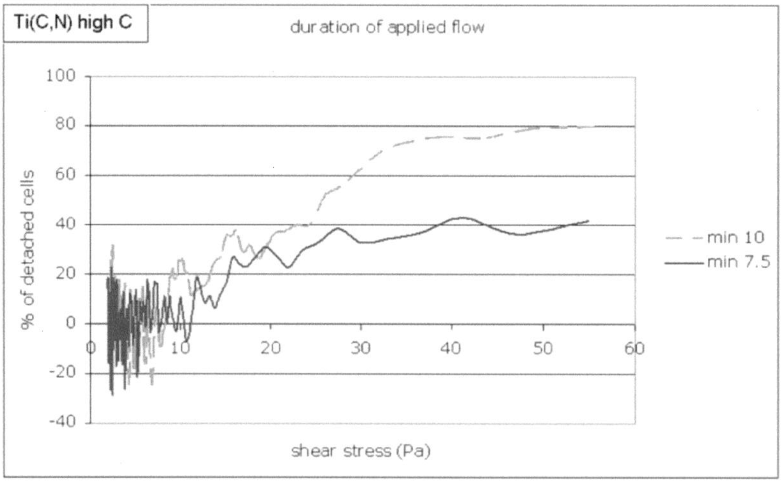

Figure 13: Cell detachment kinetics for Ti(C, N) with high carbon content coating in two exposition times.

In summary, the total values obtained in these series of experiments of the shear stress were collected and compared. Figs. **14**, **15** present a total classification of the shear stress which was detected between the coating and the biological cell.

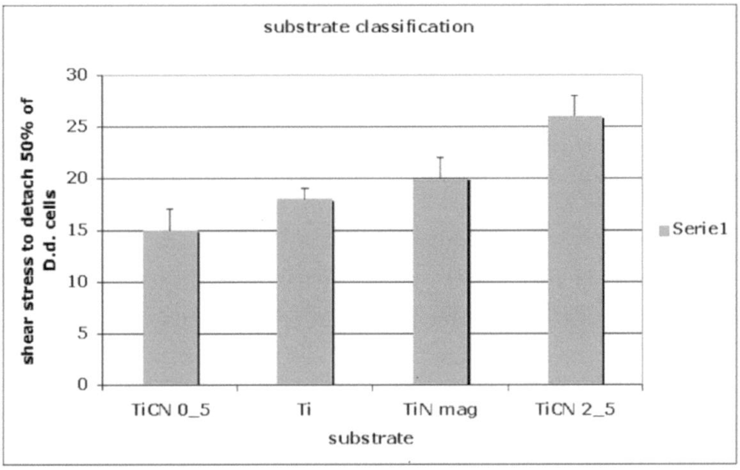

Figure 14: Shear stress classification for: Ti, TiN, Ti(C, N) low C/indication on figure TiCN 0_5/, Ti(C, N) high C/indication on figure TiCN 2_5/(series 1).

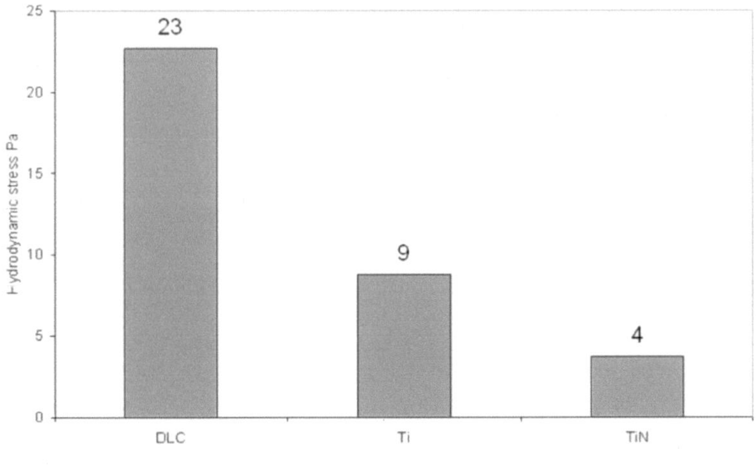

Figure 15: Shear stress classification for: DLC, Ti, TiN (mean values of a few series).

The percentage of detached cells was recalculated as a function of shear stress, since not so many cells were detached, the initial density from the distribution of remaining cells was taken. The center of the distribution was always between frames 190 and 205. The value that makes the distribution symmetrical was put. Sometimes, the distribution was not symmetrical. The center and the percentage of detached cells was shown as a function of shear stress. Data obtained for 10 min flow to data obtained after shorter flow duration were compared. In Fig. **14** there are shear stress ordered as a function of material: The less adhesive one is TiCN 0_5; the order is as follow: TiCN 0_5 < Ti = TiN magnetron < TiCN 2_5.

The radial shear flow test was used to determine the efficiency and the kinetics of the cell detachment from the plain, as well as the modified surfaces as a function of the applied shear stress [2, 3, 15]. The investigated materials were considered in the particular conditions described above. Depending on the material, a different exposition time was necessary. The results of the cell efficiency and cell rate detachment under the applied shear stress are presented in Figs. **16**, **17**. TiN

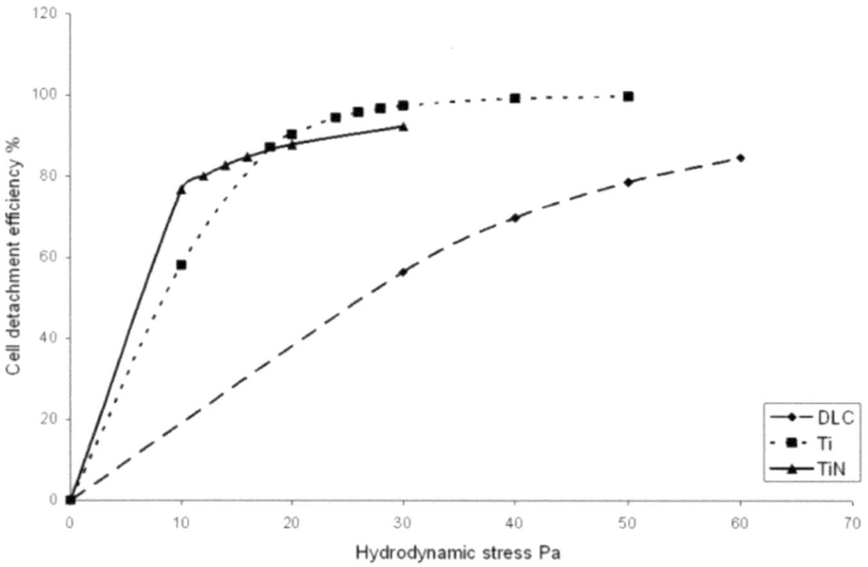

Figure 16: Cell efficiency detachment under the applied shear stress for: DLC, Ti, TiN.

exhibits a low cell material interaction. The detachment rate for titanium carbonitride with a low carbon content was over estimation, thus it was not presented in the diagram. The values of the threshold stress confirm the

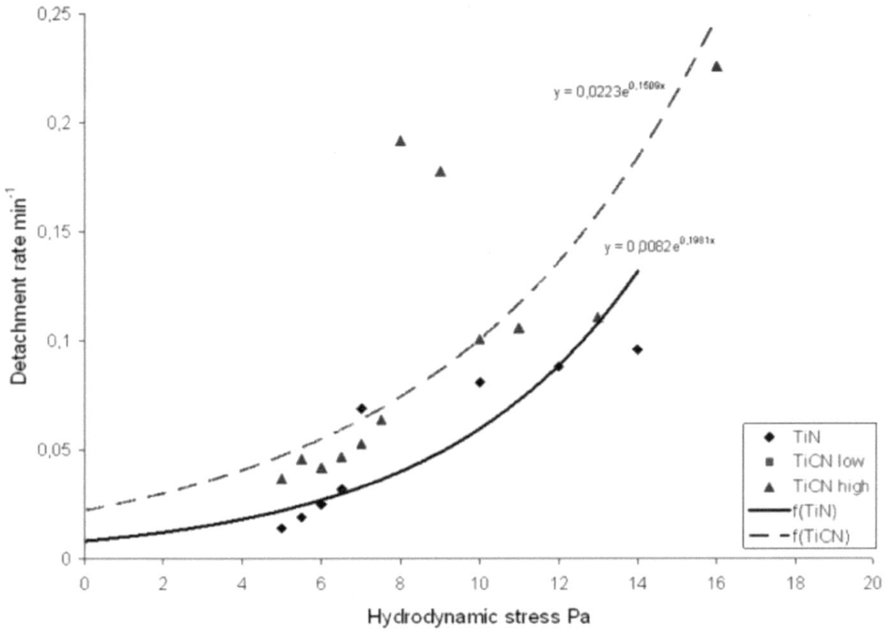

Figure 17: Cell rate detachment under the applied shear stress for TiN, Ti(C, N) low C, Ti(C, N) high C.

observations of the cell distribution. The lowest value 3Pa typical for TiN, the highest for TiCxNy(highC) with a high carbon content 7Pa. The function extrapolation for the 0 Pa hydrodynamic stress revealed a spontaneous cell detachment rate; 0.02 [1/min^{-1}] for Ti(C, N) and 0.008 [1/min^{-1}] for TiN [15]. The cell detachment test gives an overview of the cell behaviour and the shear stress which could appear between the cell and the biomaterial. The biomaterial design requires biophysical tests, as well as biocompatible tests which would illustrate the biological response of the biomaterial to the biological cell. In this case, the tests are performed in static conditions on fibroblast.

4.2. Cell-Material Interaction in Static Conditions (Biocompatibility Testing)

The presented sub-chapter describes the biocompatible tests performed on the investigated materials in static conditions. Two important aspects of the biomaterial design were considered - the stoichiometry and the phase composition modification with an introduction of carbon. To measure the biocompatibility, fibroblasts were applied.

The stoichiometric titanium nitride is known as a biocompatible material. The lack of stoichiometry could result in a negative influence on the surrounding tissue. In order to examine this problem, both groups, the biocompatible and the non-biocompatible titanium nitride, were studied. Fig. **18** presents the result of the biocompatibility tests of the example material. For this purpose titanium nitride (TiN) was considered.

The experiments revealed that the cells exhibit a high biocompatibility on the surface of the fully stoichiometric TiN (Fig. **18a**). There was a visible growth in the cells and a spread across the surface and no influence on the cell death was detected. The cells revealed individual focal adhesions.

The **c**ell function is also related to the cell shape [1]. The cells grow when spread, die when fully retracted, and differentiate into capillary tubes if maintained at a moderate degree of extension, while also forming cell-cell contacts (*e.g.*, on thin micro patterned lines).

On the non-stoichiometric titanium nitride, the cells demonstrated round shapes, typical for the cell death (Fig. **18b**). Over 50% of the deposited cells were directed to the death path. The remaining live cells did not spread on the surface.

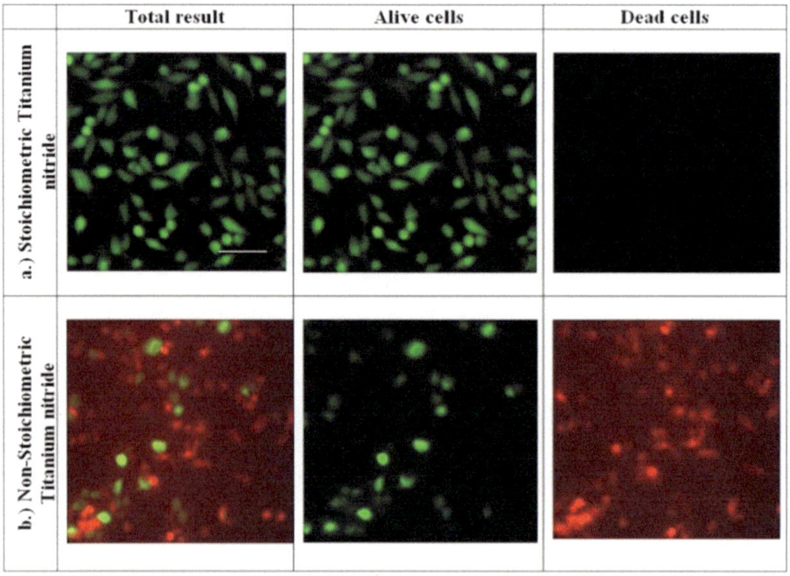

Figure 18: Material biocompatibility (fibroblasts)

a) Cells reaction on stoichiometric titanium nitride;

b) Cells reaction on non-stoichiometric titanium nitride.

As presented in Fig.**18**, the stoichiometric titanium nitride exhibits good biocompatible properties. When designing a new biomaterial, a phase composition modification by means of carbon was considered. A comparison of the biocompatible results for the stoichiometric titanium nitride, as well as for the titanium carbonitride with a low and high carbon content is presented in Fig.**19**.

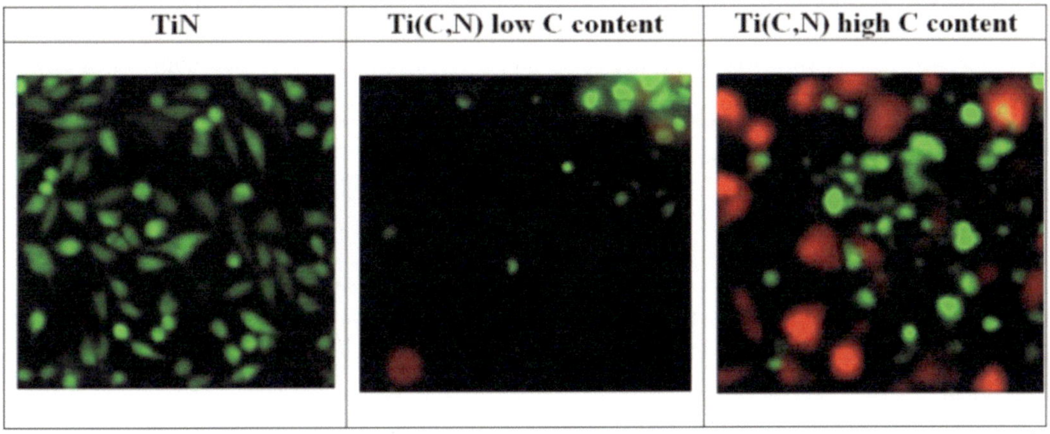

Figure 19: Biocompatibility of TiN, Ti(C, N) low C, Ti(C, N) high C (fibroblast)

5. BLOOD-MATERIAL INTERACTION

Dynamic test differs from static interactions by an introduction of shear forces activating blood cells [22-24]. The Impact-R [25] is a novel device for testing platelet function under close to physiological conditions. The device tests platelet adhesion and aggregation in an anti-coagulated whole blood (citrate buffer tubes) under arterial flow conditions [22-24]. For the performed experiments, the test was use to analyze the blood-material interaction for the modified polyurethane (PU). The PU surface was coated with diamond-like carbon (DLC), titanium (Ti), titanium nitride (TiN), titanium carbonitride (TiCN), titanium oxide (TiO), silicon doped DLC (SiDLC) and titanium doped DLC (TiDLC). The platelet- platelet aggregations caused by the activated platelets, was the main parameter which was taken under consideration. Results are presented in Fig. **20**.

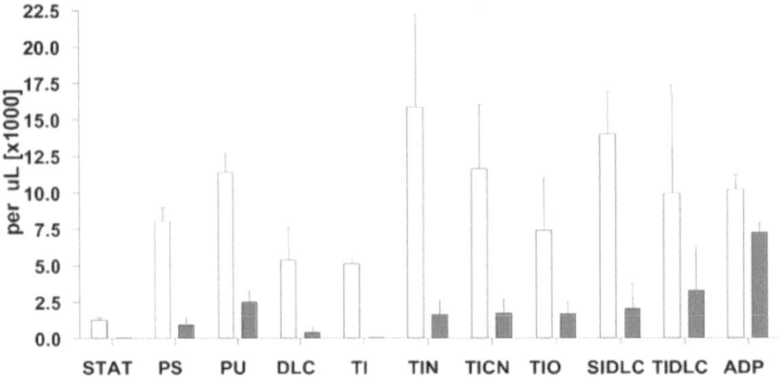

Figure 20: Platelet aggregates following the shear stress. White columns – small aggregates of a few platelets, grey columns – bigger aggregates > 2-3 platelets. Controls are: static conditions and ADP activated blood. Materials abbreviations are explained in the Methods section.

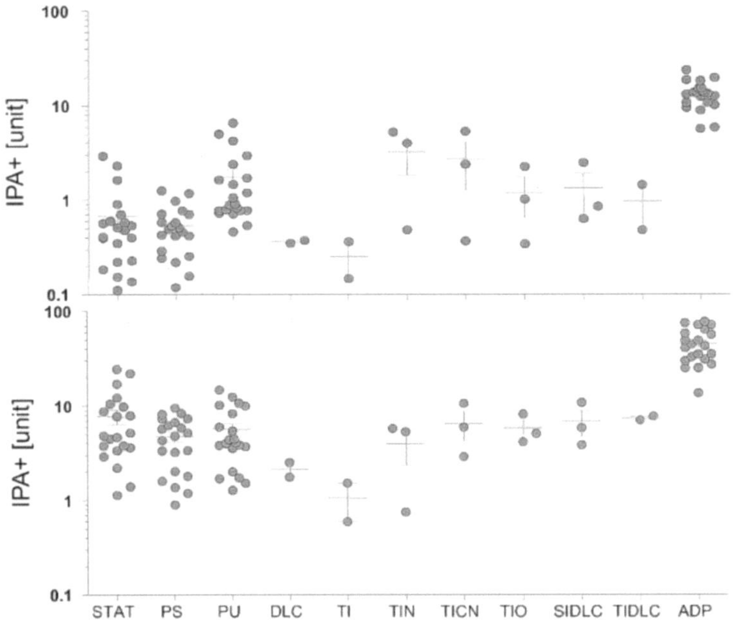

Figure 21: Platelet activation markers following the shear stress. Upper panel – integrated fluorescence of anti-selectin-P antibody. Lower panel – integrated fluorescence of PAC-1 antibody detecting activated glycoprotein IIbIIIa. Controls are: static conditions and ADP activated blood.

Platelets aggregates were distinguished as big and small ones. Big platelets aggregates were found as more than two platelets associated together and small ones as two platelets associated together. On the surface of the platelets there are the appropriate receptors, which take participation in the adhesion and aggregation. There are mainly selectins, responsible for the initial stages of the adhesion process. Selectin P which is a typical for the blood platelet, it is accumulated in the platelet granules and transported to the membrane after platelet activation [16]. The results of the platelets activation mercers are presented in Fig. **21**.

Materials abbreviations are explained in the Methods section.

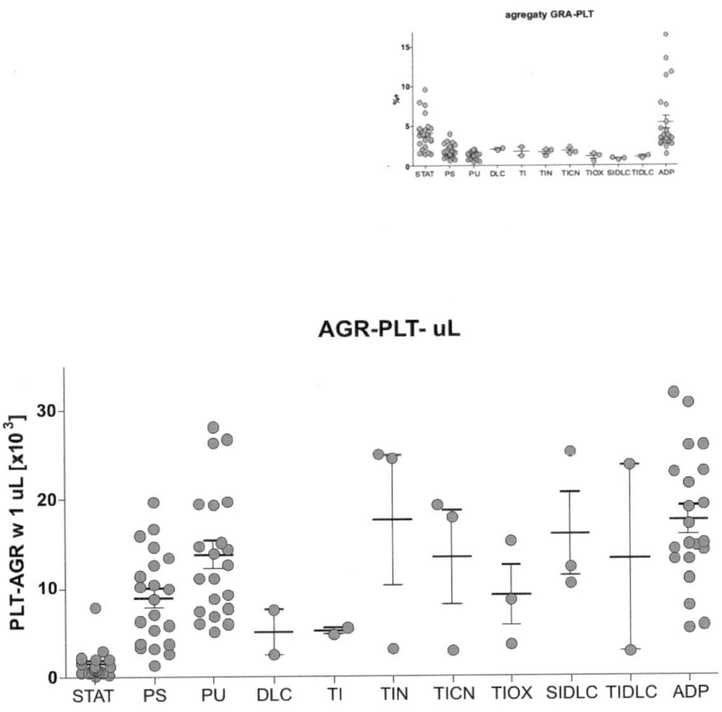

Figure 22: Granulocyte-platelet aggregates following the shear stress, as a percentage of total granulocytes. Left panel: monocyte-platelet aggregates. Right panel: neutrophil-platelet aggregates. Controls are: static conditions and ADP activated blood. Materials abbreviations are explained in the Methods section.

The other aspect in the blood- material interaction analysis, which is normally considered for the material selection deals with a “plate satelityzm” *(latine).* It consists of platelet adhesion to leukocytes neutrophils, lymphocytes, monocytes, eosinophils, and basophils. It can cause platelet pseudo-trombocytopenia, the false solutions underestimated the number of platelets [26-34]. The results of the blood-material interaction, considering to the mentioned parameters are presented in Fig. **22**.

The next aspect which was under examination was dealing with the platelet derived microparticles. Microparticles (MPs) are small membrane vesicles that are released by activated or apoptotic cells. They play a significant role in the blood clotting cascade activation. An influence of micro particles on the clots formation on the artificial surfaces is presented in Fig. **23**.

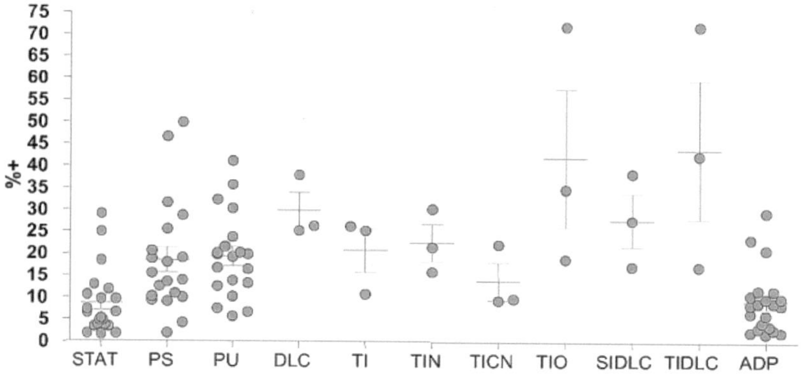

Figure 23: Microparticles thrombotic activity of blood plasma following the shear stress. Controls are: static conditions and ADP activated blood. Materials abbreviations are explained in the Methods section.

For the final material selection, the ranking plots were prepared (Fig. **24**). In general, the lower the values of aggregates or amount of activated platelets with the integrin family activated receptors and the higher amount of platelets, the better the material. Based on the ranking of PU coatings, despite of insufficient number of tested plates, examinations revealed that TiN, TiCN and possibly SiDLC have the best hemo-biocompatibility function character and should be selected for more extensive examination. There are some remaining problems of the blood-material interaction associated with the quality of the investigated material influence. Any scratches can disrupt platelets mechanically. The further parameter which could influence the results and should be concerned in discussion is standardization. In general, the better standardization for relative decrease in the platelet count the more detailed the results. Finally, the reliable method for quantification of surface bound aggregates need to be developed. It is a general problem which exists due to not fully defined blood-material interaction tests [24, 25]. Nowadays, the description of the biocompatible material is the sentence: "A hemocompatibile material must not adversely interact with any blood component". International Organization for Standardization (ISO) developed a guidance on testing medical materials that have contact with circulating blood. The guidelines does not provide the exact test methods or evaluation criteria. Instead, a list of various applicable references is suggested [22].

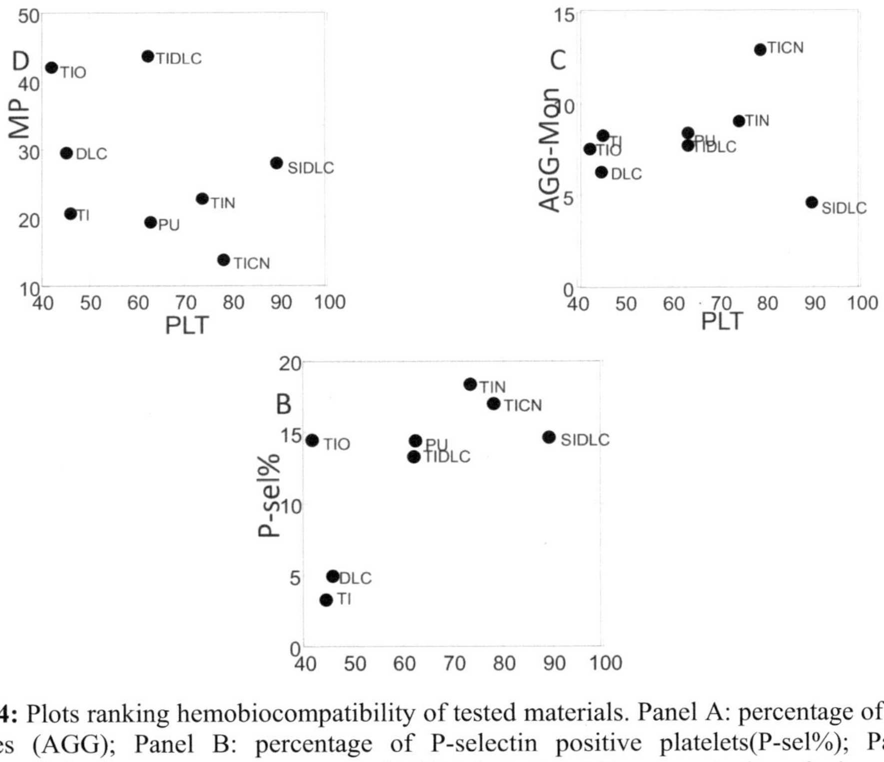

Figure 24: Plots ranking hemobiocompatibility of tested materials. Panel A: percentage of platelet aggregates (AGG); Panel B: percentage of P-selectin positive platelets(P-sel%); Panel C: percentage of platelet-monocyte aggregates (AGG-Mon); Panel D: concentration of microparticles thrombogenic activity (MP). All values are plotted against the same ordinal of percentage of platelets remained after the shear stress (PLT). Materials abbreviations are explained in the Methods section.

6. CONCLUDING REMARKS

Development of novel, low thrombogenic materials is of central importance. The presented contribution focused on the processing of the coating with an improved biocompatibility. The microstructure examinations by means of highly specialized methods tested both the quality of the surface morphology from the point of view of its homogeneity and the coating adhesion to the polymer substrate on the cross section.

The intended target of the work was to elaborate materials which would not change the properties of the PU substrate but also effectively separate it from the tissue. The polymer exhibits the change in stiffness in the long term clinical application thus the need of the surface modification. Optimisation of deposition parameter has been established from the point of view of chemical composition and surface morphology as well as satisfactory adhesion. Coatings have been fabricated on thermo-sensitive polymer substrate by application of hybrid method. It was necessary to investigate the thermal effect which could influence in the physical-mechanical parameters changes. It was found that polyurethane is thermally stable. The first effect was observed over 100°C. One can conclude that the process of layer deposition does not influence the polymer degradation. As the argument against is that the additional endothermic effect appeared and the temperature ranges of the glass transformation. This aspect is still under consideration.

A measurement of the strength of the bonds between the biomaterials and the cells is a major challenge in cellular biology, since it allows for the identification of the different species in the adhesion phenomena, thus the cell-material interaction was under examination by means of the dynamic test of the cell detachment. The biocompatibility was studied in the static test. As model cells, fibroblasts were taken under consideration. It was observed that it is highly important to focus on the proper phase composition and the stoichiometry of the material coating. The biocompatibility tests revealed the negative effect of the loss of stoichiometry, which was shown especially for titanium nitride.

The cell- material interaction was investigated using cell detachment essay. Cell detachment test gives the overview about the cell behavior and shear stress which could appear between the cell and the biomaterial. The lowest cell material interaction was found for stoichiometric titanium nitride. There is still opened question which is pointed out by the investigators over the world. Should we deal with low cell material interaction or rather consider their strong influence on the substrate? When the problem concerns the thrombus effects elimination, the answer is simple: no interaction. When one would like to elaborate more sophisticated materials then the other biological phenomena should be analyzed and the answer is not so simple. Biomaterial design requires the biophysical tests as well as biocompatible tests which would illustrate the biological response of the biomaterial into the biological cell. In this case the tests are performed in static conditions. Stoichiometric titanium nitride is known as biocompatible material. A lack of stoichiometry could result in a negative influence on the surrounding tissue which was proved by the fluorescence observations. Cell – material interaction in static conditions revealed stoichiometric titanium nitride as biocompatible material and this behaviour was damage by non-stoichiometry.

Using a simplified experimental setup, based on inexpensive commercial cone-and-plate analyzer, it was managed to reproduce the same events, which seems limiting usage of artificial materials in contact with blood. The advantage of the proposed method is low volume of donor's blood required for testing and higher throughput of the method, enabling parallel testing of materials. This method inherits the same drawbacks, as other experimental setups for hemobiocompatibility testing. Due to a high physiological variation in blood morphology, mainly platelet count, credible results can be obtained only using 8-10 replicates of the test, as was demonstrated by other authors [24, 25]. Moreover, the test is also very sensitive to the imperfect smoothness of the surface. This limitation can be evaded using a high quality test plates. Ranking of PU coatings, despite insufficient number of tested plates, revealed that TiN, Ti(C, N) and possibly Si+DLC have the best hemobiocompatibility profiles and should be selected for more extensive tests.

Even with continued advances in endovascular technology, surgical replacement of the disfunctional vessel by autologous grafts or synthetic prostheses remains the key therapy. Novel vascular grafts which provide higher patience rates, are therefore urgently required.

In recent years, new techniques have been developed to apply micro- and nanoarchitecture designs mimicking the three-dimensional morphology of the vessel wall, including bioactive surface structuring and *in vitro* manufacturing of cell formations and tissue engineered grafts.

Using these new available technologies, the research activity is performed in two main directions by the authors of this contribution:

- Nanoscaled functionalization of porous scaffold biomaterials for cardiovascular prostheses

- Nanostructural materials for implants and cardiovascular biomedical devices

The main objective of the research activity is the modification of polymer implants and tubes (cannula, conduits) on the external and internal surface for improving biocompatibility of tissue contact surfaces, especially in contact with blood. New, tissue contacting materials with main focus on blood contact aspects, for external and internal blood circulation support devices are planned to be elaborated. The surfaces of the implant devices will be covered with non-cytotoxic and anti-thrombogenic coatings biomimeticly behaved to its surrounding (*e.g.* vessel surface), which means to influence the endothelial cells to cover. Depending on the topography of the vessel inner-simulating surface and the chemical nature of the cells nonthrombogenic bio-surface, basing on cells covered polymer would be formed. It is planned to apply carbon, titanium and biopolymer based materials. Therefore, it will be used plasma pre-treatment by vacuum coating techniques. State-of-the-art devices are polymer tubes for forced external and internal blood circulation internally modified with heparin bond, preserving against the clot formation in blood contact. By lowering of the thrombogenicity, the increasing risk of bleeding is the main disadvantage of these systems. Thus, one concept is to improve the surface-blood cell interaction by using biomimetic designs. In the case of implants in contact to muscle tissue the prevention of fibrosis is main topic.

Problems with current prosthesis systems (*e.g.* bypasses) are connected to intimal hyperplasia (occlusion of vessels by thick fibroblast deposits) after surgical implantation. Generally, the biomaterial acceptance and integration is dependent and controllable by the chemical composition and topography of the surface as well as by the mechanical properties of the bulk. Although cells are on the micrometer scale, surface modifications on the nanometer scale are of high interest, because all cell functions (protein adsorption, cell adhesion/orientation/migration) are based on interactions on the nanometer range (*e.g.* focal adhesions). For artificial small vascular prosthesis, the most advanced and versatile technique for fabrication of the scaffold (support for tissue) is the *electrospinning process*, while large diameter prostheses used in clinics are knitted, woven or expanded today. Electrospinning allows an optimized control of microstructure (flat scaffolds with inherent 3D structures and macroscopic 3D shapes (tubes)) and fiber diameters in the micron range and below, all mimicking the structural properties of the biological extracellular matrix (Fig. **25**).

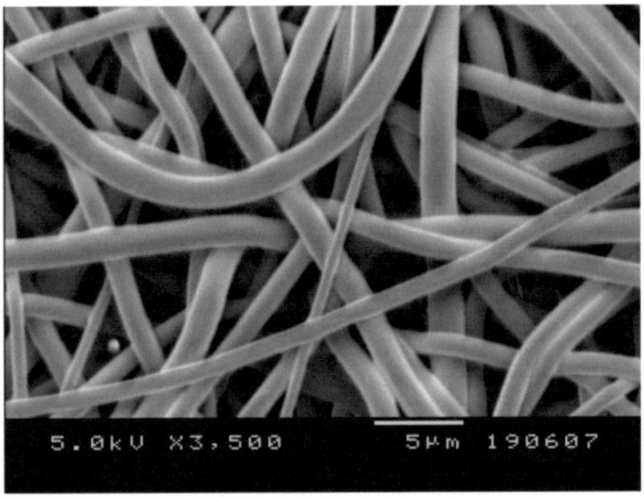

Figure 25: Polymer scaffold for precursors of tissue analogs fabricated using electrospinning.

The excellent mechanical properties of electrospun fibres impose micro-stresses necessary for maintaining the right cell phenotype. Control of pore size and distribution in the micron range allocate for space for extracellular matrix and cell metabolite production. Technological problems in electrospinning are connected to production reliability and performance, which must be optimized by jet control for future large scale industrial application. However, main task will be the reduction of thrombosis and intimal hyperplasia during clinical use of vascular grafts (which is currently the predominant restriction in clinical application of all vascular scaffolds). Attempts are both manufacturing improvements in materials (non-degradable, bioresorbable), and fiber alignment of scaffold fabrics (mechanical properties/stress) on the micro- and surface treatment on the nano-scale.

Surface treatments (physical and chemical surface modification) are effectively done by plasma techniques to shape the surface for cell-specifically focal adhesion spots of several nanometers diameter. The formation of this biomimetic nano-topography and surface chemistry is planned to be performed on the whole scaffold surface (including the open porosity), which is a novelty in plasma technology. We plan plasma etching, plasma deposition of organic and inorganic films (*e.g.* Si/Ti doped diamond like carbon) and plasma grafting with functional molecules (*e.g.* amine, carboxyl groups), respectively. Both, full chemical/topographical homogeneity and gradients are expected for targeted protein as well as endothelial and smooth muscle cell adhesion (colonization) in and outside the tube.

One of the important examination directions within precursors of tissue analogs is cell migration. A goal of such examination is targeted on contribution of an artificial surface to adhesion and direction of cell migration. A possible solution of this problem could be found in preparation of migration channels (Figs. **26-27**). They have been produced using the laser ablation technology. For an initial model analysis of cell migration, *Dictyostelium discoideum* cells were applied. Experiments were performed in static conditions in a vacuum flow chamber.

Figure 26: Migration channels fabricated using laser ablation in TiN deposited on PU substrate.

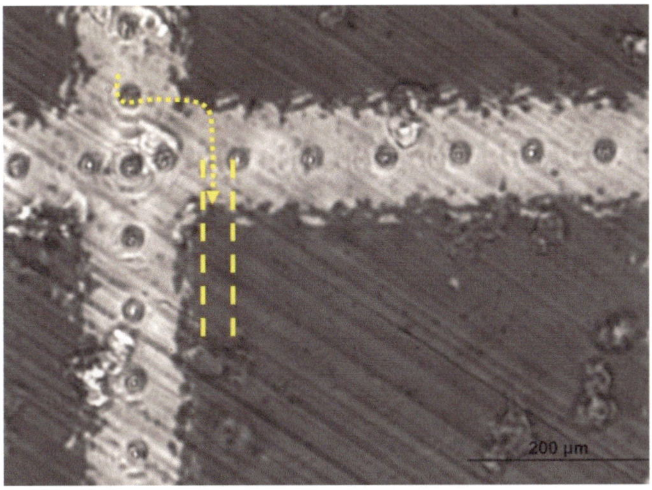

Figure 27: Dictyostelium discoideum cells migration in channels; scheme of migration.

Both strategies of material development require advanced material and biomedical *in vitro* and *in vivo* testing. Materials testing will include both the development of quick, reliable, simple test methods for future manufacturing control and the scientific description of topographical and chemical changes on the fibre and fabrics surfaces. Biomedical *in vitro* testing methods will focus both on short- to long-term protein adsorption and cell adhesion (single cell phenotypes and whole blood). *In vivo* testing is planned to be performed in animal studies in the final stage.

ACKNOWLEDGEMENTS

The research has been supported by the project CardioBioMat Era-Net-MNT/15/2009 “Nanostructural materials for implants and cardiovascular biomedical devices” and the Polish-French exchange project: Polonium “New gradient materials fabricated by laser method for blood contact application”.

Many thanks are given to Prof. H. Schima (Med. Univ. Vienna, Austria), Prof. F. Bruckert (Grenoble Institute of Technology Minatec, France), Prof. M. Sanak (Jagiellonian Medical College, Poland), Dr. J. M. Lackner (Joanneum Research, Forschungsges. mbH. Laser Center Leoben, Austria),

Dr. P. Wilczek (Foundation for Cardiac Surgery Development Zabrze, Poland) for cooperation and stimulating discussions.

CONFLICT OF INTEREST

The author(s) confirm that this chapter content has no conflict of interest.

REFERENCES

[1] Ratner B D, Hoffman. S, Schoen F J, Lemons JE. *Biomaterials Science*, Copyright Elsevier Inc. 2004.

[2] Bozzaro S, Ponte E. Cell adhesion in the life cycle of Dictyostelium, Experientia 1995; 51: 1175-81.

[3] Ashworth JM, Watts DJ. Metabolism of the cellular slime mould Dictyostelium discoideum grown in axenic culture. Biochem J 1970; 119: 175.

[4] Morio T, Urushihara H, Saito T, Ugawa Y, Mizuno H, Yoshida M, *et al.* Dictyostelium discoideum developmental cDNA project: generation and analysis of expressed sequence tags from the first-finger stage of development. DNA Res 1998; 5: 335.

[5] Decave E, Brechet Y, Bruckert F, Fourcade B., Garrivier D. Peeling model for cell detachment. Eur Phys J 2002: E 8: 79–97.

[6] Décavé E, Garrivier D, Bréchet Y, Fourcade B, Bruckert F. Shear Flow-Induced Detachment Kinetics of *Dictyostelium discoideum.* Cells from Solid Substrate Biophysical Journal 2002; 82: 2383–2395.

[7] Bruckert F, De Cave E, Garrivier D, Cossion P, Brechet Y, Fourcade B *et al.* Dictyostelium discoideum adhesion and motility under shear flow: experimental and theoretical approaches. Journal of Muscle Research and Cell Motility. 2002; 23: 651–658.

[8] Bruckert F, Demily M, Brecht Y, Boulangé L. Kinetics of yeast detachment from controlled stainless steel surfaces. Colloids Surf B Biointerfaces 2006; 51: 71-79.

[9] Bäuerle D. Laser Processing and Chemistry, Springer–Verlag, Berlin, Heidelberg 2000.

[10] Lackner J.M, Waldhauser W, Schwarz M, Mahoney L, Major L, Major B. Polymer pre-treatment by linear anode layer source plasma for adhesion improvement of sputtered TiN coatings. Vacuum 2009; 83: 302-307.

[11] Lackner J.M. Industrially- scaled hybrid Pulsed Laser Deposition at Room Temperature. Published by Orekop sc., Krakow 2005.

[12] Major B, Bruckert F, Lackner J.M, Ebner R, Kustosz R, Lacki P. Coating on TiN and Ti(C, N) basis for biomedical application to blood contact and TiN/CrN multilayered tribological systems produced by pulsed laser deposition. Arch Metal Mater 2008;53:39-48.

[13] Lackner JM, Waldhauser W, Alamanou A, Teicher C, Schmied F, Major L, Major B. Mechanisms for self-assembling topography formation in low-temperature vacuum deposition of inorganic coatings on polymer surfaces. Bull Pol Ac Tech 2010; 58(2): 281-294.

[14] Sarna J, Kustosz R, Major R, Lackner J.M, Major B. Polish Artificial Heart – new coatings, technology, diagnostics. Bull Pol Ac Tech 2010; 58(2): 329-336

[15] Major R, Bruckert F, Lackner J.M, Waldhauser W, Pietrzyk M, Major B. Kinetics of eucariote cells adhesion under shear flow detachment on the PLD deposited surfaces. Bull Pol Ac Ser Techn 2008; 56: 223-228.

[16] Sanak M, Jakiela B, Wegrzyn W. Assesment of hemocompatibility of materials with arteria blond flow by platelet functional tests. Bull Pol Ac Tech 2010; 58(2): 317-322.

[17] Thornton JA. Influence of apparatus geometry and deposition conditions on the structure and topography of thick sputtered coatings. J Vac Sci Technol 1974;11(4), 666-670.

[18] WinSAM Vario I to III, and SAM 2000, KSI, Kraemer Scientific Instruments, Lerchenweg 16-18, 35745 Herborn, Germany.

[19] Ultrasonic testing, J.Krautkraemer, H.Krautkraemer, Springer Verlag 1986.

[20] Acoustic microscopy, Andrew Briggs (Monographs on the physics and chemistry of materials, Vol. 46), Oxford University Press 1992.

[21] EVOLUTION I-III, SAM TEC GmbH, P.O. 3111, 73641 Aalen, Email: info@samtec-germany.com.

[22] Seyfert UT, Biehl V, Schenk J. *In vitro* hemocompatibility testing of biomaterials according to the ISO 10993-4. Biomol Eng 2002;19(2-6): 91-96.

[23] Otto M, Klein C.L, Koehler H, Wagner M, Roehrig O, Kirkpatrick C.J. Dymanic blood cell contact with biomaterials: validation of a flow chamber system according to international standards. J Mater Sci Mater Med 1997; 8(3):119-129.

[24] Streller U, Sperling C, Huebner J, Ranke R, Werner C. Design and evaluation of novel blood incubation systems for *in vitro* hemocompatibility assessment of planar solid surfaces. J Biomed Mater Res B Appl Biomater 2003; 66B (1): 379-390.

[25] Schaub RD, Kameneva MV, Borovetz HS, Wagner W.R. Assessing acute platelet adhesion on opaque metallic and polymeric biomaterials with fiber optic microscopy. J. Biomed Mater Res 2000; 49 (4): 460-468.

[26] Sbrana S, Della Pina F, Rizza A, Buffa M, De Filippis R, Gianetti J *et al.* Relationships between optical aggregometry (type born) and flow cytometry in evaluating ADP-induced platelet activation. Cytometry B Clin Cytom 2008; 74 (1): 30-39.

[27] Nomura S, Shouzu A, Tamoto K, Togane Y, Goto S, Uchiyama S *et al.* Assessment of an ELISA kit for platelet-derived microparticles by joint research at many institutes in Japan. J Atheroscler Thromb 2009;16(6):878-87

[28] Harrison P. Progress in the assessment of platelet function. Br J Hematol 2000; 111(3): 733-744.

[29] Shenkman B, Savion N, Dardik R, Tamarin I, Varon D. Testing of platelet deposition on polystyrene surface under flow conditions by the cone and plate(let) analyzer: role of platelet activation, fibrinogen and von Willebrand factor. Thromb Res 2000; 99 (4): 353-362.

[30] Gemmel CH. Activation of platelets by *in vitro* whole blood contact with materials: increases in microparticle procoagulant activity, and soluble P-selectin blood levels. J Biomater Sci Polym Ed 2001; 12 (8): 933-934.

[31] Varon D, Lashevski I, Brenner B, Beyar R, Lanir N, Tamarin I *et al.* Cone and plate(let) analyzer: monitoring glycoprotein IIb/IIIa antagonists and von Willebrand disease replacement therapy by testing platelet deposition under flow conditions. Am Heart J 1998; 135 (5 Pt 2 Su)S:187-193.

[32] Nomura S, Ozaki Y, Ikeda Y. Function and role of microparticles in various clinical settings. Thromb Res 2008; 123 (1): 8-23.

[33] Blann A, Shantsila E and Shantsila A. Microparticles and arterial disease. Semin Thromb Hemost 2009; 35 (5): 488-496.

[34] Gemmel CH, Ramirez SM, Yeo EL et Sefton M.V. Platelet activation in whole blood by artificial surfaces: identification of platelet-derived microparticles and activated platelet binding to leukocytes as material-induced activation events. J Lab Clin Med 1995; 125 (2): 276-287.

CHAPTER 12

Tailoring Surfaces of Nanocarriers for Cancer Therapy and Diagnosis

Gürer G. Budak*, Tolga Çamli, Fatih Büyükserin and Selman Yavuz

Gazi University NanoMedicine & Advance Technology Research Center, Ankara-Turkey

Abstract: In this chapter, the possibilities to improve *in vitro* diagnostic tests by providing more sensitive detection technologies and better nano-labels devices will be reviewed.

New therapeutic approaches based on the miniaturization of devices and targeted delivery systems are discussed as well.

Particular attention has been devoted to the development and use of "theragnostic" platforms and molecular imaging technologies which allow to systematically monitor the care process.

Keywords: Nano-labels devices, diagnosis, therapy, bio-imaging, nanomedicine.

INTRODUCTION

Nanomedicine, the application of nanotechnology to healthcare, is an essential tool to determine many unmet clinical needs of today and in the future. It exploits the improved and often novel physical, chemical and biological properties of materials at the nanometer scale. These aspects hold the promise to provide breakthroughs in nanomedicine, leading to clinical solutions within preventive medicine, diagnosis, therapy and follow-up care.

Preventive Medicine

New diagnostic tests making use of nanotechnology to quantify cancer-related biomarkers could offer an earlier and more personalized risk assessment before symptoms show up. In general, these analyses are cost-effective, sensitive, and reliable. The test itself inflicts only minimal discomfort on the patient.

***Address correspondence to Gürer G. Budak:** Gazi University NanoMedicine & Advance Technology Research Center, Ankara-Turkey; Tel: +90 312 484 6270; Fax: +90 312 484 6271; E-mail: gurer.budak@gazi.edu.tr

Nanotechnology improves *in vitro* diagnostic tests by providing more sensitive detection technologies or by providing better nano-labels that can be detected with high sensitivity once they bind to cancer-specific molecules present in the sample. Nanotechnology also improves the ease-of-use of *in vitro* diagnostic tests done by untrained users or even by patients at home. Tumor tissue with no secretion of biomarkers into blood or urine will require imaging procedures of high specificity for their early detection.

Diagnosis

If a medical check-up finds an indication or a hint of symptoms for a cancer type, it is important that more specific diagnostic procedures. These can be more laborious and expensive as they are applied to a smaller number of patients. In this case, molecular imaging, which makes use of specific targeted agents, plays a crucial role for localization and staging of a tumor.

Miniaturized imaging systems make it possible to perform image-based diagnostics everywhere and not only in research centers. Conceptually novel methods, combining biochemical techniques with advanced imaging and spectroscopy provide insight to the behavior of single cancer cells and their microenvironment for the individual patient. This leads to personalized treatment and medication tailored to the specific needs of a patient.

Therapy

In many cases, therapy is not being restricted to medication only but requires more severe therapeutic action such as surgery or radiation treatment. Planning of therapeutic interventions is based on imaging, or is performed under image guidance. Here, nanotechnology leads to a miniaturization of devices that enable minimally invasive procedures and new ways of treatment. Targeted delivery systems and nanotechnology-assisted regenerative medicine play the central role in future therapy. Targeted delivery agents allow a localized therapy which targets only the cancer cells, thereby increasing efficacy while reducing unwanted side effects. Imaging and biochemical assay techniques are used to monitor drug release or to follow the therapy progress.

Follow-Up Monitoring

Medical reasons call for an ongoing monitoring of the patient after completing the acute therapy. This is a regular check for reoccurrence, or, in the case of chronic diseases, a frequent assessment of the actual cancer status and medication planning. *In vitro* diagnostic techniques and molecular imaging play an important role in this part of the care-process, as well. Biomarkers could be systematically monitored to pick up early signs of reoccurrence, complemented by molecular imaging where necessary.

SILICA NANOCARRIERS

Among the various types of nanocarriers employed in biomedical research, silica-based nanoplatforms are gaining increasing attention due to their novel properties involving ease of preparation and modification, stability, due to biocompatibility and stealthness [2-4]. In particular, when compared with highly popular organic nanocarriers (*e.g.* liposomes and micelles) that require a poly ethylene glycol (PEG) coating to escape the reticuloendothelial system (RES), silica is intrinsically hydrophilic which allows it to escape from the opsonin proteins, resulting in high blood stability [2]. Majority of the silica related applications utilize spherical NPs which are typically synthesized by the reverse microemulsion, Stober or surfactant templated co-condensation method [4]. By varying the reaction conditions, the particle size and porosity can be finely tuned. The ability to obtain monodisperse particles with varying pore sizes (*e.g.* microporous or mesoporous silica) emerge them as ideal tools for controlled drug delivery [2, 5]. Moreover, they can be rendered fluorescent by encapsulating a fluorophore within the silica matrix during the particle formation or alternatively, by covalent attachment of the imaging agent following the production of the particles. The former approach presents several advantages compared with the use of conventional organic dyes including higher sensitivity due to the coencapsulation of several molecules into one carrier platform, enhanced photostability as well as the potential for multiplexed bioanalysis through multifluorophore encapsulation [3]. Finally, silica coating on magnetic nanocores (*e.g.* iron or manganese oxide) paved the way for the creation of multifunctional core-shell NPs that can be used for multifaceted bioimaging together with therapeutic delivery [6, 7].

Targeted diagnosis and therapy of cancer generally involves tailoring of the nanocarrier surface with various biomolecules (*e.g.* proteins, DNA or receptor molecules) through covalent coupling approaches or physical adsorption [2, 4, 5, 8]. The former is more preferred due to the ability to control the number and orientation of the biorecognition sites. Carbodiimide, schiff-base, disulfide coupling and succinimidyl ester hydrolysis are some protocols used for the covalent attachment of biomolecules to the particle surface [4, 9]. Particularly for the *in vivo* cancer applications, one can intentionally omit the modification with a targeting moiety due to the presence of "Enhanced Permeation and Retention, EPR". This passive targeting effect stems out from the increased endothelial fenestrations and architectural anarchy and results in the accumulation of injected particles in the tumor tissue even without a targeting element [10]. In this context, despite its rare use on Si NPs, PEG modification can be considered as a passive targeting strategy, which was shown to enhance the incorporation of core-shell Si NPs to MCF-7 breast cancer cells [5] and to improve their non-toxic behavior in peripheral tissue [11].

Numerous examples are present in the literature that involve a coupling of a targeting moiety to Si NP surface, [4, 5, 8] and since it is impossible to cover all of them, representative examples of DNA, receptor molecule and protein coupling to target cancer cells and tissues for imaging and therapeutic applications will be presented. Short strands of DNA/RNA called aptamers are successfully used to target acute leukemia cells [12]. Like antibodies, these aptamer molecules specifically recognize proteins and complex samples but with less cost of production and reduced immunogenicity. Fluorescent SiNPs modified with aptamers specific to acute leukemia cells were used to target and visualize acute leukemia cells *et al.* flow cytometry and fluorescence microscopy. Following this work, aptamer-modified SiNPs with magnetic cores were used to selective extract CCRF-CEM leukemia cells from whole blood samples which were then incubated with fluorescent analogs for imaging [13]. The process can be completed in less than 5 minutes and is applicable to several other cell types at a corresponding aptamer sequence modification.

Folic acid (FA) is a vitamin molecule and is widely applied for targeting NPs for both *in vitro* and *in vivo* applications due the overexpression of FA receptors in several cancer types as well as its high stability, low cost and modifiable chemical properties [4, 5]. Fluorescent SiNPs with magnetic manganese oxide cores were modified with FA for the bimodal imaging of human cervical carcinoma cell line (HeLa) [7]. Both magnetic resonance (T_1 contrast) and fluorescence imaging demonstrated the selective localization of such core-shell particles when compared with the limited internalization for the MCF-7 cells that do not overexpress FA receptors. In a recent study, FA modified mesoporous fluorescent SiNPs with a camptothecin payload successfully accumulated in human cancer xenografts and suppressed tumor growth in mice [14]. This excellent study by Tamanoi and coworkers is the first to demonstrate antitumor efficacy of SiNPs *in vivo*. It also shows biocompatibility, biodistribution and selective *in vitro* targeting of FA- modified silica nanocarriers. Fluorescence and ICP-MS was used to confirm tumor accumulation which revealed that although FA-conjugated drug-containing SiNPs accumulate more in the tumor tissue compared with non-targeted drug-containing counterparts, they cause very similar tumor suppression results. This surprising observation is explained by the high dose injection of nanocarriers in both cases as well as the use of MCF-7 cells that do not overexpress FA receptors for the xenograft growth [14].

Peptide and antibody modification on NP surfaces are also widely applied strategies for protein-based targeting studies [4, 8, 10, 15]. TAT and cRGD are commonly used peptides to target different cancer cell types [4, 15]. For instance, TAT-labeled fluorescent SiNPs were used to successfully bind to human lung adenocarcinoma cells (A 549) and rat brain tissue [16]. Similarly, when silica-coated magnetic NPs are coupled with cRGD peptide which recognizes the overexpressed $\alpha_v\beta_3$ integrins of the angiogenic tumors, they can be specifically used to target and image human colon cancer cells [15, 16]. Parallel to the peptide modifications, the high specificity associated with antibody-antigen interactions is widely explored for the selective and efficient binding of Si nanocarriers to different cancer lines. For example, luminophore-doped SiNPs modified with mouse anti-human CD10 antibody showed specific detection of mononuclear leukemia cells *et al.* fluorescence microscopy. The control experiments with non-target cells or antibody-deficient NPs proved the effective recognition of the antibody-modified NPs by the leukemia cells [17].

The majority of the silica-based nanocarrier applications utilize spherical NPs because the ease of preparing multicomponent systems is an alternative synthetic routes. Recent research, however, is revealing remarkable biological properties of one dimensional NPs such as increased blood circulation times and enhanced tumor accumulation when compared with spherical counterparts [18, 19]. This has caused increasing attention for the employment of one dimensional silica nanoparticles, namely silica nano tubes and test tubes (SiNTs) [3, 9]. SiNTs can be easily prepared by a versatile method called template synthesis [20] which allows absolute control over the resultant particle dimensions (Fig. **1**). Anodized alumina membranes are typically used as templates and silica is deposited generally by surface sol-gel technique. The template synthesis also allows the preparation of multifunctional SiNTs through differential modification of the inner *et al.* the outer surface of the silica structures.

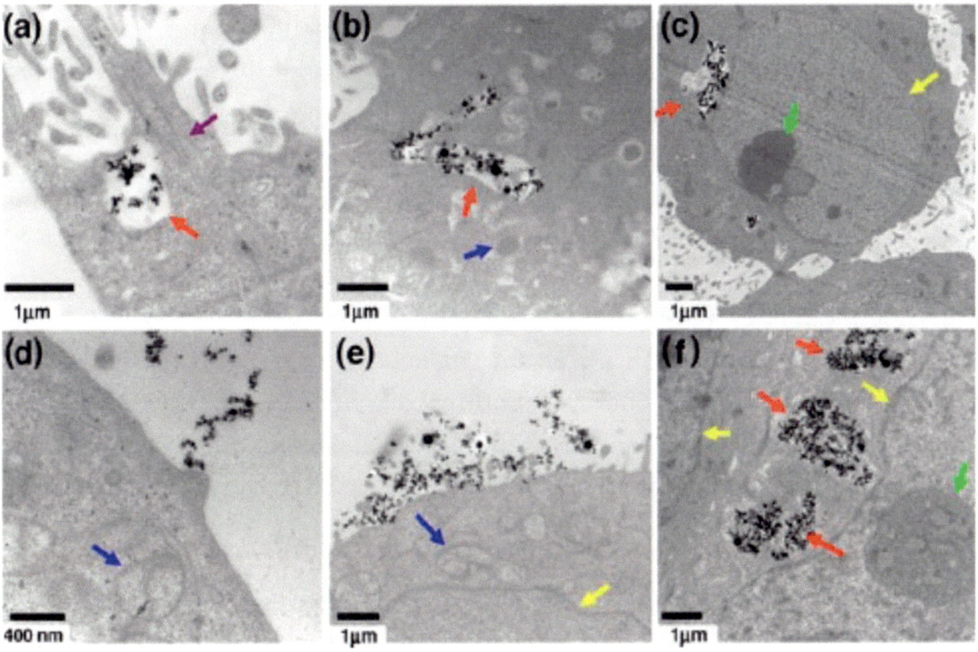

Figure 1: Transmission Electron Micrographs of MCF10A and MCF7 breast cancer cells that have been incubated with iron oxide nanoparticles for 30 min. (a-d), 4 h (b-e) and 24h (c-f) at 37°C. All slices treated with uranyl acetate to stain membranes and lead citrate to stain the nuclear body. *Colored arrows* represent selected cell organelles; nuclei *(yellow)*, neucleolus *(green)*, mitochondria *(blue)*, and vesicles with iron oxide nanoparticles inside *(red)*, filaments *(purple)* [1].

Researchers successfully applied this strategy to create fluorescently-labeled SiNTs that can specifically recognize breast cancer cells. While the test tubes were embedded in the template, the inner surfaces are decorated with a fluorophore. Upon liberation of the SiNTs with template dissolution (Fig. **1**), the outer surface are covalently coupled with IGF-IRα antibody which targets overexpressed IGF-I receptors present on the membrane of MDA-MB-231 breast cancer cells. Both confocal fluorescence and microplate reading results confirmed the specific recognition of the targeted SiNTs and minimal binding was observed in the case of a control antibody modification [9].

POLYMERIC NANOCARRIERS

The surface characteristics of nanocarriers for cancer therapy and diagnosis as well as their size are very important for the fate of the particles. In general, a small size below 100 nm scale and hydrophilic surfaces are the desired properties of nanoparticles which were used in order to avoid their subsequent clearance by macrophages [21]. Hydrophilic nanoparticles can escape from the macrophage capture. Nanoparticles could be synthesized with hydrophilic surfaces in two ways: *i)* coating the particles surface with a hydrophilic polymer, *ii)* synthesizing the polymer from hydrophilic and hydrophobic copolymer domains [22]. Decorating the nanoparticle surface with hydrophilic moieties is the first step of the preparation of nanocarriers used to tumor targeting *in vivo* applications.

Poly(methylmethacrylate) (PMMA) nanoparticles were used in several tumor distribution models in mice [23]. Different surfactants were used to modify the PMMA nanoparticles and also the distributions of uncoated nanoparticles were investigated. Prolonged half lives of coated PMMA nanoparticles were achieved and accumulation of these nanoparticles in B-16 and MaTu tumors were observed.

In order to synthesize sub-100 nm scale monodisperse PMMA nanoparticles, surfactant-free emulsion polymerization were used in the literature [24]. Cross-linked PMMA nanoparticles with the average hydrodynamic diameter range of 40 – 103 nm were produced with amidine functionality. The polymerization method uses an acetone - water mixture as the continuous phase. The effects of continuous phase composition, monomer and initiator concentrations were studied to optimize average particle hydrodynamic diameter and polydispersity index of the final particles. A linear relationship ($R = 0.995$) was obtained between the particle size and monomer concentration which allows fine tuning of the final particle size. The thermal decomposition of the 2, 2'-azobis (2-methylpropionamidine) hydrochloride initiator causes a net positive charge on nanoparticle surface that reduces with increasing solution pH due to the decomposition of the amidine moiety. The amidine group also renders the resulting PMMA latex as functional nanoparticles that can be synthesized by a single step polymerization. The presence of the amidine moiety was confirmed by Rhodamine B isothiocyanate conjugation that results fluorescence nanoparticles with bright emission.

Poly(ethylene glycol) (PEG) polymers were used as hydrophilic surface modifiers for different nanocarriers for reducing the systemic toxicity and increasing the therapeutic efficacy of the drug delivery system. PEG modification increases the nanocarrier retention time in circulatory system which the probability of accumulation at the target increases. PEG modified poly(ε-caprolactone) (PCL) and poly(α-amino ester) nanoparticles were used as drug carriers in MDR ovarian cancer cell line. The investigations showed that the combined therapy which paclitaxel (PTX) loaded PEG-PCL and ceramide loaded PEG-PCL nanoparticles were utilized decreases in tumor weight and increases in apoptotic activity [25].

PTX loaded nanocarriers produced from linear PEGlyated two-arm oligomer of cholic acids in aqueous solution were given in the literature for the ovarian cancer xenograft model. The length of linear PEG chains and configuration of cholic acid oligomer can control the size and drug loading capacity. The extreme drug loading capacities were achieved up to 12.0 mg/mL which equivalents 37.5% (w/w) of the total mass of nanoparticles [26].

In addition to PEG modification, coupling of a specific ligand on to nanocarriers which will be recognized by the cells present at the disease site is defined as active targeting. Since the tumor cells are rapidly proliferating, they over-express certain receptors to uptake of nutrients which includes folic acid, sugars and vitamins. Nanocarriers which were decorated with folic acid moieties could be targeted to cancer cells that have over-expressed folate receptors. Fig. **2** illustrates the intracellular delivery of folate attached nanocarrier and its endocytosis following a release of drug with an external stimuli process.

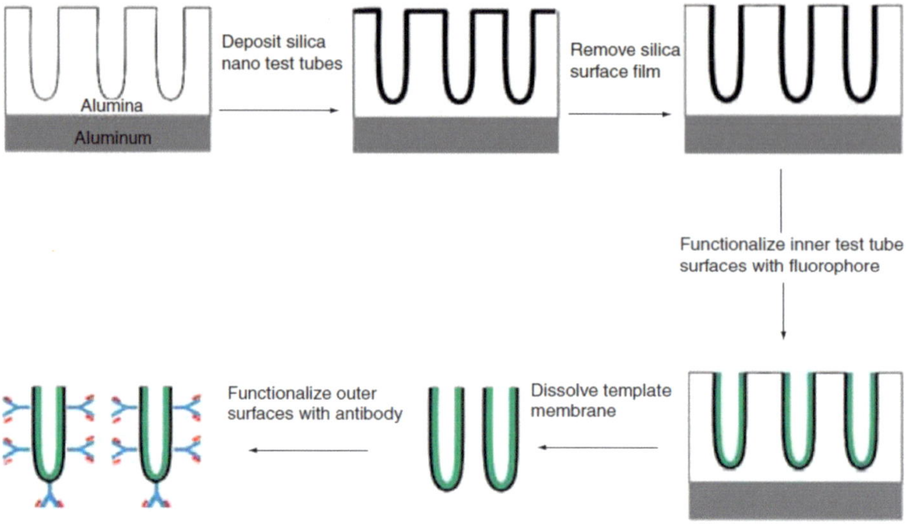

Figure 2: Schematic of SiNT synthesis and functionalization [9].

Tumor and capillary endothelial cells also express specific integrin receptors, such as $\alpha_v\beta_5$ or $\alpha_v\beta_3$ that can bind to arginine–glycine–aspartic acid (RGD) tripeptide sequence. RGD-modification has been used to orient nanocarriers to tumor cells and capillary endothelial cells of the angiogenic blood vessels [27].

GOLD NANOCARRIERS

Metal nanostructures have gained significant attention in biomedical research due to their unique optical and chemical properties. The strong optical absorption characteristics of gold nanomaterials with their outstanding biocompatibility suggest their potential applications in biotechnology, especially cancer treatments [28].

Cancer is the leading disease in the world to cause one in every four human deaths [29]. Despite many technological and clinical advances in the treatment of many types of cancers, the effective therapies are still underway. The main treatments to these cancers are chemotherapy, radiation therapy and surgical extirpation. Although chemotherapy is effective to kill the cancer cells, it also kills the healthy cells [30]. Surgery is not applicable to most of the cancers, but still it is very efficient for known tumors [31]. Radiotherapy can damage healthy cells and tissues in the path of beam during the radiation process [32]. For all these general treatments, there is a desperately need to the rapid advances which is very effective for specific cancers without killing or damaging the healthy cells.

Recently, the photothermal therapy has attracted the interest, mainly the use of lasers for thermally destruction of tumors or cancers [32]. However, the healthy tissues in the path of the light may absorb the energy and therefore reduce the speed and effectiveness of the light and increasing side injury in the body. Therefore, the laser light energy should be as low as possible to not damage the healthy cells but it should be enough both to penetrate to the deeper tumor tissues and to be absorbed by the photosensitizers. Injectable photosensitizers such as dyes, nanoshells, nanocages, and nanorods have been applied to as cancer therapy agents to absorb the low energy laser light and convert it into heat which will destruct selectively and efficiently the tumor cells [33-35]. Despite of the powerful photothermal cancer therapy, one could not control the dosage of the drug released from the nanodevices by any means.

Thermosensitive hydrogels such as poly(N-isopropylacrylamide) (polyNIPAAm) show a reversible volume phase transition from expanded form to shrunken one upon heat treatment. Yavuz *et al.* were incorporated polyNIPAAm with gold nanocages which were designed to strongly absorb near-infrared light [36]. Upon irradiation of laser light, the gold nanocages absorbed the light and then converted to heat which initiates the volume phase transition process when temperature change. Yavuz *et al.* demonstrated a laser-triggered controlled release of polymer supported dye and doxorubicin (Dox), a cancer drug for breast cancer, from polyNIPAAm-covered gold nanocages. Depending on power density of the laser and the duration of laser light irradiation, the amount of dye and Dox releases can be controlled. Moreover, the number of breast cancer cells were killed by doxorubicin can be controlled by regulating the power density of laser beam and irradiation time.

Gold nanocages are a novel class of nanostructures, which can be prepared from simple, and efficient galvanic replacement reaction of silver atoms in silver nanocubes with gold salt ($HAuCl_4$) [37]. The gold nanocages can represent a tunable surface plasmon resonance (SPR) from high energy (400nm) to the low energy, near-infrared region (1000nm). The SPR peak value of gold nanocages in a range of 900-950 nm can strongly absorb the laser light energy which does not damage to the healthy cells. The gold nanocages can have such pores having a diameter between 15 and 30 nm. The gold nanocages were covered by polyvinylpyrrolidone (PVP) which is usually around 3-5nm. On the other hand, the PVP-coated gold nanocages were applied an exchange process in poly(*N*-isopropylacrylamide) (polyNIPAAm) solution for a few days. This process gave a thicker layer of polyNIPAAm layer and the layer was around 10-15nm shown inset image.

One can easily predict that the gold nanocages can be used as capsules if the pores are closed in a someway after the loading the species such as; cancer drugs, dyes, small species, polymers; to inside of the gold nanocages. In order to control the release of the species from the gold nanocages, the coverage materials should be reversibly open-close mechanism with an external factor. This requirement can be adjusted by attaching thermosensitive polymers such as polyNIPAAm in a way that it can serve as a "door" for the pores of the gold nanocages, and permit the release of the species from inside to outside media whenever it is needed.

The near infrared light at 800-1200 nm is strongly absorbed by the gold nanocages and then converted to heat, thus heating the surroundings. If the heating the surroundings is above the lower critical solution temperature (LCST) of the polyNIPAAm, the polyNIPAAm on the gold nanocages surfaces have a phase transition from expanded form to shrunken form which open the blocking of the pores [38]. This process causes a release of the species held within the gold nanocages.

Disulfide- and thiol-based molecules are well-known for their strong binding to the gold surfaces. Disulfide containing polyNIPAAm was synthesized by using disulfide initiators which result one disulfide group in every chain of polyNIPAAm [39]. The polyNIPAAm was prepared *et al.* atom transfer radical polymerization technique which tolerates solvents of different polarity and a range of monomers with various functional groups, and also enables the control in molecular weight of the polymers. Excess amount of polyNIPAAm was mixed with gold nanocages in order to replace the polyvinylpyrrolidone (PVP). After exchanging process, free PVP and extra polyNIPAAm were discarded during washing steps. The polyNIPAAm covered gold nanocages gives a 13 nm red shift for the SPR peak which can be a proof of the polyNIPAAm binding to the gold nanocages. Furthermore, the polyNIPAAm covered gold nanocages can be dispersed in water at RT and then heated to 43°C to give a precipitate due to insoluble shrunken form of polyNIPAAm above lower critical solution temperature (LCST) around 33 °C in water.

Initially, the gold nanocages can be prepared with different sizes. These nanocages can have different sizes of pores depending on the amount of gold salt addition to the silver nanocubes. Different sizes of nanocages and pores were tried with various dyes and polymer supported dyes but the best results were obtained form 15-30 nm pore sizes with 100-120 nm gold nanocages. In order to detect the dyes by UV-Vis-NIR spectrophotometer, initially, just plain dyes were used, but, unfortunately, the release was not enough to detect by UV instrument. Therefore, polymer supported dyes were synthesized which can be easily held inside the gold nanocages.

To demonstrate the release of the species from inside of the gold nanocages by the help of polyNIPAAm coverage, PEG supported alizarin dye (Alizarin-PEG) was prepared by coupling the alcohol part of PEG (MW:~4, 300) with the acid function of alizarin yellow acid in the presence of common esterification reagents, DCC and DMAP [40]. A concentrated solution of Alizarin-PEG was mixed with polyNIPAAm covered gold nanocages and then heated to 43°C in order to load these dyes into the nanocages. After given time, the mixture was quickly cooled in ice to lock the pores by expanding the polyNIPAAm in order to keep the loaded dyes inside the nanocages. The loaded dyes were released from the nanocages by simply heating the solution above LCT of polyNIPAAm. The dye loaded gold nanocages were exposed to a constant heat (43°C) with different time scale.

The peak absorbance of Alizarin-PEG dye was observed at 357 nm, therefore, the release of dyes from gold nanocages can be easily observed in this wavelength using UV-Vis-NIR spectrophotometer. The amount of dyes released from nanocages increases with longer duration of heating process as the peak absorbance goes up. However, the amount of dyes released in each half minute lowers, and then eventually the release was very small or insignificant as the dyes were all out to the solution after a certain time. The exact concentration and amount of dyes released from nanocages can be calculated from a known concentration of dye solution by using Beer`s law.

The Alizarin-PEG dyes can be also released by the aid of laser beam. The gold nanocages are excellent absorbers of the light energy due to their optical properties. This absorbed energy can be converted to heat which makes the polyNIPAAm on the surface of nanocages to form shrunken phase. This means the pores are opened and then the dyes from inside of gold nanocages can get out by the help of the solvent. For this purpose, the gold nanocages were exposed to laser beam with different power densities. The increasing power density of laser beam makes the laser more powerful with having more energy of light. In a constant time, the gold nanocages can absorb higher energy from higher power density of laser, and therefore, more heat was released to the surroundings. Higher power density of laser was caused more release of alizarin-PEG dyes from gold nanocages due to more heating which shrinks the polyNIPAAm faster. There is a jump in peak absorbance when the power density increases from 20 mW to 40 mW. This is because of the melting of the gold nanocages and forming solid spheres which is observed by using TEM [41]. Furthermore, the concentration of dyes released from gold nanocages can be calculated by using Beer`s law.

Furthermore, to test the time dependent release of Alizarin-PEG dyes, the power density of the laser beam was kept constant at 5 mW/cm^2, but the irradiation time was changed. As observed in previous results, the first 15 seconds was the most release of Alizarin-PEG dye from gold nanocages. Upon further irradiation, the total amount of dyes released increases but the differences between each steps decrease due to remaining fewer amounts of dyes inside the nanocages.

Up to now, the release process was demonstrated with a polymer supported dye which can be easily detected by UV-Vis-NIR spectrophotometer. These exciting preliminary results encourage us to continue further investigation in a real cancer treatment, breast cancer cells, by using Doxorubicin (Dox), well-known commercial chemotherapy cancer drug. A nanomolarity of Dox can kill the breast cancer cells which can be easily observed with biological treatments [42]. In this study, the breast cancer cells were transferred to 24-well plates reached 80% confluence. Before addition of Dox loaded gold nanocages to the well plates, the well plates were kept in outside to cool down below 33°C which is the LCST of polyNIPAAm. Dox loaded gold nanocages solution was injected to the rows of well plates except two rows which was used as controls. Gold nanocages added cancer cells were irradiated by laser beam with power densities of 5, 10 and 20 mW in constant 15 seconds time period. After the irradiation experiments were done, the solutions from the well plates were replaced with MTT assay and medium solution. The plates were incubated for a day and then the solution was removed, thus leaving the precipitate behind which was used for the OD values and cell viability measurements.

The breast cancer cells without both adding gold nanocages and irradiation of laser beam were used as a reference. A small number of deaths of cancer cells after treating with DOX loaded gold nanocages without irradiation were observed. This can be due to remaining very small amount of extra Dox in the solution after washing steps. On the other hand, the laser exposure without gold nanocages gave very small deaths of cells could be due to a small heat generation during the laser process. The breast cancer cells were exposed to three different power densities of laser beam which were 5, 10 and 20 mW per cm^2. As increasing the power of laser beam, more deaths of cancer cells were observed. This can be again due to more releasing Dox to the culture media which kills more cancer cells.

Moreover, irradiation duration dependent release of DOX with the constant power density of the laser beam at 5 mW/cm^2 was demonstrated. The results from controls were consistent with the previous different laser power densities at a constant time experiment. The first fifteen minutes irradiation was the huge difference as DOX release as compared to the controls. Further irradiation with a longer time (30 and 60 seconds) gave more release of DOX from gold nanocages with a decreasing slope due to less remaining DOX inside the gold nanocages.

In summary, the gold nanocages were functionalized with polyNIPAAm which can be used as drug delivery devices with a remote control release capability. The preliminary results with PEG supported Alizarin dye show the release during heat process or laser irradiation. The heating of the dye-loaded gold nanocages solution above the low critical solution temperature of polyNIPAAm cause the shrinking of polyNIPAAm on the surface of gold nanocages which permits the release of the dyes held inside the cages to the solution. On the other hand, the gold nanocages are strong absorbers of NIR laser beam and then they converted to heat which makes again the shrinking of polyNIPAAm. Laser triggered release was demonstrated with different power densities of laser beam at constant time period, and different duration of a constant power density of laser beam. In both experiments, the increasing release of dyes was observed with longer irradiation or higher power density of laser beam. Furthermore, breast cancer destruction by using DOX (anti-cancer drug) loaded gold nanocages, a real life experiment, was demonstrated. Depending on the duration of irradiation or laser power density, the deaths of cells were observed which were treated with Dox loaded gold nanocages. These results suggest that gold nanocages are a promising new class of drug delivery devices for cancer treatment.

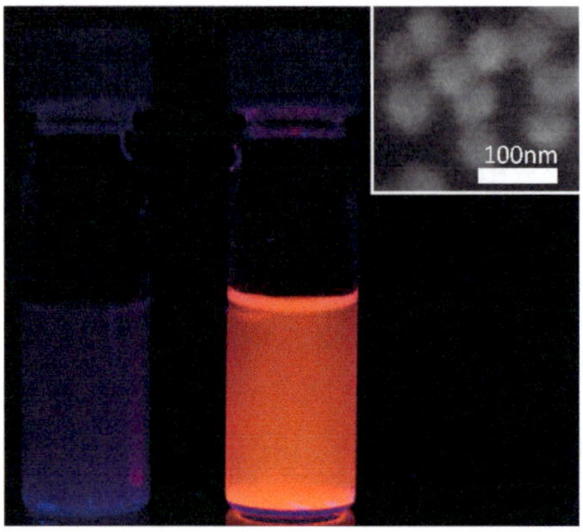

Figure 3: Fluorescence emission of naked (left) and Rhodamine B modified (right) nanoparticles under UV illumination. Inset: SEM image of cross-linked PMMA nanoparticles [24].

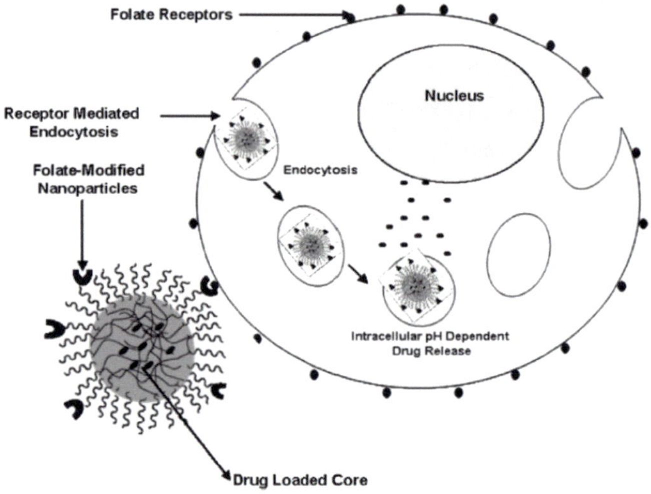

Figure 4: Illustration of active drug targeting with surface modified nanocarriers [27].

ACKNOWLEDGEMENTS

Declared none.

CONFLICT OF INTEREST

The author(s) confirm that this chapter content has no conflict of interest.

REFERENCES

[1] Zhang Y, Yang M, Portney N, Cui D, Budak G, Özbay E, Özkan M, Özkan C. Zeta potential: a surface electrical characteristic to probe the interaction of nanoparticles with normal and cancer human breast epithelial cells. Biomed Microdevices 2008; 10: 321-328.
[2] Barbe C, Bartlett J, Kong LG, Finnie K, Lin HQ, Larkin M, Calleja S, Bush A, Calleja G. Silica particles: a novel drug-delivery system. Adv Mater 2004; 16: 1959-1966.
[3] Buyukserin F, Martin CR. The use of reactive ion etching for obtaining "free" silica nano test tubes. Appl Surf Sci 2010; 256: 7700-7705.
[4] Wang L, Zhao WJ, Tan WH. Bioconjugated silica nanoparticles: development and applications. Nano Res 2008; 1: 99-115.
[5] Slowing, II, Trewyn BG, Giri S, Lin VSY. Mesoporous silica nanoparticles for drug delivery and biosensing applications. Adv Func Mater 2007; 17: 1225-1236.
[6] Lu CW, Hung Y, Hsiao JK, Yao M, Chung TH, Lin YS, Wu SH, Hsu SC, Liu HM, Mou CY, Yang CS, Huang DM, Chen YC. Bifunctional magnetic silica nanoparticles for highly efficient human stem cell labeling. Nano Lett 2007; 7: 149-154.
[7] Yang H, Zhuang YM, Hu H, Du XX, Zhang CX, Shi XY, Wu HX, Yang SP. Silica-Coated Manganese Oxide Nanoparticles as a Platform for Targeted Magnetic Resonance and Fluorescence Imaging of Cancer Cells. Adv Func Mater 2010; 20: 1733-1741.
[8] De M, Ghosh PS, Rotello VM. Application of nanoparticles in biology. Adv Mater 2008; 20: 4225-4241.
[9] Buyukserin F, Medley CD, Mota MO, Kececi K, Rogers RR, Tan WH, Martin CR. Antibody-functionalized nano test tubes target breast cancer cells. Nanomed Nanotechnolol Biol Med 2008; 3: 283-292.
[10] Ferrari M. Cancer nanotechnology: opportunities and challenges. Nat Rev Cancer 2005; 5: 161-171.
[11] Blumen SR, Cheng K, Ramos-Nino ME, Taatjes DJ, Weiss DJ, Landry CC, Mossman BT. Unique uptake of acid-prepared mesoporous spheres by lung epithelial and mesothelioma cells. Am J Respir Cell Mol Biol 2007; 36: 333-342.
[12] Herr JK, Smith JE, Medley CD, Shangguan D, Tan W. Aptamer-conjugated nanoparticles for selective collection and detection of cancer cells. Anal Chem 2006; 78: 2918-2924.
[13] Smith JE, Medley CD, Tang ZW, Shangguan D, Lofton C, Tan WH. Aptamer-conjugated nanoparticles for the collection and detection of multiple cancer cells. Anal Chem 2007; 79: 3075-3082.
[14] Lu J, Liong M, Li Z, Zink JI, Tamanoi F. Biocompatibility, biodistribution, and drug-delivery efficiency of mesoporous silica nanoparticles for cancer therapy in animals. Small 2010; 6: 1794-1805.

[15] Lin W, D. Rocca. Nanoscale metal–organic frameworks: magnetic resonance imaging contrast agents and beyond. Eur J Inorg Chem 2010; DOI: 10.1002/ejic.201000496

[16] Santra S, Yang H, Dutta D, Stanley JT, Holloway PH, Tan WH, Moudgil BM, Mericle RA. TAT conjugated FITC doped silica nanoparticles for bioimaging applications. Chem Commun 2004;24: 2810-2811.

[17] Santra S, Zhang P, Wang KM, Tapec R, Tan WH. TAT conjugated FITC doped silica nanoparticles for bioimaging applications. Anal Chem 2001; 73: 4988-4993.

[18] Park JH, von Maltzahn G, Zhang LL, Schwartz MP, Ruoslahti E, Bhatia SN, Sailor MJ. Magnetic iron oxide nanoworms for tumor targeting and imaging. Adv Mater 2008; 20: 1630-1635.

[19] Geng Y, Dalhaimer P, Cai SS, Tsai R, Tewari M, Minko T, Discher DE. Shape effects of filaments *versus* spherical particles in flow and drug delivery. Nat Nanotechnol 2007 ; 2 : 249-255.

[20] Martin C. R., Nanomaterials: a membrane-based synthetic approach. Science 1994; 266: 1961-1966.

[21] Brigger I, Dubernet C, Couvreur P. Nanoparticles in cancer therapy and diagnosis. Adv Drug Delivery Rev 2002; 54: 631-651.

[22] Cho KJ, Wang X, Nie SM, Chen Z, Shin DM. Therapeutic nanoparticles for drug delivery in cancer. Clin Cancer Res 2008; 14: 1310-1316.

[23] Lode K, Fichtner I, Kreuter J, Berndt A, Diederichs JE, Reszka R. Influence of surface-modifying surfactants on the pharmacokinetic behavior of 14C-Poly (methylmethacrylate) nanoparticles in experimental tumor models. Pharm Res 2001; 18: 1613-1619.

[24] Camli ST, Buyukserin F, Yavuz MS, Budak GG. Fine-tuning of functional poly(methylmethacrylate) nanoparticle size at the sub-100 nm scale using surfactant-free emulsion polymerization. Colloids Surf A 2010; 366: 141-164.

[25] Jabr-Milane L, van Vlerken L, Devalapally H, Shenoy D, Komareddy S, Bhavsar M, Amiji M. Multi-functional nanocarriers for targeted delivery of drugs and genes. J Control Release 2008; 130: 121-128.

[26] Li YP, Xiao K, Luo JT, Lee J, Pan SR, Lam KS. A novel size-tunable nanocarrier system for targeted anticancer drug delivery. J Control Release 2010; 144: 314-323.

[27] Ganta S, Devalapally H, Shahiwala A, Amiji M. A review of stimuli-responsive nanocarriers for drug and gene delivery. J Control Release 2008; 126: 187-204.

[28] Chen J, McLellan JM, Siekkinen A, Xiong Y, Li Z-Y, Xia Y. Facile synthesis of gold-silver nanocages with controllable pores on the surface. JACS 2006; 128: 14776-14777.

[29] Jemal A, Tiwari RC, Murray T, Ghafoor A, Samuels A, Ward E, Feuer EJ, Thun MJ. Cancer Statistics, 2004. Ca-a Cancer J Clinicians 2004; 54: 8-29.

[30] Hamilton AL, Eder JP, Pavlick AC, Clark JW, Liebes L, Garcia-Carbonero R, Chachoua A, Ryan DP, Soma V, Farrell K, Kinchla N, Boyden J, Yee H, Zeleniuch-Jacquotte A, Wright J, Elliott P, Adams J, Muggia FM. Proteasome inhibition with bortezomib (PS-341): a phase I study with pharmacodynamic end points using a day 1 and day 4 schedule in a 14-day cycle. J Clin Oncol 2005; 23: 6107-6116.

[31] Miller AB, Hoogstraten B, Staquet M, Winkler A. Reporting results of cancer treatment. Cancer 1981; 47: 207-214.

[32] Patridge AH, Burstein HJ, Winer EP. Side effects of chemotherapy and combined chemohormonal therapy in women with early-stage breast cancer. J Natl Cancer Inst. Monogr 2001; 30: 135.

[33] Vogl TJ, Mack MG, Muller PK, Straub R, Engelmann K, Eichler K. Interventional MR: interstitial therapy. Eur Radiol 1999; 9: 1479-1487.

[34] Loo C, Lowery A, Halas NJ, West J, Drezek R. Immunotargeted Nanoshells for Integrated Cancer Imaging and Therapy. Nano Lett 2005; 5: 709-711.

[35] Hirsch LR, Stafford RJ, Bankson JA, Sershen SR, Rivera B, Price RE, Hazle JD, Halas NJ, West JL. Nanoshell-mediated near-infrared thermal therapy of tumors under magnetic resonance guidance. P Natl Acad Sci 2003; 100: 13549-13554.

[36] Yavuz MS, Cheng Y, Chen J, Cobley CM, Zhang Q, Rycenga M, Xie J, Kim C, Song KH, Schwartz AG, Wang LV, Xia Y. Gold nanocages covered by smart polymers for controlled release with near-infrared light. Nat Mater 2009; 8: 935-939.

[37] Skrabalak SE. Gold Nanocages: synthesis, properties, and applications. Acc Chem Res 2008; 41: 1587-1595.

[38] Hoffman AS. Hydrogels for biomedical applications. Adv Drug Deliver Rev 2002; 54: 3-12.

[39] Tsarevsky NV, Matyjaszewski K. Combining atom transfer radical polymerization and disulfide/thiol redox chemistry: A route to well-defined (bio)degradable polymeric materials. Macromol 2005; 38: 3087-3092.

[40] Li HY, Jerome R, Lecomte P. Amphiphilic sun-shaped polymers by grafting macrocyclic copolyesters with PEO. Macromol 2008; 41: 650-654.

[41] Link S, Wang ZL, El-Sayed MA. How does a gold nanorod melt? J Phys Chem B 2000; 104: 7867-7870.

[42] Wu EC, Mikhailovsky A, Khant HA, Fu C, Chiu W, Zasadzinski JA. Oxidation-Triggered Release of Fluorescent Molecules or Drugs from Mesoporous Si Microparticles. ACS Nano 2008; 2: 2401-2409.

CHAPTER 13

Drug Delivery from Ordered Mesoporous Matrices for Bone Tissue Engineering

Renato Mortera* and Barbara Onida

Institute of Chemistry, Department of Applied Science and Technology - DISAT - Politecnico di Torino, Corso Duca degli Abruzzi 24, 10129, Torino, Italy

Abstract: The improvement of the sol-gel processes has provided a new generation of silica-based ordered mesoporous materials (OMM) for biomedical applications and bone tissue engineering. These materials, indeed, have been suggested as matrices for sustained drug release, showing that both small and large molecular drugs can be entrapped and released from the mesopores through several different processes. Many emerging biotechnologies can benefit from these OMM-based drug delivery systems. For instance, bone tissue engineering is a growing area directed towards the design of materials able to improve the bone regeneration capacity by recovering both its structure and function. In this area, controlled drug delivery from biocompatible and bioactive mesoporous materials with sufficient mechanical strength could favor the cellular growth and bone regeneration. Ordered mesoporous silica (OMS) present a high drug loading capacity and a controlled sustained release which could be helpful for this purpose, however their poor bioactive behaviour and mechanical properties drove to the development of new bioactive systems. On one hand mesoporous bioactive glasses (MBGs) have been extensively investigated, on the other hand the combination of OMS and bioactive macroporous scaffolds has been proposed in order to obtain hierarchical systems for bone substitution.

Keywords: Silica-based materials, mesoporous materials, bone tissue engineering, drug release, bioactive glasses.

1. INTRODUCTION

One of the most important problem that arises as a result of severe pathologies or in consequence of surgical treatments is the administration of large quantities of drugs to the patients with a significant lengthening of hospital stays, recovery time and in some cases with the occurrence of several side effects.

***Address correspondence to Renato Mortera:** Institute of Chemistry, Department of Applied Science and Technology - DISAT - Politecnico di Torino, Corso Duca degli Abruzzi 24, 10129, Torino, Italy; Tel: +39 011 0904631; Fax: +39 011 0904699; E-mail: renato.mortera@polito.it

Lia Rimondini, Claudia Bianchi and Enrica Vernè (Eds)

For this reason the field of implantable devices able to release in a controlled and targeted way drug molecules, specific hormones, growth factors, chemotherapeutic and antiestrogens agents is a research area in continuous growth [1].

Controlled drug delivery systems have to achieve precisely spatial and/or temporal delivery of therapeutic agents to the target site. Generally, a temporally and spatially-controlled drug delivery system should maintain the concentration of drug in the precise sites of the body within the optimum range and under the toxicity threshold, which improve the therapeutic efficacy and reduce toxicity. In the past years, controlled drug delivery has been developed in polymer-based systems and in novel inorganic materials-based systems [2].

Since the pioneer work reported by Vallet-Regì *et al.* in 2001 [3], ordered mesoporous silica (OMS) have occupied a significant position in the literature on drug delivery, due to their narrow pore size distribution and their high specific surface area [4], in combination with the possibility to modulate pore diameter in a wide range by varying synthesis conditions.

Several studies have been carried out on the system proposed in 2001, and several others ordered mesoporous materials (OMM) have been suggested as matrices for sustained drug release, showing that both small and large molecular drugs can be entrapped and released from the mesopores through several different processes [5-7].

Many emerging biotechnologies can benefit from these OMM-based drug delivery systems. For instance, bone tissue engineering is a growing area directed towards the design of materials able to improve the bone regeneration capacity by recovering both its structure and function [8]. In this area, controlled drug delivery from biocompatible and bioactive mesoporous materials with sufficient mechanical strength could favor the cellular growth and bone regeneration [9].

In the present contribution we shortly resume a few examples of ordered mesoporous matrices for drug delivery in bone tissue engineering. Indeed several comprehensive reviews have been recently published [5-7, 9, 10] and may be referred to for a wider perspective.

2. ORDERED MESOPOROUS MATERIALS FOR BONE TISSUE ENGINEERING

The improvement of the sol-gel processes has provided a new generation of silica-based ordered mesoporous materials for biomedical applications and bone tissue engineering [11-13]. These new materials exhibit a high capability to host active agents that contribute to the tissue's healing processes. In this strategy, the use of structure-directing agents is essential for obtaining highly ordered structures. Indeed, under appropriated synthesis conditions, surfactants self-organize into micelles which link the hydrolyzed silica precursors through the hydrophilic component and self-assembly to form an ordered mesophase, according to several factors, *i.e.* surfactant chemistry, organic:inorganic phase ratio, surfactant concentration, temperature, pH.

Once the product is dried and the surfactant removed (calcined or extracted with solvents), a mesoporous structure is obtained, exhibiting a high surface area and an ordered porosity. The applications of these materials range from many technological purposes to coatings and membranes in the biomaterials field and in tissue engineering [9].

Ordered mesoporous silica (OMS), having pore sizes in the range of 2-50 nm and inner surface silanol and siloxane reactive groups, have been proposed for the first time as implantable systems by Vallet-Regì and co-workers in 2001 [3]. Taking into account the good biocompatibility of SiO_2-derived materials with bone tissue and their porous features, several mesoporous SiO_2 structures such as MCM-41, MCM-48 and SBA-15 were proposed as local drug delivery systems to be implanted in bone [9]. Indeed, the presence of silanol groups and porosity seems to be crucial in the apatite layer formation. In this sense, the layer formation is enhanced by the presence of pores larger than 2 nm, and a direct relation between pore size and volume and the nucleation rate has been drawn [14].

Pure silica MCM-41 and SBA-15 mesoporous materials are able to develop an apatite-like layer under *in vitro* conditions [14, 15]. However, the kinetic formation is too slow to consider them as bioactive bone grafts [8]. Therefore in the last years the synthesis of bioactive mesoporous glasses (MBGs) belonging to different multicomponent systems has been extensively investigated [9, 16, 17]. The *in vitro* bioactive behaviour of these MBGs turned out to be much faster and intense than in conventional sol-gel glasses due to the much higher surface area and porosity [9], however their mechanical resistance resulted lower [18].

As a whole, the possibility to combine bioactivity, controlled drug release and mechanical properties represents a marked improvement in the field of bone tissue engineering. This could be achieved combining the release properties shown by the ordered mesoporous systems with the bioactivity of macroporous materials employed to fabricate implants [19-21].

3. DRUG DELIVERY FROM ORDERED MESOPOROUS SILICA

MCM-41 was the first ordered mesoporous material to be employed as drug delivery matrix and nowadays it still occupies a significant position in the literature, due to its easy synthesis and versatility. Other groups of mesoporous materials with larger pore size such as SBA, hollow mesoporous spheres and MSU were also used for drug delivery, but their use increased only in the last two years, particularly regarding the controlled release of large molecules [5, 7, 10].

In the pioneer work of Vallet-Regì and her colleagues [3], two kinds of MCM-41 with a pore size of 2.5 nm and 1.8 nm respectively were tested using ibuprofen as drug. An adsorption of about 30% by weight was observed for both samples.

Later on, Horcajada *et al.* have conducted an investigation focusing on the influence of pore size of MCM-41 materials on drug adsorption and delivery. This study revealed that the amount of ibuprofen adsorbed is directly related with the pore size and the delivery rate in a simulating body fluid (SBF) solution decreases as the pores size decreases in the range of 3.6-2.5 nm [15]. It is worth noting the presence of a peculiar discontinuity between 24 and 27 hours of release, which is observed only for pore sizes below 2.7 nm.

Qu and co-workers [22] analyzed a series of MCM-41 mesoporous silica materials with similar pores diameter but different morphologies, showing that the drug-loading amount is directly correlated to the Brunauer-Emmett-Teller surface area and pore volume, while the drug release profiles is dependent on the morphologies of mesoporous silica carriers M. Manzano *et al.* confirmed these results comparing the long-term ibuprofen release behavior of irregular-shaped MCM-41 with MCM-41 synthesized in form of spheres [23]. However, in all the above reported cases the authors have examined MCM-41 materials compacted into disks, therefore effects due to diffusion phenomena inside the tablets may not be ruled out.

In order to exclude effects of particle size and particle morphology, Cauda *et al.* reported the use of systems without significant external surface, *i.e.* ordered mesoporous silica synthesized within the channels of anodic alumina membranes [24]. They confirmed the data reported by Horcajada *et al.* on MCM-41 tablets [15].

In 2004 Cavallaro *et al.* [25] investigated the release of four anti-inflammatory agents, *i.e.* diflunisal, naproxen, ibuprofen and its sodium salt, from MCM-41. Drug release studies were performed at pH 1.1 and 6.8 mimicking gastrointestinal fluids and showed clearly the protective role played by the mesoporous carrier, especially regarding the diflunisal administration.

Organic modification of silica permits precise control over the surface properties and pore sizes for specific applications. Both reactive and passive organic groups can be incorporated in the porous solids by grafting methods or by co-condensation under surfactant control [26]. For drug delivery based on mesoporous materials, several investigations using organic modified mesoporous silica have been reported, starting from the work of Tourné-Peteilh and co-workers in which the functionalization of the hexagonal mesostructure MCM-41 with 3-glycidoxypropylsilane in order to obtain a covalent bond between the drug and the silica walls was proposed [27].

Horcajada and colleagues compared the ibuprofen uptake and release capacity from MCM-41-type mesoporous materials superficially modified by a post-synthesis method with different organic groups [28]. The adsorbed ibuprofen amount resulted around 19 wt% for the materials modified with polar groups and about 11 wt% for the materials modified with non-polar groups. This work showed the possibility to choose the adsorbed drug content and the delivery rate according to the host-guest chemical interaction by selecting an adequate modification with an organic group.

Among the others mesoporous materials, the hexagonal SBA-15 is becoming more and more used as drug carrier thanks to its large pore size (above 6 nm) [7, 10]. Izquierdo-Barba and colleagues have reported the delivery of ibuprofen from SBA-15 disks to SBF, as comparison with MCM-41 [29], pointing out a SBA-15 behavior similar to the kinetics observed for MCM-41 with pores larger than 2.7 nm.

Andersson *et al.* [30] prepared a series of mesoscopically ordered silicas with different pore sizes, pore connectivity, and pore geometry including MCM-41, SBA-3 and SBA-1 as carriers for controlled drug delivery systems. They reported that the degree of ibuprofen loading is dependent on the specific surface area and the pore diameter of the host matrix and showed that the release process in SBF is mainly diffusion-controlled. However clear differences were observed, ascribed not only to the differences in the pore connectivity and pore geometry of the materials, but also to the aqueous stability of the matrix. It has been also suggested that this latter phenomenon could influence the release kinetics.

In 2009 Gao and co-workers [31] developed a SBA-15-based drug delivery system for the storage and controlled release of a drug with a cell-penetrating peptide (CPP) as a vector, by using a one-pot liquid-crystal templating method. They did not remove the surfactant from the SBA-15 mesopores and the hydrophobic poly(propyl oxide) shell of the pore-filling Pluronic F127 acts as the diffusion-limiting factors of the drug-CPP conjugate. A sigmoidal *in vitro* release into phosphate buffered saline (pH 7.4) was observed and the typical release duration was 5 days.

Zhu and co-workers reported a facile route for preparation of irregular hollow mesoporous spheres (HMS) and employed them for drug storage and delivery using ibuprofen [32]. The comparison with MCM-41 pointed out that the HMS exhibit much more storage capacity than MCM-41 (up to 110 wt% with respect to 30 wt%, typical for MCM-41).

4. DRUG DELIVERY FROM MESOPOROUS BIOACTIVE GLASSES

Since the first synthesis procedure reported by Yan *et al.* in 2004 [17], also mesoporous bioactive glasses (MBGs) have been proposed for local drug delivery [9].

Xia and colleagues [33] have synthesized well-ordered MBGs with high specific surface area, encapsulating gentamicin into the mesopores by adsorption method. Results showed that the amount of drug loading of MBG was three times more than that of conventional sol-gel glasses. The outcomes of drug release in distilled water and in SBF showed that MBGs effectively decreased the initial burst. During the release period, gentamicin was released from the MBGs at a much lower release rate as compared to that from conventional glasses. Furthermore, the drug release was sensitive to the pH and ionic concentration of the release medium suggesting possible controls of the release rate.

In 2010 Lin and co-workers [34] functionalized MBGs with photoactive coumarin, obtaining a reversible gate operation. Coumarin-modified MBG was used as a drug delivery carrier to investigate drug storage/release characteristics using phenanthrene as a model drug. Irradiation with UV light (λ>310 nm) induced photo-dimerization of the coumarin-modified MBGs, which led to the pores' closing with cyclobutane dimers and trapping of the guest phenanthrene in the mesopores. However, irradiating the dimerized-coumarin-modified MBGs with shorter wavelength UV light (λ~250 nm) regenerated the coumarin monomer derivative by the photo-cleavage of cyclobutane dimers, such that trapped guest molecules were released from the mesopores. The system showed great potential in light-sensitive intelligent drug delivery systems and bone tissue engineering.

5. DRUG DELIVERY FROM HIERARCHICAL POROUS SYSTEMS

Pore dimensions of OMM are in the 2-50 nm range, whereas the size of cells is in the 10-200 μm range. As a consequence, mesopores are too small to allow cell uptake, which needs dimensions in the order of microns. Therefore, bone porosity, which ranges between 1 and 3500 μm, is necessary for several physiological functions [35]. In fact, bone tissue engineering requires the design of 3D scaffolds with interconnected macroporosity within the 20-1000 μm range and sufficient mechanical strength (higher than 2 MPa) [36]. Such macroporosity is essential to allow bone cell penetration, adherence, growth and proliferation, leading to bone tissue in-growth and eventually vascularisation on implantation. As described above, OMM allow drug release systems to be designed, with the purpose of incorporating drugs to be subsequently released in a localized and controlled way [9].

In order to combine the drug up-take and release properties of MCM-41 and SBA-15 mesoporous silica with the bone regeneration capacity and the mechanical properties of bioactive glass-ceramic macroporous scaffolds belonging to two different systems (the SiO_2-CaO-K_2O (SCK) and the CaO-Na_2O-P_2O_5-MgO-K_2O-CaF_2 (Fa-GC)), different composite systems with hierarchical porosity have been developed [19-21, 37, 38].

The incorporation of OMS has been achieved by impregnation of the scaffold, upon dipping, with the OMS synthesis batch. The OMS synthesis procedure has been modified to use more gentle pH values, so to minimize the scaffold degradation, in terms of bioactivity, morphology and mechanical properties, due to the impregnation procedure.

The amount of OMS incorporated was estimated through three different methods, resulting between 10 % and 30 % for all systems and the morphology of the OMS incorporated inside the macropores of the scaffolds resulted to be the same of the OMS synthesized in powder form.

The mechanical properties turned out to be better by using an alkaline OMS synthesis batch [20]. Indeed, the presence of MCM-41 spheres, anchored on scaffolds struts, strengthen the structure and the compressive strength increases up to 3 times for Fa-GC scaffolds. On the contrary the SBA-15-based composites did not show any increase of the mechanical resistance [19].

All the systems showed an important bioactive behavior: hydroxyapatite agglomerates have been observed on the samples surface after 1 week in SBF. No significant differences were observed between OMS-containing scaffolds and as-made scaffolds during the bioactivity tests.

The presence of mesoporous silica yielded an increase of three times in the capacity of ibuprofen (IBU) incorporation for the SCK system (from 3.9% to 11.8%) and of ten times for the Fa-GC one (from 0.8 % to 8.0%) [19, 20, 38].

The IBU release curves of MCM-41-SCK and MCM-41-Fa-GC composites in SBF were very similar and showed the same features observed for MCM-41 spheres in powder form (Fig. **1**). Two distinct regions are observed: the former, between 0 and 8 hours, characterized by a very fast increase of drug concentration in SBF; the latter, between 8 and 120 hours, with a slow released which seems to continue over 120 hours. A step has been observed between 24 and 27 hours. The same feature has been reported in literature for MCM-41 with similar pore size [15] and seems to be related to pores occlusion phenomena [39, 40].

On the contrary the scaffold alone (herein named Ref) released the small amount of adsorbed IBU in the first 8 hours, clearly pointing out the role of MCM-41 phase in the incorporation and release of IBU from MCM-41-SCK and MCM-41-Fa-GC composite scaffolds.

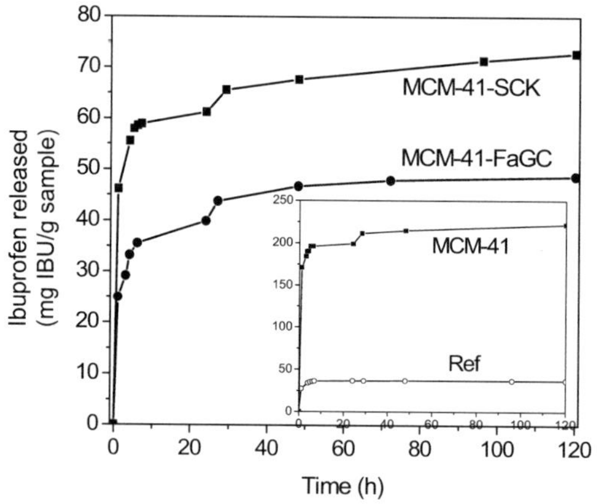

Figure 1: IBU release curves from MCM-41-SCK and MCM-41-Fa-GC composites to SBF, compared to those obtained for MCM-41 spheres in powder form and reference scaffold.

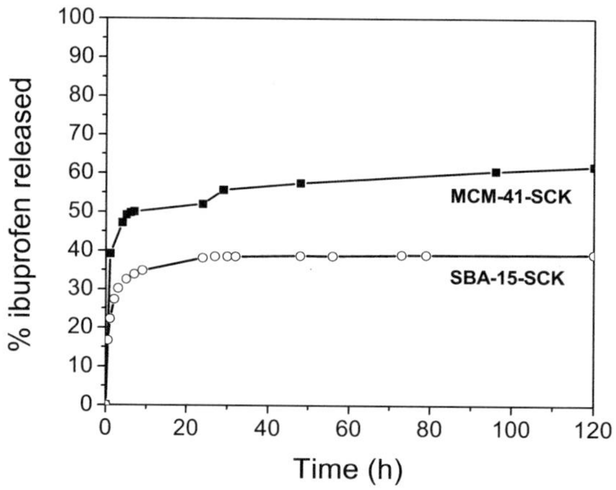

Figure 2: Percentage of released ibuprofen from SBA-15-SCK and MCM-41-SCK composites.

The percentage of released ibuprofen from SBA-15-SCK composite showed an initial burst effect, delivering around 35% of the total amount of incorporated drug in the first 10 hours; a slower release is observed between 10 and 24 hours. Then, a plateau is reached at 40%w/w.

In the case of the MCM-41-SCK system, after 120 hours around 60% of adsorbed drug has been released, *i.e.* a higher value than in the case of SBA-15-SCK (Fig. **2**).

It is worth noting that for both calcined MCM-41 and SBA-15 in powder form the release of ibuprofen in SBF in similar conditions has been reported to be complete [6, 12, 23]. The partial release from these composites might be ascribed to a pore blocking occurring during the contact time between the composite scaffolds and the SBF solution, in fact the bioactivity of the systems induces precipitation of apatite precursors on the surface in the very first hours of soaking in SBF solution. These precipitated species might partially block pores, preventing further drug diffusion out of the composite.

All the data suggested the feasibility of the proposed approach, *i.e.* the use of ordered mesoporous silica incorporated inside glass-ceramic bioactive scaffolds in view of the local drug release in bone tissue engineering.

Since it is not possible to exclude that during *in vivo* experiments some particles could be released in the body due to the more complex physiological conditions with respect *in vitro* models, the choice of an uniform morphology of the OMS may be important in the case of a possible leaching of the particles from the scaffolds, in view of the development of systems for *in vivo* applications [41]. Moreover, a homogeneous particles size is significant to model the release features in different biological conditions [42, 43].

For this reason a hierarchical system has been synthesized, combining an SCK scaffold with monosized MCM-41 silica spheres, which revealed similar properties of the above described OMS-SCK composites [37].

6. CONCLUSIONS

The combination of the sol-gel syntheses with the preparation of bioactive glasses has opened new perspectives in the chemistry of these materials [9].

It has been shown that OMS are promising for application as controlled drug delivery systems. These materials, indeed, present a high drug loading capacity and a controlled sustained release which could be helpful in bone tissue engineering [6]. However their poor bioactive behaviour and mechanical properties [8] opened up the development of new bioactive materials. On the one hand MBGs have been extensively investigated [9]. On the other hand the combination of OMS and bioactive macroporous scaffolds has been proposed in order to obtain hierarchical systems [21].

The drug release from these OMM seems to be a diffusion-controlled process, depending on pore size and functional groups on the wall. However, only a few investigations have been attempted in order to have a complete description of the real release kinetics [7]

The direct synthesis of bioactive scaffolds with hierarchical porosity and precise architectures to fit into specific bone defects also will have a deep impact in the near future [18]. In this sense, rapid prototyping techniques are excellent tools for this purpose and they may fundamental instruments for the development of a new bioceramics generation [9].

ACKNOWLEDGEMENTS

Declared none.

CONFLICT OF INTEREST

The author(s) confirm that this chapter content has no conflict of interest.

REFERENCES

[1] Ariga K, Vinu A, Hartmann M. Ordered Mesoporous Materials in the Context of Drug Delivery. In Nalwa HS, editor. American Scientific; 2005.

[2] Yang Q, Wang SH, Fan PW, Wang LF, Di Y, Lin KF, Xiao FS. pH-responsive carrier system based on carboxylic acid modified mesoporous silica and polyelectrolyte for drug delivery. Chem Mater 2005; 17: 5999-6003.

[3] Vallet-Regi M, Ramila A, del Real RP, Perez-Pariente J. A new property of MCM-41: Drug delivery system. Chem Mater 2001; 13: 308-311.

[4] Kresge CT, Leonowicz ME, Roth WJ, Vartuli JC, Beck JS. Ordered Mesoporous Molecular-Sieves Synthesized by a Liquid-Crystal Template Mechanism. Nature 1992; 359: 710-712.

[5] Hartmann M. Ordered mesoporous materials for bioadsorption and biocatalysis. Chem Mater 2005; 17: 4577-4593.

[6] Manzano M, Vallet-Regi M. New developments in ordered mesoporous materials for drug delivery. J Mater Chem 2010; 20: 5593-5604.

[7] Wang SB. Ordered mesoporous materials for drug delivery. Microporous Mesoporous Mat 2009; 117: 1-9.

[8] Vallet-Regi M. Ordered mesoporous materials in the context of drug delivery systems and bone tissue engineering. Chem-Eur J 2006; 12: 5934-5943.

[9] Arcos D, Vallet-Regi M. Sol-gel silica-based biomaterials and bone tissue regeneration. Acta Biomater 2010; 6: 2874-2888.

[10] Manzano M, Colilla M, Vallet-Regi M. Drug delivery from ordered mesoporous matrices. Expert Opin Drug Deliv 2009; 6: 1383-1400.

[11] Izquierdo-Barba I, Colilla M, Vallet-Regi M. Nanostructured Mesoporous Silicas for Bone Tissue Regeneration. J Nanomater 2008. doi:10.1155/2008/106970

[12] Vallet-Regi MA, Ruiz-Gonzalez L, Izquierdo-Barba I, Gonzalez-Calbet JM. Revisiting silica based ordered mesoporous materials: medical applications. J Mater Chem 2006; 16: 26-31.

[13] Vila M, Hueso JL, Manzano M, Izquierdo-Barba I, de Andres A, Sanchez-Marcos J *et al.* Carbon nanotubes-mesoporous silica composites as controllable biomaterials. J Mater Chem 2009; 19: 7745-7752.

[14] Izquierdo-Barba I, Ruiz-Gonzalez L, Doadrio JC, Gonzalez-Calbet JM, Vallet-Regi M. Tissue regeneration: A new property of mesoporous materials. Solid State Sciences 2005; 7: 983-989.

[15] Horcajada P, Ramila A, Perez-Pariente J, Vallet-Regi M. Influence of pore size of MCM-41 matrices on drug delivery rate. Microporous Mesoporous Mat 2004; 68: 105-109.

[16] Li N, Jie Q, Zhu SM, Wang RD. Preparation and characterization of macroporous sol-gel bioglass. Ceramics International 2005; 31: 641-646.

[17] Yan XX, Yu CZ, Zhou XF, Tang JW, Zhao DY. Highly ordered mesoporous bioactive glasses with superior *in vitro* bone-forming bioactivities. Angew Chem Int Ed 2004; 43: 5980-5984.

[18] Li X, Wang XP, Chen HR, Jiang P, Dong XP, Shi JL. Hierarchically porous bioactive glass scaffolds synthesized with a PUF and P123 cotemplated approach. Chem Mater 2007; 19: 4322-4326.

[19] Onida B, Cauda V, Fiorilli S, Verné E, Vitale-Brovarone C, Viterbo D *et al.* Incorporation of ordered mesoporous silica inside a bioactive porous scaffold in view of controlled drug release. Stud Surf Sci Catal 2005; 158: 2027-2032.

[20] Mortera R, Onida B, Fiorilli S, Cauda V, Vitale-Brovarone C, Baino F *et al.* Synthesis and characterization of MCM-41 spheres inside bioactive glass-ceramic scaffold. Chem Eng J 2008; 137: 54-61.

[21] Cauda V, Fiorilli S, Onida B, Verné E., Vitale-Brovarone C, Viterbo D *et al.* SBA-15 ordered mesoporous silica inside a bioactive glass-ceramic scaffold for local drug delivery. J Mater Sci: Mater Med 2008; 19: 3303-3310.

[22] Qu FY, Zhu GS, Lin HM, Zhang WW, Sun JY, Li SG *et al.* A controlled release of ibuprofen by systematically tailoring the morphology of mesoporous silica materials. J Solid State Chem 2006; 179: 2027-2035.

[23] Manzano M, Aina V, Arean CO, Balas F, Cauda V, Colilla M, *et al.* Studies on MCM-41 mesoporous silica for drug delivery: Effect of particle morphology and amine functionalization. Chem Eng J 2008; 137: 30-37.

[24] Cauda V, Muhlstein L, Onida B, Bein T. Tuning drug uptake and release rates through different morphologies and pore diameters of confined mesoporous silica. Microporous Mesoporous Mat 2009; 118: 435-442.

[25] Cavallaro G, Pierro P, Palumbo FS, Testa F, Pasqua L, Aiello R. Drug delivery devices based on mesoporous silicate. Drug Delivery 2004; 11: 41-46.

[26] Stein A, Melde BJ, Schroden RC. Hybrid inorganic-organic mesoporous silicates - Nanoscopic reactors coming of age. Adv Mater 2000; 12: 1403-1419.

[27] Tourne-Peteilh C, Lerner DA, Charnay C, Nicole L, Begu S, Devoisselle JM. The potential of ordered mesoporous silica for the storage of drugs: The example of a pentapeptide encapsulated in a MSU-Tween 80. Chemphyschem 2003; 4: 281-+.

[28] Horcajada P, Ramila A, Gerard F, Vallet-Regi M. Influence of superficial organic modification of MCM-41 matrices on drug delivery rate. Solid State Sciences 2006; 8: 1243-1249.

[29] Izquierdo-Barba I, Sousa E, Doadrio JC, Doadrio AL, Pariente JP, Martinez A *et al.* Influence of mesoporous structure type on the controlled delivery of drugs: release of ibuprofen from MCM-48, SBA-15 and functionalized SBA-15. J Sol-Gel Sci Technol 2009; 50: 421-429.

[30] Andersson J, Rosenholm J, Areva S, Linden M. Influences of material characteristics on ibuprofen drug loading and release profiles from ordered micro- and mesoporous silica matrices. Chem Mater 2004; 16: 4160-4167.

[31] Gao CB, Izquierdo-Barba I, Nakase I, Futaki S, Ruan JF, Sakamoto K, Sakamoto Y, Kuroda K, Terasaki O, Che S. Mesostructured silica based delivery system for a drug with a peptide as a cell-penetrating vector. Microporous Mesoporous Mat 2009; 122: 201-207.

[32] Zhu YF, Shi JL, Shen WH, Chen HR, Dong XP, Ruan ML. Preparation of novel hollow mesoporous silica spheres and their sustained-release property. Nanotechnology 2005; 16: 2633-2638.

[33] Xia W, Chang J. Well-ordered mesoporous bioactive glasses (MBG): A promising bioactive drug delivery system. J Control Release 2006; 110: 522-530.

[34] Lin HM, Wang WK, Hsiung PA, Shyu SG. Light-sensitive intelligent drug delivery systems of coumarin-modified mesoporous bioactive glass. Acta Biomater 2010; 6: 3256-3263.

[35] Sanchez-Salcedo S, Werner J, Vallet-Regi M. Hierarchical pore structure of calcium phosphate scaffolds by a combination of gel-casting and multiple tape-casting methods. Acta Biomater 2008; 4: 913-922.

[36] Hollister SJ. Porous scaffold design for tissue engineering. Nat Mater 2005; 4: 518-524.

[37] Mortera R, Baino F, Croce G, Fiorilli S, Vitale-Brovarone C, Verné E *et al.* Monodisperse Mesoporous Silica Spheres Inside a Bioactive Macroporous Glass-Ceramic Scaffold. Adv Eng Mat 2010; DOI: 10.1002/adbi.200900075.

[38] Vitale-Brovarone C, Baino F, Miola M, Mortera R, Onida B, Verné E. Glass-ceramic scaffolds containing silica mesophases for bone grafting and drug delivery. J Mater Sci: Mater Med 2009; 20: 809-820.

[39] Mortera R, Fiorilli S, Garrone E, Onida B. Structural changes of MCM-41 spheres during ibuprofen release to SBF. Stud Surf Sci Catal 2008; 174B: 1001-1004.

[40] Mortera R, Fiorilli S, Garrone E, Verné E, Onida B. Pores occlusion in MCM-41 spheres immersed in SBF and the effect on ibuprofen delivery kinetics: A quantitative model. Chem Eng J 2010; 156: 184-192.

[41] Huang SS, Yang PP, Cheng ZY, Li CX, Fan Y, Kong DY, Lin J. Synthesis and characterization of magnetic FexOy@SBA-15 composites with different morphologies for controlled drug release and targeting. J Phys Chem C. 2008; 112: 7130-7137.

[42] Amatore C. Theoretical trends of diffusion and reaction into tubular nano- and mesoporous structures: General physicochemical and physicomathematical modeling. Chem-Eur J 2008; 14: 5449-5464.

[43] Costa P, Lobo JMS. Modeling and comparison of dissolution profiles. Eur J Pharm Sci 2001; 13: 123-133.

CHAPTER 14

Methods to Study the Mechanism of Bioactivity

Antonio J. Salinas* and María Vallet-Regí

Departamento de Química Inorgánica y Bioinorgánica, Facultad Farmacia; Universidad Complutense; 28040 Madrid; Spain, and Centro de Investigación Biomédica en Red: Bioingeniería, Biomateriales y Nanomedicina, CIBER-BBN, Spain

Abstract: The formation of so-called bioactive bond in which specific compositions of glasses and glass-ceramics strongly bond with bone, is a complex process highly influenced by the material surface reactivity when in contact with the biological fluids and cells. Indeed, non bioactive materials elicit a foreign body reaction and are isolated by a non adherent fibrous capsule responsible for the failure of many prostheses. It is clear that understanding the complex mechanisms that form the bioactive bond can help to design new prostheses and improve the present, increasing their survival rates with the consequent economic and social benefits. An important role in the bioactive bond formation is played by biomolecules and cells, which can only be investigated in complex *in vitro* cell cultures studies, aided by bioreactors, or *in vivo*, in animal models. However, the reason why an implanted material is isolated from the body by a fibrous capsule or strongly bonded to osseous tissue is its surface reactivity that can be evaluated under *in vitro* conditions in artificial solutions with certain blood plasma characteristics like pH, temperature, inorganic ion concentration, *etc.* These solutions were designed to evaluate new candidates to implantable materials and have also led to great advances in the study of the mechanisms of bioactivity. This chapter presents the methods used to study the mechanism of formation of bioactive bond.

Keywords: Bioactive glasses, glass ceramics, SBF solution, bioactivity, hydroxyapatite.

1. BIOACTIVE GLASSES AND GLASS-CERAMICS

Bioactive glasses and glass-ceramics were extensively investigated in recent years for use in bone tissue replacements or to coat non bioactive biomaterials [1-6]. These bioceramics have an interesting property for clinical applications because

***Address correspondence to Antonio J. Salinas:** Departamento de Química Inorgánica y Bioinorgánica Facultad de Farmacia; Universidad Complutense; 28040 Madrid; Spain, and Centro de Investigación Biomédica en Red: Bioingeniería, Biomateriales y Nanomedicina, CIBER-BBN, Spain; Tel: +3491 3941790; Fax: +3491 3941786; E-mail: salinas@farm.ucm.es

Lia Rimondini, Claudia Bianchi and Enrica Vernè (Eds)

they bind directly to bone without being isolated by a not adherent fibrous capsule. This capsule causes relative micromovements in the *in vivo* biomaterial/tissue interface yielding to the failure of many prostheses. Instead, bioactive bioceramics provide a stable union with living tissues which should produce a significant increase in the duration of the prosthesis. For this reason, in the 1980s and 1990s, bioactive glasses and glass-ceramics were profusely investigated to obtain as high as possible bioactivity for bioglasses, mainly aimed to be used for bone grafts, and with enhanced mechanical properties for bioactive glass-ceramics that, in general, exhibited a slower bioactive response than bioglasses [7-9].

Most of the glasses and glass-ceramics investigated as biomaterials contain SiO_2 as network former, and other metallic oxides actuating as network modifiers that were included with different purposes. For instance, initial bioactive glasses, obtained by quenching of a melt, always contained an alkaline oxide, generally Na_2O, to decrease the melting temperature making cheaper the glass manufacture. In the case of the bioglasses, designed for implants, it was verified, precisely from *in vitro* studies as will be described later, that these oxides of monovalent cations also produced an increase in the glass reactivity in aqueous media, playing a leading role in the formation of the bioactive bonding. Also, the traditional glasses contained oxides of divalent and trivalent cations to increase their stability. With this objective, and taking into account the inorganic component of bone is a calcium phosphate, bioactive glasses usually contained also CaO and P_2O_5, the second one that should play a role as a network former, given the high valence of phosphorus (V). Furthermore, P_2O_5 use to be present in glasses in relatively small amounts due to the necrosis of bone described for the presence of large amounts of phosphorus [10].

In recent times most part of bioactive glasses and glass-ceramics obtained by the classic method of melting and quenching belong to the Na_2O-K_2O-CaO-MgO-P_2O_5-SiO_2 system. For glasses, the compositional limits to present bioactivity were established as 14-30 mol% of alkali oxides ($Na_2O + K_2O$), 14-30 mol% of alkaline earth oxides (CaO + MgO) and < 59 mol % of SiO_2 [11]. In addition, small amounts of other oxides have been included looking specific improvements as bactericidal activity, inhibition of bone resorption or antioxidant effects or with oxides as Ag_2O, ZnO, or even Ce_2O_3 y Ga_2O_3 [12-15]. Regarding the glass-ceramics, mainly obtained in the mentioned system, main research was focused on the thermal treatments to partially devitrify the parent glasses, analyzing the effects of the proportion crystalline phase/glassy phase and the amount and type of crystalline phases in the bioactive response and the mechanical behavior of materials [16]. Some authors also proposed the chemical etch to induce or increase bioactivity in glass-ceramics with striking mechanical properties for biomedical applications [17].

2. *IN VITRO* STUDIES: A PRELIMINARY STEP TO ASSESS BIOCERAMICS

Although the first synthesized bioactive glasses were directly implanted in experimental animals [10], it was soon found that this was an unreasonable approach since, apart from the ethical aspects associated with animal suffering, materials should be implemented fairly long periods (from several weeks to some months) before knowing if a particular glass or glass-ceramic composition exhibited a bioactive response. However, experimental data obtained in these initial *in vivo* studies brought a very interesting result which took effect in the further development of new bioactive ceramic. Indeed, analyzing the interface between the bioactive glass and bone it was found the presence in the bonding layer of hydroxycarbonate apatite (HCA) nanocrystals analogous of that present in bone. In other words, the bonding layer between the biomaterial and living tissue had a structure similar to the natural bone, which is a composite that HCA nanocrystals served as reinforcement of collagen fibers. This caused some authors submitting a prerequisite that a material present *in vivo* bioactivity was such material to have a surface reactivity such that be coated by a layer of HCA nanocrystals when soaked in a fluid simulating blood plasma.

In Fig. **1** is displayed the sequence of methods with increasing complexity followed to evaluate new bioactive ceramics showing the *in vitro* studies in acellular solutions in the first step.

Initially, because studied glasses already contained the ions calcium and phosphate that could be released to the medium to form the calcium phosphate layer, the first assay solution was an aqueous solution buffered at plasma pH of 7.40 with tris(hydroxymethyl) aminomethane and hydrochloric acid (TRIS buffer) at 37 ºC [18]. This solution was enough to evaluate glasses with high levels of bioactive response. However, when at the end of the 1980s Kokubo *et al.* tried to evaluate glass-ceramics that had shown their ability to bond to bone when implanted in animals, they checked that the HCA layer was not formed in TRIS buffer. Therefore, they decided to develop a new solution in which HCA layer be formed on the glass-ceramic surface. Result of these studies was a set of solutions being that denoted K9 which produced the best results. K9 is usually considered the original Simulated Body Fluid (SBF) [19]. SBF is an acellular solution that contains a concentration of inorganic ions almost equal to plasma (see Table **1**) buffered at physiological pH with TRIS buffer. As you can see in the Table **1** that solution was improved including sulfate ions at the expenses of chloride ions and this corrected-SBF is that most researchers nowadays called as SBF [16]. Numerous changes and improvements of the SBF solution have been proposed, as it will be seen later, but traditional assays soaking the materials in SBF under static conditions is an easy and quick way to evaluate new candidates for bioactive biomaterials.

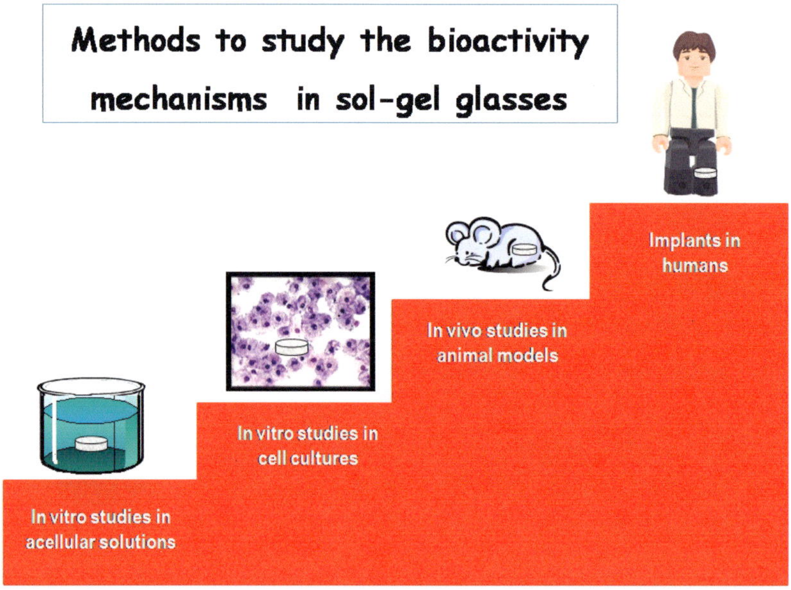

Figure 1: Sequence of studies before a new glass or glass-ceramic is implanted.

However, what interests us here is how these trials in simulated biological fluids allowed further bioactivity mechanism and also how they contributed greatly to design new bioactive glasses and glass-ceramic.

For example, these tests allowed verifying that the formation of the layer rich in calcium and phosphorus is a surface process where the new silanol groups (Si-OH) formed on the surface when the glass is soaked play a key role. For this reason, Li *et al.* [18] proposed get through a processing sol-gel obtained porous bioactive glass with high specific surface area and numerous silanol groups even before be immersed in the assay solution. As a result, an important family of bioactive sol-gel glasses can be considered a direct consequence of the results obtained in the *in vitro* bioactivity trials. These sol-gel glasses also contain a wide distribution of pores in the mesoporous region and present a textural properties has much to do with the thermal treatments and the composition of the glass, mainly the relative proportion of CaO and SiO_2 [21]. In general, greater CaO contents increase the pores size and reduce specific surface area whereas higher SiO_2 contents produce an increase in the specific surface area as well as the amount of micropores in the material. Furthermore, it were recently proposed so called bioactive "template glasses" a glass family of sol-gel glasses obtained by using templates, that act as directing structure agents, as used in the synthesis of well known materials silica with ordered mesoporosity [22].

3. *IN VITRO* STUDIES: A POWERFUL TECHNIQUE TO STUDY THE MECHANISM OF BIOACTIVITY

In vitro studies in solutions simulating human plasma have become essential tools to understand the mechanism of bioactivity in bioactive glasses and glass-ceramics. As it was said, these studies were initially designed to preselect the best materials to implant in animal models. The starting point of the studies is that bioactive bioceramics display this behavior when implanted because they are able to be coated by a layer of HCA nanocrystals, analogous to those present in bone. That way, when collagen fibers coming to encapsulate the material find the HCA layer, new bone is formed firmly bonding artificial material and living tissue. Consequently, these studies typically aimed at keeping soaked materials believed to have a compatible with the formation of a bond bioactive surface reactivity into a solution at 37 °C mimicking blood plasma. After selected times materials are extracted and, after drying, characterized by various experimental techniques to assess the possible variations experienced by the materials surface. Furthermore, changes experienced by the assay solution indicative of specific processes of materials dissolution or precipitation of certain salts during the assays are also monitored. In Fig. **2** are including some of the experimental techniques used to check the HCA layer formation: X-ray diffraction (XRD), Fourier Transform Infrared (FTIR) spectroscopy, Scanning Electron Microscopy (SEM), Energy Dispersive X-ray Spectroscopy (EDS), Transmission Electron Microscopy (TEM), Electron diffraction (ED), X-ray Photoelectron Spectroscopy (XPS), Nuclear Magnetic Resonance (NMR), *etc*.

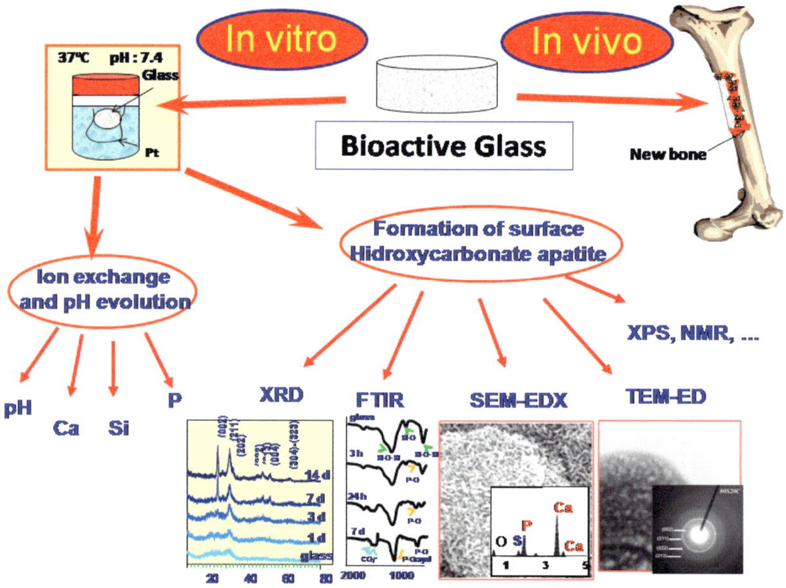

Figure 2: The bioactive glass will bond to bone when implanted. In a SBF solution it suffers surface reactions producing variations in the solution and in the glass surface that can be monitored by different experimental techniques.

The *in vitro* studies have also allowed detect the mechanism of formation of bioactive bond. Indeed, compared with the traditional *in vivo* tests where the behavior of the material in contact with the living tissue is evaluated after weeks or months, in the *in vitro* tests, simple and with not bloody nature, materials can be assessed within minutes of immersion. This has made it possible to know that although the new bone formation process need weeks, the glasses with high levels of bioactivity undergo changes on its surface after few seconds in contact with aqueous media used for the so-called *in vitro* bioactivity tests.

3.1. Ways to Perform the *In Vitro* Bioactivity Studies

There are many parameters which can be modified during *in vitro* studies in acellular solutions including the solution composition and dynamic conditions, the shape of the material or the techniques that confirms the bioactivity. Following the main aspects to consider when *in vitro* bioactivity studies are carried out.

The *In Vitro* Solution: For the initial bioactive glasses developed by Hench *et al.* it was used TRIS buffer with a pH of 7.4, identical to plasma. Evaluated glasses contained Ca, P, Si and Na oxides. The Na^+ ions were very quickly leached to the solution which allowed a layer HCA formation if the glass composition were within the bioactivity window [1]. However, when bioactive apatite-wollastonite (A-W) glass-ceramic proposed by Kokubo *et al.* was evaluated in TRIS it was not able to form the HCA layer. For this reason, Kokubo *et al.* proposed the Simulated Body Fluid (SBF) [19]. SBF use also TRIS to adjust the pH to 7.4 but also contains also a concentration of inorganic ions almost equal to human plasma (See Table **1**). As is observed, the main difference is the concentration of

Table 1: Composition of some simulated body fluids compared with blood plasma (mM).

	Na^+	K^+	Mg^{2+}	Ca^{2+}	Cl^-	HCO_3^-	HPO_4^{2-}	SO_4^{2-}	pH	Buffer
Blood Plasma	142.0	5.0	1.5	2.5	103.0	27.0	1.0	0.5	7.35-7.43	biological
Tris-buffer	-	-	-	-	-	-	-	-	7.4	Tris
Original SBF	142.0	5.0	1.5	2.5	148.8	4.2	1.0	-	7.4	Tris
Corrected SBF	142.0	5.0	1.5	2.5	147.8	4.2	1.0	0.5	7.4	Tris
CISP*	142.0	5.0	1.5	2.5	103	24-27	1.0	0.5	7.4	CO_2 / HCO_3^-

*Carbonated Simulated inorganic plasma (from Ref. [20]).

carbonate ions, 4.2 mM, significantly lower than the concentration in plasma, 27 mM. It must be take into account that it is not possible to dissolve more carbonate in the solution under the ambient conditions. To keep the electroneutrality of the solution, the carbonate anion deficit is offset by a chloride ions concentration greater than in plasma. In addition SBF is supersaturated with respect to apatite and must be prepared carefully to avoid unwanted precipitation.

When attempts were made to obtain a solution with an analogous to the plasma carbonate ions concentration, there was a trend to coprecipitate calcium carbonate with the HCA. Several authors have attempted to remove the Tris buffer for being a buffer not physiological and also because its tendency to complex calcium ions which would exert a harmful influence on the study of bioactivity had been published [20]. Also it was proposed the use of commercial solutions, like Hanks Balanced Salt Solution, HBBS, stable for longer times that SBF. In fact, one of the main drawbacks of SBF is that, because it is metastable, must be prepared when being used and stored refrigerated for a maximum of 60 days (following the initial Kokubo´s recipe) or 30 days, following his more recent advices [23]. The top of Fig. **3** shows the classic form of the *in vitro* testing in different types of solutions, under static conditions and without renewing the dissolution (integral method). The bottom of the figure shows several modifications and improvements introduced in this type of studios, some of which have been proposed by the authors of this chapter [20, 24].

Shape of the Sample: Another important issue to be considered in the *in vitro* tests is the shape in which the sample is used (powder, pieces, compacts). Some authors recommend carrying out the *in vitro* bioactivity tests with the samples as powders because in this way the surface of the solid that contacts the liquid, and the subsequent reactivity increases. However, our group most of the times studied the samples as pieces, even obtained compacting powders, because in this way, it is available a surface to determine if some new material was formed or not as a result of the SBF treatment.

Static or Stirring Conditions: Obviously, the HCA formation process is a chemical process and the agitation (magnetic or orbital) will increase the homogenization of the dissolution facilitating chemical reaction. However, an excessive stirring can hinder the deposition a phase type apatite on the surface of bioactive glasses and glass-ceramics.

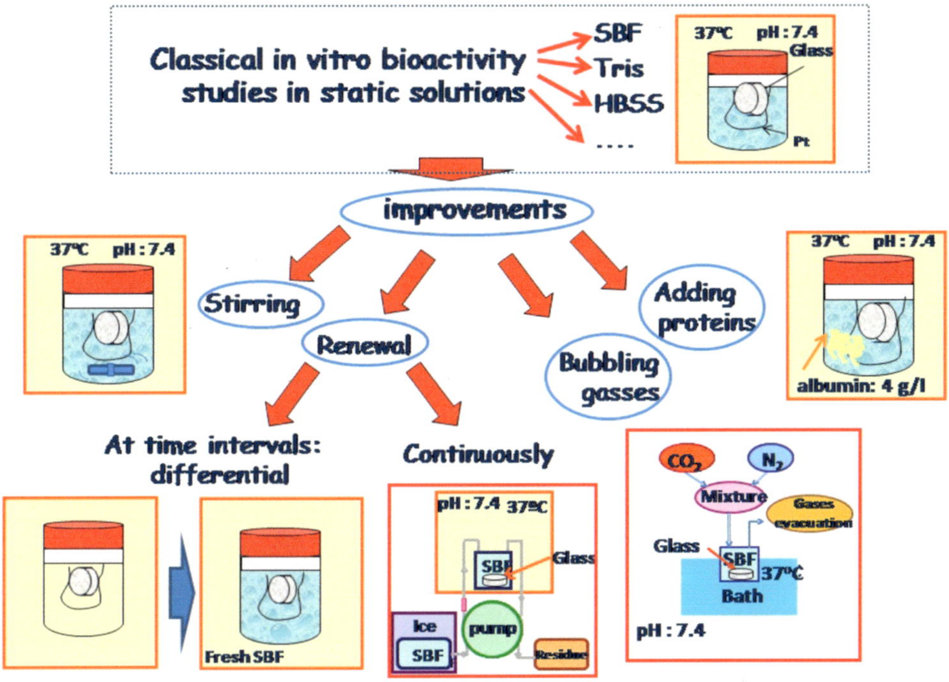

Figure 3: The classical *in vitro* studies in SBF or analogous solutions, and some improvements proposed by different authors to mimic better the physiological conditions.

Renewal or not the Dissolution: SBF featuring advantages when used to evaluate the bioactivity of bioceramics with bioactivity average as the A-W glass ceramic, showed enormous changes in the concentration of the dissolution when it was used to evaluate the highly bioactive glasses. For this reason, some authors proposed the renewal of the assay solution at time intervals (differential method). Others proposed the use of a dynamical system with the use of peristaltic pumps made a continuous renewal of the dissolution test which would be identical to the initial concentration at all times (dynamic or continuous method) [24]. It has been very informative in terms to understand aspects of the mechanism of bioactivity, to compare results obtained for highly bioactive glass studied in static and integral SBF and in dynamic.

Table **2** shows the most outstanding results for both methods when used to study a sol-gel glass of composition 80 mol% SiO_2 and 20 mol% CaO (80S20C glass). From these results, different mechanisms of formation of the apatite were proposed in each case, mainly based on the pH differences in the assay solution (see Fig. **4**).

Table 2: Comparison of a highly bioactive sol-gel glass (80 mol % SiO_2-20 mol % CaO) studied in SBF under static conditions or with continuous renewal of the test solution.

	Corrected SBF (initial)	**Static (Integral) (after 7 days)**	**Continuous renewal (after 7 days)**
Variations in solution			
pH	7.4	8.0	7.4
Ca (mM)	2.5	8.5	2.5
P (mM)	1.0	0.3	0.8
Si (mM)	0	2.1	0
Variations in the layer			
Formation a-CaP* (hours)		1.5	3
Cristal size of HCA after 7 d (nm)		>5 & <14	>14
Ca/P molar ratio (after 7 d)		1.6	1.2
Layer thickness (μm)		2.5	13
Presence of Si		Yes	No

* a-CaP: amorphous calcium phosphate.

3.2. Techniques to Analyze the New Layer Formed on Bioactive Ceramics

To characterize the glasses and glass-ceramics surface after *in vitro* tests one of the most important techniques is the Fourier Transform Infrared Spectroscopy (FTIR). New materials that form *in vitro* in the first moments exhibit small crystal size. For this reason diffraction techniques, such as X-ray Diffraction (XRD) or Electron Diffraction (ED) are very little useful until materials formed have reached enough size to be detected by ED (5 nm) or XRD (14 nm) [25].

However, the FTIR spectroscopy is very sensitive to the presence of small amounts of phosphate groups even found in the amorphous form. Also detects the presence of silanol groups on the surface of the glass which, play an important role in the formation of calcium phosphate layer. Finally, FTIR spectroscopy have two other important features when it comes to study the bioactivity of bioactive glasses and glass-ceramics: (i) it allows to detect the presence of carbonate groups in the HCA formed, and (ii) is also able to detect a certain crystallinity degree in the calcium phosphate formed. This last feature is very striking and has played an important role both the development of the mechanism of bioactivity and the study of new materials designed for implants. Indeed the FTIR spectroscopy is not a diffractometric measurement but it is very sensitive to the presence of specific functional groups. However, in the *in vitro* bioactivity studies, is particularly interesting to note that the band around 570 cm^{-1} indicative of amorphous calcium phosphate, is split into two bands at approximately 563 and 603 cm^{-1} when phosphate acquires a certain degree of crystallinity. This was confirmed when the new layer formed was studied by other experimental techniques. On the other hand, XRD is not a very appropriate to study the new layer formed *in vitro* since, except for long immersion periods of materials with high bioactivity, the size of the HCA crystals is on the edge of nanometers. As a result, XRD maxima are diffuse and overlapped making difficult the correct phase identification.

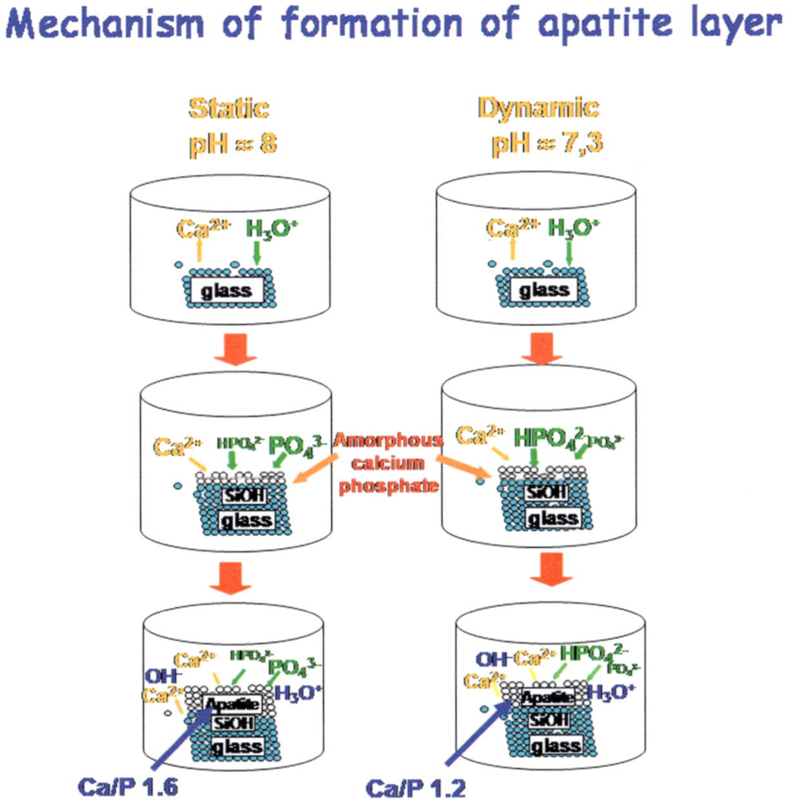

Figure 4: Comparison of the mechanism proposed to explain the differences in the calcium phosphate formed in static conditions and with continuous renewal of the solution.

A very important experimental technique is SEM that can detect the formation of the new layer formed. In addition, when a backscattered electrons detector is used it can be provide interesting information on compositional variations in the layer. It is also of interest to use the EDS analysis, frequently available as in SEM microscopes that reports semiquantitative composition variations. Very interesting is also to compare the results of EDS analysis that penetrate into the layer distances on the order of one micrometer, with XPS that analyze only the first few nanometers. This comparison allows checking the evolution in composition of the layer with the immersion time in SBF and detecting the heterogeneities with the deep in the composition of the layer.

4. MECHANISM OF BIOACTIVITY IN BIOACTIVE GLASSES AND GLASS-CERAMICS

It is necessary more than one experimental technique to propose the formation of bioactive bonding mechanism. They are a set of more in some cases provide different views of the same phenomenon and others supplement which have allowed to propose these mechanisms have slight differences depending on the type of material and its kinetic of bioactive response.

There is a mechanism that explains the bioactive glasses and glass-ceramics widely disseminated in the scientific community that was proposed in 1991 by Hench [1]. According to this mechanism, the bioactive bond is formed by a sequence of reactions that takes place on the surface of the material when it contact with the biological fluids. Table **3** shows the sequence of 11 stages of reaction occurring in a bioactive glass surface when in contact with biological fluids and leading to new bone production. The 5 early stages, which are reproduced in the *in vitro* bioactivity tests, require few days, even less, and will make the time required for the bone formation. It is also interesting to note that in less bioactive glasses or glass-ceramics, stages 1-3 overlapped with 4 and 5 and instead detecting a silica-rich layer, is directly detected the newly formed HCA layer.

A slightly different mechanism for bioactive glasses at the CaO-P_2O_5-SiO_2 system was proposed by Kokubo in 1992 [26]. The key role in the formation of the layer was played by the Ca^{2+} ions leached to the medium, because new Si-OH groups were formed and the apatite supersaturation of solution increased, both factors favoring the HCA formation.

Table 3: Sequence of 11 stages that occur at the surface of a bioactive glass, after Hench [1].

	Chemical/Biological processes	**Effect**	**Result**	**Time required**
1	Interchange Na^+ and H_3O^+	Formation of Silanol (Si-OH)	Silca rich layer	minutes
2	Loss of soluble silica	Release of $Si(OH)_4$; formation Si-OH		
3	SiO_2 condensation and repolymerization	Formation of Siloxane bonds (Si-O-Si)		
4	Migration of Ca^{2+}, $PO_4^{3-}$to the silica rich layer	Formation of amorphous calcium phosphate (a-CaP)	a-CaP	hours
5	Crystallization incorporating OH^-, CO_3^{2-}	Crystallization of hydroxycarbonate apatite (HCA)	HCA	days
6	Adsorption of biological moieties in the HCA layer	Adsorption and desorption of growth factors	Growth of Bone	weeks
7	Action of macrophages	Preparation the implant site for tissue repair		
8	Adsorption of Osteoblasts (OBs) stem cells	Attachment stem cells		
9	Differentiation and proliferations of OBs	Synchronized proliferation and differentiation of the cells		
10	Generation of matrix	Growth factors stimulating cell proliferation and formation extracellular matrix proteins		
11	Crystallization of matrix	Mature osteocytes in a collagen-HCA matrix		

In 2001, Salinas *et al.* [27] studied the differences between the calcium phosphate-rich layers after soak 7 days in SBF a bioactive sol-gel glass in two different conditions: static SBF without renewal (integral) and SBF continually renovated with the help of a peristaltic pump (continuous). Different formation mechanisms were proposed to explain the differences detected in the newly formed layer as a result of the compositional differences and pH of solution (see Fig. **4**).

In 2002 Salinas *et al.* [28] studied the effect in the bioactivity of include amounts of phosphorus in a SiO_2-CaO glass. In P-free glass, the amorphous calcium phosphate was formed in a few hours, but the HCA formation required 7 days. On the contrary, when amounts of P_2O_5 up to 5% mol were included it was becoming necessary longer times to form amorphous calcium phosphate, but the nanocrystalline HCA was detected after only 4 days of soaking. In 2005 Vallet-Regí *et al.* [29] analyzed the nanostructure of two glasses with and without by High Resolution TEM (HRTEM) checking that phosphorus bind to calcium forming Si doped β-Tricalcium phosphate (Si-β-TCP) nanocrystals. These results allow to propose different mechanisms according to which the glass containing or not, phosphorus. On the P-containing glass, calcium ions leaching were impeded, so hampered both the formation of new silanol groups and increase the supersaturation of solution explaining the more time required to form amorphous calcium phosphate. However, in the P-containing glass the Si-β-TCP nanocrystals acted as crystallization nucleii so virtually formed directly HCA without going through the stage of amorphous calcium phosphate.

On the other hand, in 2006 Lopez-Noriega *et al.* [22] proposed the manufacture of new bioactive glasses with ordered mesoporosity in the presence of surfactants acting as structure directing agents. In some of these template glasses it was found that stage 5 of Table **3**, *i.e.* the crystallization of HCA, was detected after only 8 hours of immersion in SBF. A detailed study of the process by HRTEM performed by Izquierdo-Barba *et al.* [30] showed that in this case, and as consequence of high bioactivity of the material, it was obtained octacalcium phosphate (OCP), as intermediate stage analogous to formed in the *in vivo* biomineralization processes. In this case, although the pH of SBF greatly increased because of the high reactivity of the material, the surface pH of the silica rich layer with many silanol groups in template glass was considerably below (6.7) than in conventional sol-gel glasses (7.4) and this made possible the above-mentioned OCP deposition.

The most important aspects of the bioactivity mechanisms referred in this section are shown in a comparative way in Fig. **5**.

5. CONCLUSIONS

In vitro tests are a powerful tool to determine the bioactivity mechanism of bioactive glasses and glass-ceramics. In 1991 and 1992 Hench and Kokubo proposed respectively two mechanisms highlighting the importance of the formation of a layer rich in silica, Hench, as a consequence of leaching of Na^+ ions, Kokubo, by the formation of new silanol groups and increase of the supersaturation of solution, both effects due to leaching of Ca^{2+} ions. Both mechanisms are correct for materials and conditions in which were studied. However, our group proposed specific mechanisms when the conditions of the *in vitro* studies changed, for example: static SBF *vs.* continuous SBF, Si-Ca sol-gel glass *vs.* Si-Ca-P sol-gel glass or conventional sol-gel glass *vs.* greatly-bioactive template glass. The main aspects of the above-mentioned mechanisms have been summarized in Fig. **5**.

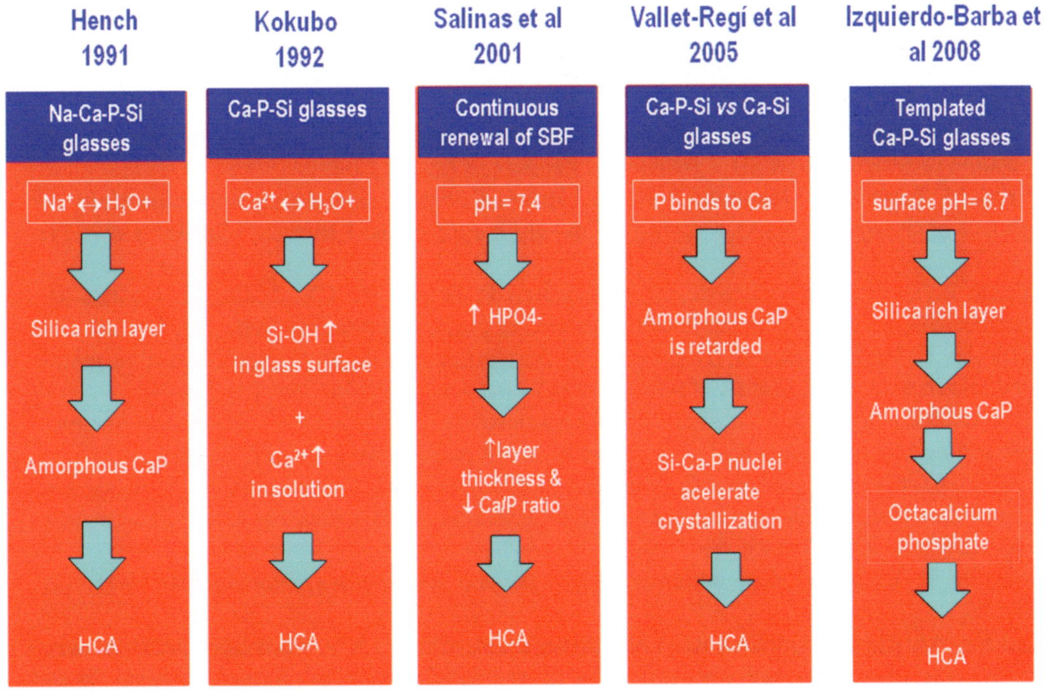

Figure 5: Some mechanisms explaining the HCA formation in bioactive glasses and glass-ceramics.

ACKNOWLEDGEMENTS

Declared none.

CONFLICT OF INTEREST

The author(s) confirm that this chapter content has no conflict of interest.

REFERENCES

[1] Hench LL. Bioceramics: from concept to clinic. J Am Ceram Soc 1991; 74: 1487-510.

[2] Vallet-Regí M. Ceramics for medical applications. J Chem Soc, Dalton Trans 2001; 97-108.

[3] Vallet-Regí M. Revisiting ceramics for medical applications. Dalton Trans. 2006; 5211-20.

[4] Salinas AJ, Vallet-Regí M. Evolution of ceramics with medical applications. Z Anorg Allg Chem 2007; 1762-73.

[5] Vallet-Regí M. Bioceramics: where do we come from and which are the future expectations. Key Eng Mater 2008; 377: 1-18.

[6] Vitale-Brovarone C, Verné E, Robiglio L, Martinasso G, Canuto RA. Biocompatible glass-ceramic materials for bone substitution. J Mater Sci: Mater Med 2008; 19: 471-8.

[7] Hench LL, Anderson O. Bioactive glasses. In: Hench LL, Wilson J, Eds. An introduction to bioceramics. Singapore, World Scientific, 1993, pp. 41-62.

[8] Vallet-Regí M, Ragel CV, Salinas AJ. Glasses with medical applications. Eur J Inorg Chem 2003; 1029-42.

[9] Kokubo T, Kim HM, Kawashita M. Novel bioactive materials with different mechanical properties. Biomaterials 2003; 24: 2161-75.

[10] Hench LL, Splinter RJ, Allen WC, Greenlee TK. Bonding mechanisms at the interface of ceramic prosthetic materials. J Biomed Mater Res 1972; 2: 117-41.

[11] Brink M, Turunen T, Happonen R-P, Yli-Urpo A. Compositional dependence bioactivity of glasses of the system Na_2O-K_2O-MgO-CaO-B_2O_3-P_2O_5-SiO_2. J Biomed Mat Res 1997; 37: 114-21.

[12] Verné E. Di Nunzio S, Bosetti M, Appendino P, Brovarone CV, Maina G, Cannas M. Surface characterization of silver-doped bioactive glass. Biomaterials 2005; 25: 5111-9.

[13] Kokubo T, Ito S, Sakka S, Tamamuro T. Formation of a high strength bioactive glass-ceramic in the system MgO-CaO-SiO_2 -P_2O_5. J Non-Cryst Solids 1986; 21: 536-40.

[14] Du RN, Chang J, Ni SY, Zhai NY. Characterization and *in vitro* bioactivity of zinc-containing bioactive glass and glass-ceramics. J Biomater Appl 2006; 20: 341-60.

[15] Salinas AJ, Shruti S, Malavasi G, Menabue L, Menziani MC, Vallet-Regí M, Synthesis and characterization of Ce_2O_3, Ga_2O_3 and ZnO containing mesoporous bioactive glasses Acta Biomaterialia 2011; 7: 3452-8.

[16] Kokubo T. Bioactive glass ceramics properties and applications, Biomaterials 1991; 12: 155-63.

[17] Salinas AJ, Román J, Vallet-Regí M, Oliveira JM, Correia RN, Fernandes MH. *In vitro* bioactivity of a glass and glass-ceramics of the $3CaO \cdot P_2O_5$-$CaO \cdot SiO_2$-$CaO \cdot MgO \cdot 2SiO_2$ system. Biomaterials 2000; 21: 251-7.

[18] Li R, Clark A E, Hench L L, An investigation of bioactive glass powders by sol-gel processing. J Appl Biomat 1991; 2: 231-9.

[19] Kokubo T, Kushitani H, Sakka S, Kitsugi T, Yamamuro T, ′Solutions able to reproduce *in vivo* surface-structure changes in bioactive glass-ceramic A-W`. J Biomed Mat Res 1990; 24: 721-34.

[20] Marques PAAP, Magalhaes MCF, Correia RN, Martin AI, Salinas AJ, Vallet-Regí M. Ceramics *In vitro* mineralization Protocols: a supersaturation problem. Key Eng Mater 2004; 254-256: pp. 143-146.

[21] Vallet-Regí M, Salinas AJ, Martínez A, Izquierdo-Barba I, Pérez-Pariente J. Textural properties of CaO-SiO_2 glasses for use in implants. Solid State Ionics 2004; 172: 441-4.

[22] López-Noriega A, Arcos D, Izquierdo-Barba I, Sakamoto Y, Terasaki O, Vallet-Regí M. Ordered mesoporous bioactive glasses for bone tissue regeneration. Chem Mat 2006; 18: 3137-44.

[23] Kokubo T, Takadama H. How useful is SBF in predicting *in vivo* bone bioactivity? Biomaterials 2006; 110; 27: 2907-15.

[24] Izquierdo-Barba I, Salinas AJ, Vallet-Regí M. Effect of continuous solution exchange on the *in vitro* reactivity of a CaO-SiO_2 sol-gel glass. J Biomed Mater Res 2000; 51: 191-99.

[25] Vallet-Regí M, Izquierdo-Barba I, Salinas AJ. Influence of P_2O_5 on crystallinity of apatite formed *in vitro* on surface of bioactive glasses. J Biomed Mater Res 1999; 46: 560-5.

[26] Kokubo T. A/W Glass-ceramic: processibng and properties. In: Hench LL, Wilson J, Eds. An introduction to bioceramics. Singapore, World Scientific, 1993, pp. 75-88.

[27] Salinas AJ, Vallet-Regí M, Izquierdo-Barba I. Biomimetic apatite deposition on calcium silicate gel glasses. J Sol-Gel Sci Techn 2001; 21: 13-25.

[28] Salinas AJ, Martín A I, Vallet-Regí M. Bioactivity of three CaO-P_2O_5-SiO_2 sol-gel glasses. J Biomed Mater Res 2002; 61: 524-32.

[29] Vallet-Regí M, Salinas AJ, Ramírez-Castellanos J, González-Calbet JM. Nanostructure of bioactive sol-gel glasses and organic inorganic hybrids. Chem Mater 2005; 17: 1874-9.

[30] Izquierdo-Barba I, Arcos D, Sakamoto Y, Terasaki O, López-Noriega A, Vallet-Regí M. High performance mesoporous bioceramics mimicking bone mineralization. Chem Mater 2008; 20: 3191-8.

CHAPTER 15

New Trends in Bone Tissue Engineering Scaffolds: Hierarchically Porous Glass and Glass-Ceramic Structures

Francesco Baino* **and Chiara Vitale Brovarone**

Institute of Materials Engineering and Physics, Applied Science and Technology Department – DISAT, Politecnico di Torino, Corso Duca degli Abruzzi 24, 10129, Torino, Italy

Abstract: Biomaterial choice and porosity features play a crucial role in scaffolds design. Bioactive glasses are ideal candidates for manufacturing bone tissue engineering scaffolds as they can bond to bone and stimulate bone cells towards osteogenesis. A 3-D network of highly interconnected macropores is essential to allow cells migration into the scaffolds, as well as the *in vivo* vascularization of the implant. An ordered structure of nanopores can impart an added value to the scaffold, enhancing its bioactive properties and providing drug uptake/release abilities to the system. This chapter highlights the great potential carried by hierarchical glass scaffolds, exhibiting pore scales from the meso- to the macro-range, and their impact in the field of bone tissue engineering.

Keywords: Scaffolds, bioactive glass, macropores, mesopores, bone repair.

1. INTRODUCTION

Over the last decade, one of the most important trends of biomaterials research has been the increasing acceptance of biologically inspired approaches. In fact, biological structures can carry new ideas and design concepts towards the development of ideal biomaterials and implants. It s known that the complex functionality of the various natural tissues includes a continuous change from one structure to another one. For instance, the fascinating architecture of bone, with a transition from a dense and stiff external structure (cortical bone) to a porous internal one (cancellous bone) clearly shows that functional gradation has been a key concept of biological adaptation: in this specific case, the structural organization of bone tissue optimizes the materials response to external loads [1].

Therefore, scaffolds mimicking the functional pores gradation typical of natural

***Address correspondence to Francesco Baino:** Applied Science and Technology Department, di Torino, Corso Duca degli Abruzzi 24, 10129 Torino, Italy; Tel: +39 011 0904668; Fax: +39 011 090 4699; E mail: francesco.baino@polito.it

Lia Rimondini, Claudia Bianchi and Enrica Vernè (Eds)

bone could be considered as ideal bone substitutes. In addition, it should be taken into account that pores belonging to different size scales can impart peculiar properties and abilities to the scaffold itself. For instance, pores in the nano-range can be useful to uptake and subsequently release therapeutic agents and drugs, thereby providing an important added value to the implant.

Future directions of research in bone tissue engineering seem to go towards the development of hierarchically porous ordered systems able not only to mimic Nature but also to fulfil multiple requirements such as bone bonding ability, cells activity guidance and *et al.* drug therapy, which all would have been regarded as impossible to be incorporated in a single device few years ago.

2. NEED AND SEARCH FOR AN IDEAL BONE TISSUE ENGINEERING SCAFFOLD

Bone tissue exhibits the unique and fascinating ability of self-repairing by continuous remodelling; however, there are several pathological conditions that require bone replacement. In fact, the substitution of bone portions of different size and shape may be necessary due to congenital affections, trauma, tumour removal and age-related diseases such as osteoarthritis and osteoporosis. Harvesting the patient's tissue from a donor site and, afterwards, transplanting it to a host site has been traditionally considered the gold standard solution in orthopaedic reconstructive surgery owing to excellent immunocompatibility [2]; however, the use of the so-called autografts carries some problems since they are available in limited amount, an additional surgical procedure is needed for the patient and there is the risk of infection and/or persistent pain at the harvesting site [3]. An optional solution is given by the transplantation of tissue belonging to cadavers (freeze-dried bone) [4]; however, the use of allografts is affected by limited availability, risk of graft rejection from the patient's immune system, need for immunosuppressant drugs and ethical/religious concerns. Finally, non-living transplants from other species can be a third option for bone replacement [4]; however, xenografts carry problems analogous to those of allografts and, in addition, the clinical outcomes after their use are quite controversial.

The alternative to transplants is the implantation of man-made devices, properly fabricated by using natural, semisynthetic or synthetic materials, that can successfully interface with living host tissues [5]. The significant advantages of implants over transplants are high – ideally unlimited – material availability, good reproducibility and major versatility of the implanted device.

Bone tissue engineering scaffolds are 3-D porous structures that should provide structural support for the cells and the newly formed tissue and promote the natural processes of bone regeneration. Bioactive glasses and glass-ceramics, due to their versatile properties that can be properly designed depending on their composition and/or synthesis method, are very attractive materials for producing scaffolds devoted to bone regeneration [6]. The first bioactive glass, belonging to the SiO_2-Na_2O-CaO-P_2O_5 system (Bioglass®), was synthesized by Hench and co-workers in the early 1970s [7]; since then, many other glasses have been proposed by materials and medical researchers for bone tissue engineering applications; however, most of such glasses has been produced and tested in form of powders, particles, granulates or dense bulk of various shape and size, and only a limited group of them has been chosen to produce porous scaffolds.

An ideal scaffold for bone tissue engineering applications should fulfil a complex set of requisites [8, 9]; essentially, it should be (i) biocompatible, (ii) produce non-toxic products of degradation, (iii) resorb at the same (or slower) rate as the bone is repaired, (iv) exhibit a 3-D structure of highly interconnected macropores (above 50 %vol.), (v) elicit a bioactive response, *i.e.* the formation of a hydroxyapatite (HA) layer on its surface in order to bond to the host bone (HA has a compositional and crystallographic similarity to bone mineral phase [5-7]), (vi) promote cells adhesion, spreading and proliferation, (vii) have mechanical properties matching those of the host bone, (viii) have the potential to be easily produced and sterilized.

The features of scaffold porosity, *i.e.* pores size, shape, distribution and interconnection, strongly affect several important characteristics of the scaffold itself. Table **1** summarizes the seven levels of pores size that should be exhibited by an ideal bone tissue engineering scaffold. It should be carefully taken into account that cells play a key role in bone tissue engineering applications. In the light of this consideration, as reported in Table **1** the importance of the five porosity levels at the macro-scale (levels III-VII) is really crucial, since pores ranging within 10-3500 μm, corresponding to the porosity of natural bone, are necessary for many physiological functions and can directly interact with cells [10]. For instance, interconnected macropores within the 10-500 μm range are essential to allow bone cells penetration, adhesion, growth and proliferation, thereby leading to bone tissue in-growth and eventual vascularisation of the implant. Furthermore, pores within 1-10 μm, generating a diffused surface roughness, favour proteins adhesion and osteoblasts attachment on the scaffold trabeculae [11].

It should be also kept in mind that bone porosity is strongly organized according to the concept of functional gradation. In fact, it is well known that bone architecture exhibits non-uniform porosity distribution; this is particularly evident, for instance, in the longitudinal cross-section of long bones, in which the bone at the ends (epiphyses) shows a sponge-like appearance (cancellous or trabecular bone) whereas the bone in the central part (diaphysis) is rather dense with low pores content (cortical bone) [1]. This important lesson from Nature should be taken into account for developing bone tissue engineering scaffolds with gradients of porosity, since heterogeneity of pores features can result in optimized structural, mechanical and biological properties in comparison to homogeneously porous biomaterials.

Table 1: Hierarchically porous structure of an ideal glass-based scaffold for bone tissue engineering.

Level of pores size	Pores size range	Function
I	2-50 nm	Uptake and release of drugs, growing factors, organic moieties. The nano-textural features of the glass can enhance its bioactive properties.
II	50 nm-1 μm	Useful to further increase the specific surface area available for the ion-exchange phenomena responsible of bioactivity.
III	1-10 μm	Useful for promoting cells adhesion (surface roughness).
IV	10-100 μm	Useful for cells penetration.
V	100-500 μm	Useful for bone in-growth and implant vascularisation.
VI	500-1000 μm	Useful to lower the Young's modulus of the implant, thereby reducing the stress-shielding phenomenon.
VII	1000-3500 μm	Useful for fixation through wiring-like sutures during surgical implantation.

Pores in the meso-range, corresponding to the level I in Table **1**, are too small to allow cells uptake, which needs dimensions in the order of microns. However, mesopores can be very useful in imparting an added value to the scaffold [12]. For instance, mesopores can provide the functionality of encapsulating biological agents or drugs that can be subsequently delivered *et al.* according to properly designed release kinetics. Furthermore, it has been recently demonstrated that the textural properties of mesoporous glasses play a key role in enhancing the bioactivity of the material with respect to melt-derived bioactive glasses [13].

Table 2: Key advancements in the development of hierarchically porous glass/glass-ceramic (GC) scaffolds.

Year (first synthesis)	Material/composition (%mol.)	Type[a]	References[b]
2002	58S ($60SiO_2$-$36CaO$-$4P_2O_5$)	SGGS	[30-34]
2003	70S30C ($70SiO_2$-$30CaO$)	SGGS	[31-35]
2007	80S15C ($80SiO_2$-$15CaO$-$5P_2O_5$)	MBGS	[37-42]
2008	GC-SCK (glass system: $50SiO_2$-$44CaO$-$6K_2O$) + SBA15 spheres	HPCS	[18]
2008	GC-SCK (glass system: $50SiO_2$-$44CaO$-$6K_2O$) + MCM41 spheres	HPCS	[20, 21]
2008	GC-70S26C ($70SiO_2$-$26CaO$-$4P_2O_5$)	SGGS	[36]
2008	GC-Bioglass® (glass system: $46.1SiO_2$-$26.9CaO$-$24.4Na_2O$-$2.6P_2O_5$)	PGS	[15]
2009	GC-FaGC (glass system: $50SiO_2$-$18CaO$-$7Na_2O$-$6P_2O_5$-$7K_2O$-$3MgO$-$9CaF_2$) + MCM41 spheres	HPCS	[23]
2010	GC-CEL2 (glass system: $45SiO_2$-$26CaO$-$15Na_2O$-$3P_2O_5$-$4K_2O$-$7MgO$)	PGS	[16]

[a] PGS = pore-graded scaffold; HPCS = hierarchically porous composite scaffold; SGGS = sol-gel glass scaffold; MBGS = mesoporous bioactive glass scaffold.
[b] References refer to numbering in the main text.

Many evidences seem to indicate that the future of scaffolds goes towards multi-scale systems with porosity from the meso- to the macro-range in order to impart to biomaterials advanced properties over the traditional ones, such as enhanced bone-bonding properties or drug uptake/release abilities. Drug release from mesoporous systems will not be treated extensively here, as this topic is reviewed in detail by Mortera and Onida in another chapter of the present volume.

In the last decade, essentially two major classes of hierarchically porous glass-derived scaffolds have been developed by several research groups: (i) glass-ceramic scaffolds with gradients of porosity at the macro-scale mimicking the functional pores gradation of bone tissue and (ii) multi-level porous glass-based scaffolds with pores size from the meso- to the macro-scale. Table **2** chronicles the development of these innovative smart structures that will be described in detail in the next sections.

3. PORE-GRADED SCAFFOLDS AT THE MACRO-SCALE

As recently reviewed by Miao and Sun [14], HA, β-tricalcium phosphate (β-TCP) and other calcium-phosphate ceramics, as well as their composites with some biocompatible polymers, have been widely adopted for making graded scaffolds able to mimic the pore-graded structure of natural bone. Among bioactive glasses, however, to the best of the authors' knowledge only Bioglass® [15] and CEL2 [16] have been proposed for the this purpose in two recently published studies (Table **2**). Two examples of pore-graded CEL2-derived scaffolds are shown in Fig. **1**.

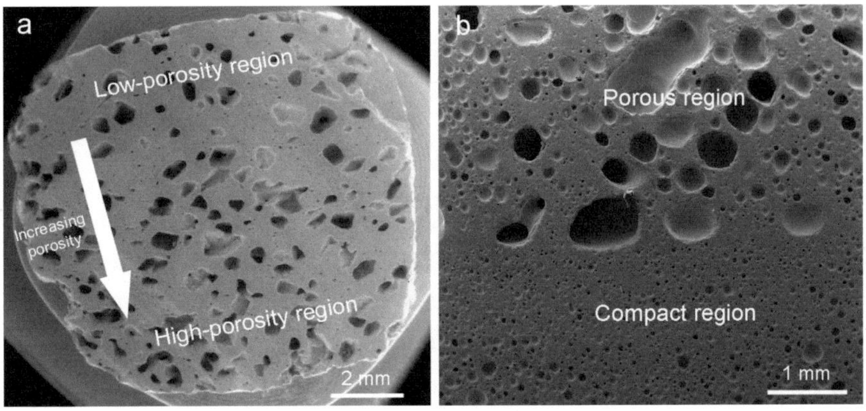

Figure 1: Glass-ceramic (GC-CEL2) scaffolds with gradients of porosity at the macro-scale: (a) pores gradation in the cross-section of a scaffold prepared through PE burning-out method and (b) interface between porous and compact layers of a scaffold prepared by sponge replication/enamelling and mimicking the cancellous/cortical bone system.

Bretcanu *et al.* [15] described an interesting and versatile approach for manufacturing 3-D highly porous (> 80%vol.) foam-like glass-ceramic (GC) Bioglass®-derived scaffolds. The authors used pre-formed polyurethane (PU) sponges with tailored gradient of porosity as sacrificial templates for manufacturing scaffolds through the replication technique. The PU sponge was pre-formed by compressing it in metal mould and then by thermally treating it at low temperature

(200 °C for 1 h in air). By varying the shape and size of the mould, PU templates with different porous features/gradients could be successfully obtained. It should be underlined that the pore-graded architecture could also carry the potential of reducing the dramatic brittleness of GC-Bioglass® scaffolds (the compressive strength of homogenously porous GC-Bioglass® scaffolds ranged within 0.3-0.4 MPa [17]), but no data about the actual mechanical strength of these graded scaffolds has been reported in the literature up to now.

Very recently, Vitale-Brovarone *et al.* [16] used CEL2 particles to fabricate graded glass-ceramic scaffolds by means of different processing methods, *i.e.* sponge replication, polyethylene (PE) burning-out and enamelling together with various combinations of such techniques. GC-CEL2 scaffolds able to mimic the porosity gradient of cancellous bone (Fig. **1a**) or to reproduce the trabecular/cortical bone system (Fig. **1b**) were successfully obtained. It was possible to design the final scaffold – in terms of structural similarity to bone, pores features and mechanical strength – by varying or coupling together the different fabrication methods. This remarkable versatility can be important and useful to fulfil specific criteria depending on the end use, such as high-strength properties required for the substitution of load-bearing bone portions.

4. SCAFFOLDS WITH DIFFERENT SCALES OF POROSITY

4.1. Glass-Ceramic/Mesoporous Silica Composite Scaffolds

In the last few years, some attempts for fabricating glass-based multi-scale porous structures have been carried out by using glass-ceramic macroporous scaffolds as carriers for mesoporous silica spheres. The mesopores size and arrangement in the mesophase can be carefully controlled and designed according to some synthesis parameters such as temperature, pH and type of templating agent (surfactant). The purpose of such composite scaffolds was twofold, as they could combine the properties of traditional glass-derived scaffold, *i.e.* mechanical support in the defect zone, bone bonding ability and bone tissue regeneration, with the unique features supplied by mesoporous materials, such as enhanced bioactivity and controlled drug adsorption/release useful for drug therapy *in situ*.

The first prototype of such composite system was developed by Cauda *et al.* [18] and it involved the incorporation of SBA-15 mesoporous silica particles inside a bioactive GC-SCK scaffold (Table **2**) produced through the PE burning-out method [19]. The authors' purpose was to obtain a composite scaffold able to promote the successful integration of the graft and the local drug (ibuprofen) delivery in the implant surroundings. This composite system showed a drawback related to the SBA-15 synthesis conditions (strongly acidic, $pH < 1$), which led to the GC-SCK scaffold degradation during the incorporation of the mesophase by dipping route. Afterwards, the study was extended to MCM-41 silica spheres, which posses narrower pores size (2-3 nm) in comparison with SBA-15 (5-8 nm), and the synthesis conditions were optimized in order to avoid the scaffold damage (mild pH ~9) [20]. Finally, a highly bioactive GC-FaGC scaffold (Table **2**) [22] was proposed as a carrier for MCM-41 mesoporous spheres (Fig. **2**) that were found to play a key role in enhancing the drug adsorption ability of the whole composite system [23].

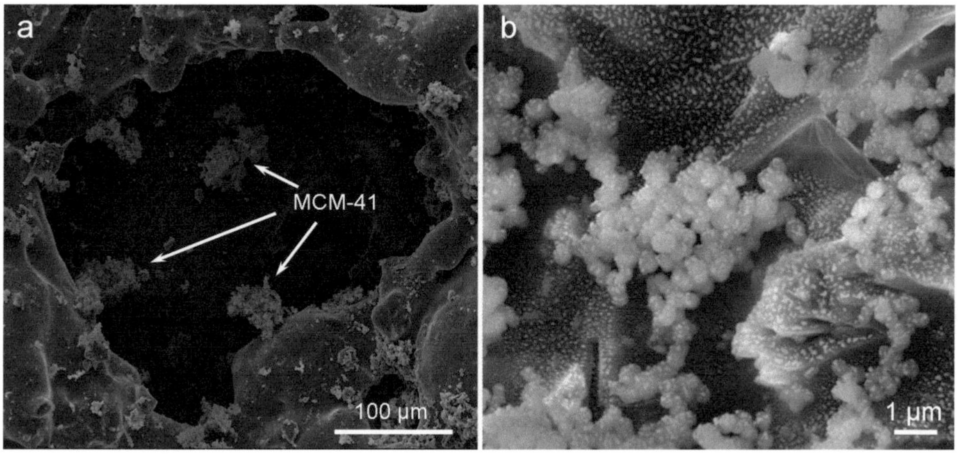

Figure 2: MCM-41/GC-FaGC composite scaffold: (a) MCM-41 agglomerates anchored on scaffold walls and (b) high-magnification SEM micrograph of the MCM-41 agglomerates, showing the spherical morphology of the mesophase particles.

Although the obtained results were promising and encouraging, however some remarks have to be taken in account: (i) the proposed systems (bioactive glass-ceramic scaffold + silica mesophase) were composite, *i.e.* constituted by two different materials that can present problems of interfacial bonding; (ii) from a methodological viewpoint, the synthesis of the whole system was rather easy and cheap but required long time of processing; (iii) although some authors [24] reported the bioactivity of some pure-silica mesophases, this issue is currently under debate in the literature and, specifically, several evidences seem to demonstrate the lacking of MCM-41 and SBA-15 bioactivity [25, 26].

An attractive solution to overcome the last problem (lacking of bioactivity in pure silica mesophases) is the manufacturing of monomaterial (non-composite) bioactive glass scaffolds with multi-scale porosity. This different approach led to the fabrication of sol-gel glass scaffolds and mesoporous glass scaffolds, that will be discussed in the next sections.

4.2. Sol-Gel Glass Scaffolds

Hench and co-workers demonstrated that silica content in melt-derived glass should be less than 60 %mol. to allow the material to bond to bone as a consequence of the bioactive process [27, 28]. However, HA formation and bone bonding ability can be also achieved in glasses with up to 90 %mol. silica if the glass is obtained by a sol-gel process. Sol-gel bioactive glasses were synthesized for the first time in the early 1990s [29]; in comparison to melt-derived glasses, sol-gel glasses can be prepared at relatively lower processing temperatures, exhibit higher bone bonding rates and excellent resorption properties. The synthesis methods and the potential applications of sol-gel glasses in the biomedical field have been extensively reviewed elsewhere [30].

Sol-gel process was recently combined with foaming methods, thereby leading to a new generation of hierarchically porous biomaterials for biomedical applications. In 2002, for the first time Sepulveda *et al.* [31] foamed sol-gel glasses to obtain macroporous scaffolds mimicking the trabecular architecture of the cancellous bone. The synthesis method involved the sol foaming with the aid of a surfactant; on gelation, the spherical bubbles formed in the sol after vigorous stirring became permanent in the gel and led to a 3-D sponge-like highly porous structure similar to that of cancellous bone. Sol-gel glass scaffolds exhibited an excellent bioactive behaviour due to the high surface area that is provided by the nanoporous network inherent to the sol-gel process, and is available for ion-exchange phenomena with biological fluids. Sol-gel foaming led to scaffolds exhibiting a bimodal porous structure, constituted by large pores up to 500 μm connected by pores windows (10-100 μm), together with a random-like mesoporous texture (10-20 nm).

Jones and Hench demonstrated that many variables of sol-gel processing, such as surfactant agent and glass composition, can be used to control the final pores network structure of the scaffold [32, 33]. Sol-gel glass foams were also tested *in vitro* with osteoblasts cultures and were found potentially suitable for bone repair [34].

Jones *et al.* [35] showed that the compressive strength of sol-gel glass scaffolds can be improved by carefully adjusting the thermal treatments necessary for scaffold processing; however, a further optimization of scaffold mechanical properties, especially for improving fracture toughness, is mandatory before foreseeing real clinical applications. Scaffolds derived from sol-gel process usually remain glassy, but recently Rainer *et al.* [36] fabricated bioactive glass-ceramic scaffolds *et al.* foaming starting from sol-gel glass powders. The glass-loaded foams underwent severe shrinkage (>75%) during sintering and, therefore, were characterized by relatively low porosity (48 %vol.) in comparison with that of cancellous bone. The scaffolds were bioactive and the proposed fabrication method is very interesting and versatile for producing patient-tailored grafts, but no indications about scaffolds mechanical properties were reported.

4.3. Mesoporous Bioactive Glass Scaffolds

The sol-gel glass scaffolds proposed in the literature are characterized by both macro- and nano-pores, but the mesoporous texture is not organized according to a well-defined symmetry but rather exhibits a random or worm-like arrangement. In 2004 a fundamental work of Yan *et al.* who first reported the combination between supramolecular chemistry and sol-gel process for synthesizing mesoporous bioactive glasses (MBGs) [37], opened the doors of a new and challenging field of research for biomaterials scientists. Since then, the synthesis methods and potential biomedical applications of MBGs have been extensively investigated by several research groups, with particular reference to the drug release ability and excellent bioactivity of MBGs [13]. Fig. **3** shows the typical nanotextural arrangement of ordered mesopores in a MBG belonging to the SiO_2-CaO-P_2O_5 system (80S15C).

MBGs have been mainly investigated in form of powders, but since 2007 some research works have reported the fabrication of 3-D scaffolds using MBGs as starting materials. Yun *et al.* [38] synthesized hierarchically porous 3-D MBG scaffolds by using a combination of sol-gel, double polymers templating and rapid prototyping techniques. The resulting scaffold was characterized by three scales of porosity: mesopores (average size ~5 nm) deriving from the use of $EO_{100}PO_{65}EO_{100}$ (F127) as a template, macropores in the 10-30 μm range deriving from methyl cellulose and giant pores (30-1000 μm) created by rapid prototyping. *In vitro* tests showed that a HA layer formed on scaffold trabeculae after soaking for 24 h in simulated body fluid (SBF).

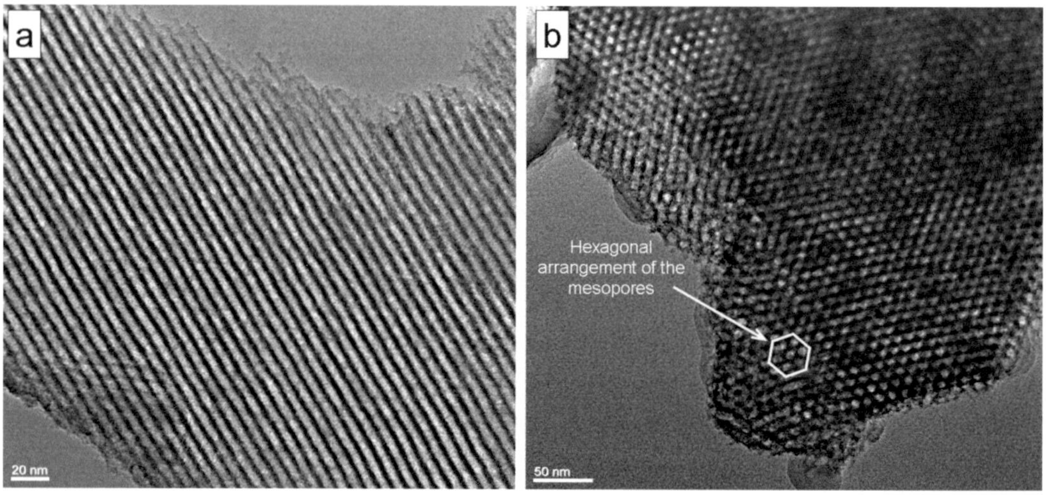

Figure 3: TEM pictures of a typical MBG (80S15C): (a) longitudinal arrangement of the mesopores (parallel channels) and (b) transversal cross-section of the mesopores, that are ordered according to a hexagonal symmetry.

Li *et al.* [39] reported the synthesis of multi-scale porous MBG scaffolds by using the block copolymer $EO_{20}PO_{70}EO_{20}$ (P123) and a PU macro-porous sponge as co-templates and demonstrated that a HA layer can form on scaffold surface after soaking in SBF for 4 h. The scaffold exhibited a 2-scale hierarchical porosity including macropores in the 200-700 μm range and mesopores with average size of ~3.7 nm.

Zhu *et al.* [40] successfully prepared 3-D porous MBG scaffolds by a combination of PU sponge and P123 surfactant as co-templates and evaporation-induced self-assembly (EISA) process. Studies of biological compatibility showed that human bone-derived cells cultured on the scaffolds for 1, 3 and 7 days presented a good degree of attachment and spreading. In addition, hierarchical porous MBG scaffold exhibited a greatly enhanced bone-forming bioactivity in comparison with traditional bioactive glass (BG) scaffold of same composition due to its high surface area and pore volume. Drug release studies on these scaffolds by using gentamicin have been also performed [41]. The drug uptake ability of MBG scaffolds was over twofold higher than that of BG scaffold; in addition, as far as drug delivery is concerned during the whole release period in SBF gentamicin was released from the MBG scaffold at a much lower release rate in comparison with that from BG scaffolds.

At present, MBG scaffolds with multi-scale porosity are at a preliminary stage of investigation and a lot of research work has to be done in the future. For example, almost no indications about the mechanical strength of MBG scaffolds have been reported in the literature. Very recently, Wu *et al.* [42] reported an interesting research work about this crucial issue: firstly the authors tested as-such MBG scaffolds in compression (60 kPa), and afterwards they made an attempt to reinforce scaffold structure by using a polymeric coating. Specifically, they soaked MBG scaffolds in silk solution and found that silk-induced modification improved the uniformity and continuity of scaffold pore network, maintained a high porosity content (~94 %vol.) as well as large pore size (200-400 μm) and increased the mechanical strength up to 250 kPa. It can not be ignored that the mechanical resistance of MBG-based scaffolds is over one order of magnitude lower than that of cancellous bone: this is a dramatic drawback affecting any actual clinical applications, as the intrinsic mesoporous texture featuring the MBGs imparts high brittleness to scaffold structure, thereby causing problems related to sample manipulation and its safe implantation in the patient's bone.

5. SUMMARY AND OUTLOOK

Nature demonstrates that functional gradation is an optimized design rule for living tissues. In the last decade, the bio-mimicry principle combined with the concepts of tissue engineering has led to fascinating and promising results.

Next generation scaffolds for bone tissue engineering should be designed as hierarchically ordered porous structures not only to provide the structural integrity required for bone support and the physico-chemical environment suitable for bone regeneration – that, however, are still the main scaffold functions – but also to carry the added value of further smart functionalities, such as improved bioactivity, high strength properties or the ability of acting as vehicles for controlled drug release *in situ*.

For instance, graded scaffolds mimicking pores gradation of natural bone exhibit higher strength than that of homogeneously porous scaffolds fabricated with the same material and, therefore, they may be a helpful resource for replacing load-bearing bone portions [16].

Furthermore, biomaterials with hierarchical bimodal pores size distribution in the meso- (2-50 nm) and macro- (> 50 nm) range can offer interesting properties in comparison with materials exhibiting monomodal porosity. A typical system featured by hierarchical porosity is composed by a 3-D macroporous structure serving as a support inside which a different material having smaller pores can be incorporated or deposited. Ideally, the macroporous network aims to ensure the chemical and mechanical stability of the system as well as good mass transport properties owing to the high pores interconnection, whereas the mesoporous coating can provide the functionality for a given applications, such as the encapsulation of biological or pharmaceutical agents. In the field of bone tissue engineering, glass-derived scaffolds are optimal candidates as substrate for the development of porous bodies with hierarchical porosity, as their highly interconnected pores network can allow the access of mesophase precursors also in the inner region of the structure. The purpose of such systems is twofold, since they are expected to combine the properties of the glass-based scaffold, *i.e.* mechanical support in the defect zone and bone bonding ability, with the unique features supplied by mesoporous materials, such as local drug release and enhanced bioactive properties.

The use of pure silica mesoporous materials (SBA15, MCM41) [18, 20-23] and MBGs [31-35] in the manufacturing of tissue engineering scaffolds is still at a preliminary stage of research, but the results achieved up to now are promising and very challenging.

Sol-gel glass and MBGs scaffolds exhibit several attractive features but also dramatic limitations concerning mechanical properties; therefore, a great effort should be addressed in the next years towards the fabrication of hierarchically structured glass scaffolds with adequate mechanical strength, at least comparable to that of cancellous bone.

As suggested in some recent research works [43, 44], non-destructive high-resolution (~1 μm) X-ray computed microtomography (micro-CT) could be an useful tool for acquiring important data in order to improve the structural and mechanical properties of the scaffolds.

In addition, MBGs scaffolds are undoubtedly worthy of investigation also from another viewpoint. The ordered texture of mesopores was shown to impart high bioactivity properties to MBGs and was also found useful for drug uptake/delivery application. It should be very interesting to analyze in detail whether the kinetics of bioactive process and drug release are in competition with each other. This essentially means to assess whether the formation of HA crystals on mesopores walls can have an antagonistic effect on the material's drug release ability.

Finally, the tailoring of hierarchically porous scaffolds with precise shape, size and 3-D architecture to fit into specific and often irregular bone defects will be a great challenge in the near future. The development and setting of standard methods for fabricating complex scaffolds with multiple functionalities will have an ever increasing impact in biomaterials research, also in order to reduce the gap existing between "research products" on a Lab-scale and "industrial products" intended for an actual clinical use.

ACKNOWLEDGEMENTS

Declared none.

CONFLICT OF INTEREST

The author(s) confirm that this chapter content has no conflict of interest.

REFERENCES

[1] Boskey AL. Mineralization of bones and teeth. Elements 2007; 6: 385-92.
[2] Reichert JC, Saifzadeh S, Wullschleger ME, Epari DR, Schutz MA, Duda GN, Schell H, Van Griensven M, Redl H, Hutmacher DW. The challenge of establishing preclinical models for segmental bone defect research. Biomaterials 2009; 30: 2149-63.
[3] Younger EM, Chapman MW. Morbidity at bone graft donor site. J Orthop Trauma 1989; 3: 192-5.
[4] Schlickewei W, Schlickewei C. The use of bone substitutes in the treatment of bone defects – The clinical view and history. Macromol Symp 2007; 253: 10-23.

[5] Hench LL, Polak J. Third-generation biomedical materials. Science 2002;295:1014-7.
[6] Hench LL. Biomaterials: a forecast for the future. Biomaterials 1998; 19: 1419-23.
[7] Hench LL, Splinter RJ, Allen WC, Greenlee TK. Bonding mechanisms at the interface of ceramic prosthetic materials. J Biomed Mater Res 1972; 2: 117-41.
[8] Karageorgiou V, Kaplan D. Porosity of 3D biomaterial scaffolds and osteogenesis. Biomaterials 2005; 26: 5474-91.
[9] Jones JR, Gentleman E, Polak J. Bioactive glass scaffolds for bone regeneration. Elements 2007; 3: 393-9.
[10] Simske SJ, Ayers RA, Bateman TA. Porous materials for bone engineering. Mater Sci Forum 1997; 250: 151-82.
[11] Schwartz Z, Boyan BD. Characterisation of microrough bioactive glasses: surface reactions and osteoblast responses. J Cell Biochem 1994; 56: 340-7.
[12] Mourino V, Boccaccini AR. Bone tissue engineering therapeutics: controlled drug delivery in three-dimensional scaffolds. J R Soc Interface 2010; 7: 209-27.
[13] Vallet-Regi M. Ordered mesoporous materials in the context of drug delivery systems and bone tissue engineering. Chem Eur J 2006; 12: 5934-43.
[14] Miao X, Sun D. Graded/gradient porous biomaterials. Materials 2010; 3: 26-47.
[15] Bretcanu O, Samaille C, Boccaccini AR. Simple methods to fabricate Bioglass®-derived glass-ceramic scaffolds exhibiting porosity gradient. J Mater Sci 2008; 43: 4127-34.
[16] Vitale-Brovarone C, Baino F, Verné E. Feasibility and tailoring of bioactive glass-ceramic scaffolds with gradient of porosity for bone grafting. J Biomater Appl 2010; 24: 693-712.
[17] Chen QZ, Thompson ID, Boccaccini AR. 45S5 Bioglass®-derived glass-ceramic scaffolds for bone tissue engineering. Biomaterials 2006; 27: 2414-25.
[18] Cauda V, Fiorilli S, Onida B, Verné E, Vitale-Brovarone C, Viterbo D, Croce G, Milanesio M, Garrone E. SBA-15 ordered mesoporous silica inside a bioactive glass-ceramic scaffold for local drug delivery. J Mater Sci: Mater Med 2008; 19: 3303-10.
[19] Vitale-Brovarone C, Verné E, Appendino P. Macroporous bioactive glass-ceramic scaffolds for tissue engineering. J Mater Sci: Mater Med 2006; 17: 1069-78.
[20] Mortera R, Onida B, Fiorilli S, Cauda V, Vitale-Brovarone C, Baino F, Verné E, Garrone E. Synthesis of MCM-41 spheres inside bioactive glass-ceramic scaffold. Chem Eng J 2008; 137: 54-61.
[21] Mortera R, Baino F, Croce G, Fiorilli S, Vitale-Brovarone C, Verné E, Onida B. Monodisperse mesoporous silica spheres inside a bioactive macroporous glass-ceramic scaffold. Adv Eng Mater 2010; 12: 256-9.
[22] Baino F, Verné E, Vitale-Brovarone C. 3-D high strength glass-ceramic scaffolds containing fluoroapatite for load-bearing bone portions replacement. Mater Sci Eng C 2009; 29: 2055-62.
[23] Vitale-Brovarone, Baino F, Miola M, Mortera R, Onida B, Verné E. Glass-ceramic scaffolds containing silica mesophases for bone grafting and drug delivery. J Mater Sci: Mater Med 2009; 20: 809-20.
[24] Izquierdo-Barba I, Ruiz-Gonzalez L, Doadrio JC, Gonzalez-Calbet J, Vallet-Regi M. Tissue regeneration: A new property of mesoporous materials. Solid State Sci 2005; 7: 983-989
[25] Mortera R, Fiorilli S, Garrone E, Onida B. Structural changes of MCM-41 spheres during ibuprofen release to SBF. Stud Surf Sci Catal B 2008; 174: 1001-4.
[26] Mortera R, Fiorilli S, Garrone E, Verné E, Onida B. Pores occlusion in MCM-41 spheres immersed in SBF and the effect on ibuprofen delivery kinetics: a quantitative model. Chem Eng J 2010; 156: 184-92.

[27] Hench LL, Wilson J. Introduction to Bioceramics. World Scientific Publishing. Singapore, 1993.

[28] Hench LL. The story of Bioglass®. J Mater Sci: Mater Med 2006; 17: 967-78.

[29] Li L, Clark AE, Hench LL. An investigation of bioactive glass powders by sol-gel processing. J Appl Biomater 1991; 2: 231-9.

[30] Avnir D, Coradin T, Lev O, Livage J. Recent bio-applications of sol-gel materials. J Mater Chem 2006; 16: 1013-30.

[31] Sepulveda P, Jones JR, Hench LL. *In vitro* dissolution of melt-derived 45S5 and sol-gel derived 58S bioactive glasses. J Biomed Mater Res 2002; 61: 301-11.

[32] Jones JR, Hench LL. Effect of surfactant concentration and composition on the structure and properties of sol-gel-derived bioactive glass scaffolds for tissue engineering. J Mater Sci 2003; 38: 1-8.

[33] Jones JR, Hench LL. The effect of processing variables on the properties of bioactive glass foams. J Biomed Mater Res B 2004; 68: 36-44.

[34] Valerio P, Guimaraes MHR, Pereira MM, Leite MF, Goes AM. Primary osteoblast cell response to sol-gel derived bioactive glass foams. J Mater Sci: Mater Med 2005; 16: 851-6.

[35] Jones JR, Ehrenfried LM, Hench LL. Optimising bioactive glass scaffolds for bone tissue engineering. Biomaterials 2006; 27: 964-73.

[36] Rainer A, Giannitelli S, Abbruzzese F, Licoccia S, Traversa E, Trombetta M. Fabrication of bioactive glass-ceramic foams mimicking human bone portions for regenerative medicine. Acta Biomater 2008; 4: 362-9.

[37] Yan X, Yu C, Zhou X, Tang J, Zhao D. Highly ordered mesoporus bioactive glasses with superior *in vitro* bone-forming bioactivities. Angew Chem Int Ed 2004; 43: 5980-4.

[38] Yun HS, Kim SE, Hyeon YT. Design and preparation of bioactive glasses with hierarchical pore networks. Chem Comm 2007: 21; 2139-41.

[39] Li X, Wang X, Chen H, Jiang P, Dong X, Shi J. Hierarchically porous bioactive glass scaffolds synthesized with a PUF and P123 cotemplated approach. Chem Mater 2007; 19: 4322-6.

[40] Zhu Y, Wu C, Ramaswamy Y, Kockrick E, Simon P, Kaskel S, Zreiqat H. Preparation, characterization and *in vitro* bioactivity of mesoporous bioactive glasses (MBGs) scaffolds for bone tissue engineering. Microp Mesop Mater 2008; 112: 494-503.

[41] Zhu Y, Kaskel S. Comparison of the *in vitro* bioactivity and drug release property of mesoporous bioactive glasses (MBGs) and bioactive glasses (BGs) scaffolds. Microp Mesop Mater 2009; 118: 176-82.

[42] Wu C, Zhang Y, Zhu Y, Friis T, Xiao Y. Structure-property relationships of silk-modified mesoporous bioglass scaffolds. Biomaterials 2010; 31: 3429-38.

[43] Jones JR, Poologasundarampillai G, Atwood RC, Bernard D, Lee PD. Non-destructive quantitative 3D analysis for the optimisation of tissue scaffolds. Biomaterials 2007; 28: 1404-1413

[44] Jones JR, Atwood RC, Poologasundarampillai G, Yue S, Lee PD. Quantifying the 3D macrostructure of tissue scaffolds. J Mater Sci: Mater Med 2009; 20: 463-71.

CHAPTER 16

Surface Modification of Bioactive Glasses

Sara Ferraris* **and Marta Miola***

Applied Science and Technology Department – DISAT, Politecnico di Torino, C.so Duca degli Abruzzi 24, 10129, Torino, Italy

Abstract: Bioactive glasses are widely studied in the biomaterial field because of their ability to chemically bind to living tissues and to positively stimulate cells through the release of specific ions. They are extremely versatile materials, in fact their composition can be tailored in order to improve particular properties (*e.g.* bioactivity, resorbability, mechanical resistance ...). In this research work, different strategies aimed to modify the surface of bioactive glasses have been reviewed. The main idea of these treatments is to combine the conventional properties of bioactive glasses (bioactivity, ion release ...) with more specific ones, by the addition of particular ions or the grafting of biologically active molecules (*e.g.* proteins or drugs). At first a brief introduction on bioactive glasses is reported, then surface modification by the introduction of active ions is described and finally the surface functionalization with biomolecules (proteins or drugs) is discussed.

Keywords: Bioactive glass, osteoinductive, osteoconductive, surface functionalize-tion, antimicrobial.

1. BIOACTIVE GLASSES

Bioactive glasses belong to surface reactive glass-ceramics that are able to chemically bond to living tissues, through the formation of an hydroxyapatite layer on their surface. They have been described as osteoproductive materials [1], in fact they are able not only to provide a biocompatible interface along which bone migrates, as simply osteoconductive materials do, but also to promote osteogenic stem cells colonization and elicit both an intracellular and an extracellular response at their interface, by the release of specific ions (Si^{4+}, Na^{+}, Ca^{2+} ...).

Bioactivity mechanism of bioactive glasses foresees at first the ion exchange of network modifying Ca^{2+} and Na^{+} ions from the glass with H_3O^{+} ions from the

***Address correspondence to Sara Ferraris and Marta Miola:** Applied Science and Technology Department – DISAT, Politecnico di Torino, C.so Duca degli Abruzzi 24, 10129, Torino, Italy; Tel: +39 011 0904717; Fax: +39 011 090 4699; E-mails: sara.ferraris@polito.it; marta.miola@polito.it body fluid, the development of silanols (Si-OH) at the surface *via* the partial dissolution of silica at the interface and the condensation and re-polymerisation of the silica-rich surface layer. Subsequently phosphate and calcium ions from the body fluid are incorporated in the silicate layer to form an amorphous calcium phosphate layer that crystallizes into biologically equivalent hydroxyapatite. In *in vivo* environment these first steps are followed by the adsorption of biological molecules, the action of macrophages, the attachment and differentiation of stem cells, the generation and crystallization of matrix and finally the formation of new bone [1].

The basic components of bioactive glasses are SiO_2, Na_2O and CaO. Different oxides, such as P_2O_5, MgO, K_2O, B_2O_3 or even Al_2O_3, can be introduced in the glass composition in order to tailor its properties.

Surface reactivity strongly depends on material composition. In particular, for traditionally prepared (melt and quenching) glasses, a silica content up to about 53% mol allows hydroxyapatite crystallization within 2h, with an SiO_2 increase between 53% mol and 58% mol 2-3 days are required, while compositions with silica content higher than 60% mol do not form a crystalline apatite layer even after four weeks in SBF [1]. The substitution of silica with P_2O_5 results in completely resorbable bioactive glasses. Phosphate based glasses can dissolve completely in aqueous media and this dissolution behaviour may be easily altered *via* the chemistry of the glasses to give materials with a variable degree of solubility [2, 3]. Al_2O_3 addition can significantly improve glass mechanical properties [4], but on the other hand, an amount higher than 3% mol can completely inhibit the material bioactivity [1]. Fluoride additions can promote the formation of the apatite layer while reducing the rate of network dissolution.

The effect of bioactive glasses on cell behaviour has been widely investigated in the scientific literature. In particular it has been observed that ions released from bioactive glasses can stimulate osteoblast activity and differentiation.

In 2001 Gao *et al.* [5] studied the response of Saos-2 osteoblastic cells to two bioactive glasses with different Si contents. They observed that both glasses can stimulate mRNA levels of proα1(I) collagen, alkaline phosphatase and BMP-2 at 4, 7 and 14 days. Finally they suggest that cell behaviour could be affected by Si content.

Recently Hench [6] reviews new theories on the ability of bioactive glasses to genetically stimulate bone regeneration. It has been observed that new bone had colonized the surface of an array of 45S5 Bioglass® particulate placed in a surgically created defect in a monkey model. New bone was formed around the particles and created a regenerated structure of bone around the bioactive glass particles. Moreover it has been noticed that not only the glass but also the ionic dissolution products released from 45S5 Bioglass® (mainly Si and Ca ions) influenced and controlled the cell cycle of osteogenic precursor cells and ultimately controlled the differentiated cell population. In particular it has been observed that the ions produced osteostimulation when present at a particular concentration range (15–30 ppm Si and 60–90 ppm Ca).

Thanks to their structure, it is easy to tailor bioactive glass composition in order to reach specific functions (*e.g.* antibacterial or cell stimulating properties). It is possible to introduce active ions both in the bulk material composition, adding specific oxides, and just on the surface, for example by means of ion exchange processes.

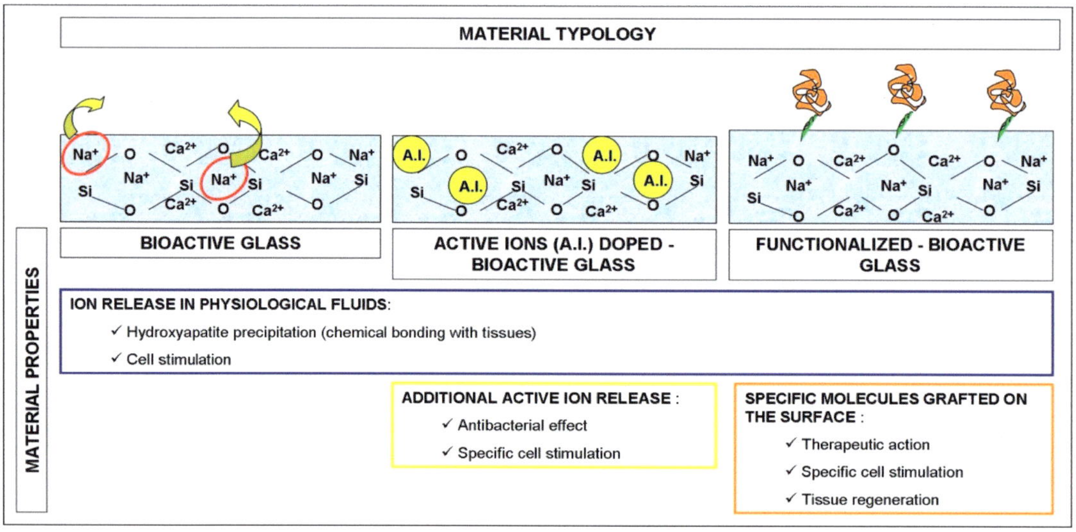

Figure 1: Properties of traditional and modified bioactive glasses.

Moreover, the high surface reactivity of bioactive glasses allows to expose reactive hydroxyls groups that can be employed for the grafting of specific molecules. In this context it is possible to introduce on material surface drugs, in order to reach a localized therapeutic effect, or biomolecules (proteins or growth factors) to improve tissue regeneration process.

The addition of active ions, drugs or biomolecules offers the opportunity to combine the traditional properties of bioactive glasses (briefly described above), with specific functionalities useful for different applications.

Fig. **1** summarizes opportunities and properties of traditional and modified bioactive glasses, that are briefly described below.

2. ACTIVE IONS DOPED BIOACTIVE GLASSES

Ion exchange process has been successfully employed in order to introduce silver only in the outermost surface layer of bioactive glasses and glass-ceramics in order to confer antibacterial properties. Vernè *et al.* [7-12] applied a patented [13] ion exchange process for the surface enrichment of two different bioactive glasses: the first one (SCNA - 57% SiO_2, 3% Al_2O_3, 34% CaO and 6% Na_2O – mol%) presents a simple composition and a low degree of bioactivity, while the second one (Fa-GC – 50% SiO_2–18% CaO–9% CaF_2–7% Na_2O–7% K_2O–6% P_2O_5–3% MgO) is a fluoroapatite containing glass-ceramic with a more complex composition and an higher degree of bioactivity. Both materials have been studied and surface modified in bulk form (glass and glass-ceramic slices) [7, 9] and also as plasma spray coatings on titanium substrates and AISI316L steel [8, 10]. Good results have been obtained also on glass-ceramic macroporous scaffolds [11, 12].

It has been observed that silver introduction does not alter the amorphous or partially crystalline structure of the pristine material. The staring glass bioactivity is also maintained after ion-exchange process. Silver is gradually released in simulated body fluid up to 1 month. The quantity of silver introduced onto material surface can be tailored in order to confer effective antibacterial action against *S. aureus* (2 fold reduction of colony forming units) and biocompatibility for osteoblast cells.

The amount of Ag^+ ions introduced onto the glass surface depends on its composition, in particular it raises increasing the exchangeable ions (mainly Na^+) content; so the amount of silver added to Fa-GC resulted higher then the one added to SCNA, in the same process conditions. As far as process parameters are concerned, the quantity of Ag^+ introduced in the glass depends on temperature, silver concentration in the bath and duration of the treatment, in particular it raises increasing the cited variables.

Finally an increase in the material specific surface area allows the addition of an higher amount of silver, so plasma spray coatings and macroporous scaffolds result richer in Ag^+ than plan glass slices. So, the surface topography and the shape of material (bulk, coatings, powders, macroporous structure...) have to be considered to tailor the ion-exchange process parameters.

Ag-doped glasses can be also obtained using the melt and quenching technique, by introducing silver in the reactants, both in silica and phosphate-based glass [14, 15], or by sol-gel method [16, 17]. These research works investigate the bioactivity and the antibacterial effect using the most frequent bacterial species involved in biomaterial-related infections.

Bioactive glasses have been also doped with other metal ions such as zinc, copper and gallium. Zinc has been introduced since it is an essential trace element in the human body that has shown a stimulatory effect on bone formation [18, 19]; moreover it has antimicrobial properties [20]. The bioactivity of zinc-doped glasses has been deeply investigate, nevertheless the Zn effect on bioglasses reactivity is still a debatable argue.

Recently, the need for new antimicrobial agents suggests some authors [21, 22] to introduce gallium onto phosphate-base glass and overwork the degradation of this material to obtain the delivery of gallium ions in a controlled way. The antimicrobial effect of gallium has been confirmed towards tests on several pathogenic bacteria.

Finally, Knowles *et al.* [23, 24] investigate the antibacterial effect of glass and glass fibre doped with a variable amount of copper. Also in this works, the antibacterial ions release occurs as the glass degrades; so, by tailoring the material dissolution rate it is possible to obtain a controlled ions delivery.

3. FUNCTIONALIZED BIOACTIVE GLASSES

The goal of surface functionalization is to immobilize particular molecules (proteins, enzymes, peptides or drugs, such as antibiotics or antineoplastic agents) on biomaterials for the purpose of inducing specific cell and tissue responses (e. g. adhesion or differentiation) of reducing infection risk or even remove bacteria from the surgical site. In this way the idea is obtain specific functionalities by means of proper molecules delivered directly to the tissue-biomaterial interface.

From an industrial and commercial point of view, the introduction of biomolecules or drugs onto material surface can change the classification of the final medical device (*e.g.* according to 93/42 directive concerning medical devices). In fact, whether the biomolecule act as the principal function of the product, it can be shifted from conventional medical device to drug, and requires different and more complex procedures for certification and commercialization.

Surface functionalization of biomaterials with the aim of improve their bone integration is now an extremely active field of scientific research, so a number of solution has been explored all over the world. Metallic and polymeric surfaces are the most widely investigated [25-29] and few attempts have been performed also on bioactive glasses and ceramics. As said before the surface functionalization of bioactive glasses allows the combination of peculiar properties of the starting material (bioactivity and release of ions able to stimulate cells) with specific one resulting from the chosen molecule.

At first silanization of bioactive glass-ceramic scaffolds (Bioglass®) has been explored for further grafting of biologically active molecules. It has been successfully carried out in aqueous media with [30] and without [31] glutaraldehyde employment. It has been observed that the treatment that involves material exposition to an aqueous solution at 80°C for up to 4h can favour the exposition of the crystalline phase improving bioactivity despite of the presence of silane layer.

Nano-particles of sol-gel derived bioactive glasses have been surface modified with amminopropyltriethoxysilane (APTES) by wet-chemical method in a dynamic inert nitrogen atmosphere, without glutaraldehyde employment [32]. Silanization allows the surface enrichment in amino groups that has been suggested as useful way to improve cell proliferation and viability in addition to improved dispersibility of nano-particles.

Mesoporous bioactive glasses (MBGs) have been surface modified through the introduction of amino- (N-MBGs) and carboxylic- (C-MBGs) groups using 3-amminopropyltriethoxysilane and triethoxysilypropyl succin anhydride respectively as functionalising molecules [33]. No glutaraldehyde has been used for grafting. It has been observed that the ordered mesoporous structure is partially destroyed after the grafting procedure and in particular that surface area is significantly reduced. *In vitro* bioactivity tests indicate that the morphology, the nucleation and growth rate of carbonate-hydroxyapatite (CHA) is affected by functionalization. In fact N-MBGs promote the formation of spherical CHA particles (different from the rod-like CHA grown on conventional MBGs), and the bioactivity does not decrease with the decreasing surface area. In contrast the nucleation and growth rate of CHA on C-MBGs decrease to a large extent with increasing carboxylic groups content.

The opportunity to conjugate osteoproductive characteristics of bioactive glasses with periodontal regeneration potentialities of enamel matrix proteins has been explored [34]. Bioglass® granules as such or coated with Emdogain®(commercial mixture of enamel matrix proteins isolated from old pigs), by incubation in carbonate buffer containing the proteins, have been compared with a less reactive glass in *in vitro* tests with mouse pre-osteoblastic cell line (MC3T3-E1). It has been observed that Bioglass® alone or combined with Emdogain®, has the ability to support the growth of osteoblast-like cells *in vitro* and to promote osteoblast differentiation by stimulating the expression of major phenotypic markers. In addition, it has been noticed that the bioactive granules coated with Emdogain® reveal significantly higher protein production than the bioactive granules alone after 20 days of cell culture.

The effect of bioactive glass (Bioglass®) coating and of vascular endothelial growth factor release on *in vitro* cellular response and *in vivo* angiogenesis and bone regeneration has been studied [35]. Polymeric scaffolds (poly(lactide-co-glycolide)) have been loaded with VEGF and/or coated with Bioglass®. In the presence of a bioactive glass coating, no pronounced differences in the differentiation of human mesenchymal stem cells *in vitro* have been observed. However, it has been noted a significantly enhanced mitogenic stimulation of endothelial cells in the presence of the bioactive glass coating, with an additive effect with VEGF release. This trend has been maintained *in vivo*, where coated VEGF-releasing scaffolds demonstrate significant improvements in blood vessel density at 2 weeks *versus* coated control scaffolds. At 12 weeks, bone mineral density has been significantly increased in coated VEGF-releasing scaffolds *versus* coated controls, while only a slight increase in bone volume fraction has been observed.

Recently, Vernè *et al.* immobilized alkaline phosphatase (ALP) enzyme on bioactive glasses surface [36, 37]. Two different glasses have been investigated, the first one (SCNA - 57% SiO_2, 3% Al_2O_3, 34% CaO and 6% Na_2O – mol%) presents a simple composition and a low degree of bioactivity, while the second one (CEL2 – 45% SiO_2, 3% P_2O_5, 26% CaO, 7% MgO, 15% Na_2O, 4% K_2O) presents a more complex composition and a higher bioactivity. Silanization with 3-aminopropyltriethoxysilane (APTES), without glutaraldehyde employment has been considered for functionalization together with direct linking to surface exposed hydroxyls groups. ALP has been grafted maintaining its activity on both glasses (despite of the high surface reactivity of CEL2 one) with the two considered techniques. The enzyme can be released in solution slowly, by dipping sample in simulated body fluid, or faster by ultrasonic washings in physiological solutions; in all cases the enzyme maintain its activity in solution. ALP presence onto material surface does not inhibit the starting bioactivity of glasses, moreover it enhances and quickens hydroxyapatite precipitation kinetics. In fact the physiological role of this enzyme is the hydrolysis of organic phosphates, so its active site presents an high affinity for phosphate groups; in a simulated body fluid, rich in phosphate groups and calcium ions, ALP can act as nucleation centre for hydroxyapatite.

Surface functionalization can be employed also for the introduction of therapeutic molecules onto material surface. In this case the bone bonding and stimulating properties of bioactive glasses can be coupled with specific treatment requirements.

One of the most frequent complication of prosthetic surgery is infection development. In order to overcome or fight this situation several solutions have been proposed both in clinical and research fields.

In order to treat infections, antibiotic loaded polymethylmethacrylate (PMMA) beads and temporary prostheses are widely employed together with antibiotic loaded PMMA-based bone cements [38-41]. The opportunity to combine bioactive properties of bioactive glasses with antibacterial ones of antibiotics in bone cements has been explored adding both drug and glass powders into the cement composition [42]. Biodegradable systems (such as collagen sponge, polylactic and polyglicolic acids and calcium phosphates based ceramics) have also been considered for the localized delivery of drugs [38].

Antibiotic (isepamicin sulphate - ISP and cefmetazole - CMZ) loading into porous apatite-wollastonite glass-ceramics (A-W GC) has been performed under vacuum [43]. Drug absorption and release ratios almost proportional to the material porosity have been observed. The most porous A-W GC allowed ISP release at bactericidal concentration up to 42 days. Lower amounts have been obtained for CMZ.

A vacuum infiltration method has also been employed in order to introduce the antibiotic cefuroxime acetate into porous blocks of another bioactive glass (containing SiO_2, CaO, P_2O_5, B_2O_3 and Na_2O) [44]. *In vitro* drug release in amount higher than *S. aureus* MIC up to 6 days has been observed, *in vivo* drug loaded implants show the ability to reduce osteomyelitis in rabbit and induce new bone formation.

A borate bioactive glass powder (< 50 μm) has been used to prepare antibiotic (Teicoplanin) loaded pellet by the addition of chitosan, citric acid and glucose [45]. Drug release has been observed up to 30 days *in vitro* and up to 9 days *in vivo*. In physiological fluids pellets progressively degrade and transform into porous hydroxyapatite. Faster reactivity kinetics have been observed *in vivo*.

Moreover, a degradable and bioactive borate glass has been used as a carrier of vancomycin in the treatment of osteomyelitis [46]. Vancomycin was mixed with borate glass powders and phosphate buffered curing solution in order to create pellets for *in vivo* test on infected rabbit. The implants show excellent biocompatibility, they are very effective in eradicating osteomyelitis and simultaneously stimulate bone regeneration.

Cancer therapy is one of the main problems and challenges of the modern medicine. Several solutions have been explored in order to reduce the side effects of conventional systemic chemotherapy and radiotherapy, the majority of them are based on localized drug delivery. A lot of literature works employ surface functionalization of magnetic or polymeric nanoparticles in order to perform magnetic or immunological drug targeting devices [47, 48, 49]. Moreover also hydroxyapatite has been considered as support material for antineoplastic agents loading [50]. Recently surface grafting of chemotherapeutic drugs to a bioactive glass-ceramic has been carried out by Vernè *et al.* [51]. A bioactive and ferrimagnetic glass-ceramic (named SC45), previously developed in the same research group [52, 53] has been considered as base material. Due to iron oxides presence among starting reagents and to the particular temperature controlled melting route, the cited glass contains magnetite crystals embedded in an amorphous matrix analogous to Bioglass® one. Magnetite confers ferrimagnetic properties to the glass-ceramic that can be employed in order to obtain material heating under alternating magnetic field. This properties can be interestingly employed for the local application of heat in hyperthermic therapy of cancer. Thanks to the bioactive glass matrix this glass ceramics can be considered for bone substitution after solid tumours removal. The surface functionalization of this material with proper antineoplastic agents allows the combination of localized hyperthermia and chemotherapy.

A proper water-based pre-treatment has been developed in order to expose reactive –OH groups onto material surface avoiding silica gel formation. Two different chemotherapeutic agents (Cisplatinum and Doxorubicin) have been grafted directly to the hydroxylated surface. After one day of uptake at about the 80 % of doxorubicin and the 40% of Cisplatinum have been loaded onto the glass-ceramic samples. Drug release in solutions (distilled water for Doxorubicin and physiological saline for Cisplatinum) have been investigated, any trend has been observed but, for both drugs, the maximum release have been observed after one day and has been quantified as at about 15% for Doxorubicin and 50% for Cisplatinum. It must be considered that the active portion of Doxorubicin remains exposed after grafting so it is able to exploit its therapeutic function also from material surface, on the other hand drug release is fundamental for Cisplatinum effectiveness.

4. CONCLUSION

Surface can be defined as the visiting card of a material in the biological environment, in fact the interaction between an artificial object and the biological components happens mainly at their interface. Surface functionalization represents an useful tool for tailoring surface properties for specific applications.

Bioactive glasses as such are able to positively affect cellular and tissular behaviour thanks to their composition (bioactivity and ions release). Glass structure is particularly favourable for the introduction of various ions for different applications (antibacterial agents, cell stimulating ones…). Moreover bioactive surfaces reactivity can be employed for the exposition of active groups for biomolecular grafting. On the other hand it must be taken into account that this reactivity has not to excessively interfere with anchoring procedure, and so a compromise must be reached. Finally the maintenance of the starting bioactivity upon modification must be always verified.

The surface modification of bioactive glasses is a challenging opportunity to couple the basic potentialities of bare materials with new one for specific applications.

ACKNOWLEDGEMENTS

Declared none.

CONFLICT OF INTEREST

The author(s) confirm that this chapter content has no conflict of interest.

REFERENCES

[1] Cao W, Hench LL. Bioactive materials Ceram Int 1996; 22: 493-507

[2] Franks K, Abrahams J, Knowles C. Development of soluble glasses for biomedical use Part I: *In vitro* solubility measurement J. Mater Sci Mater Med 2000; 11:609-14

[3] Knowles JC, Phosphate based glasses for biomedical applications J Mater Chem 2003; 13: 2395–2401

[4] Demirkesen E, GoG. ¨ Filler Effect of Al_2O_3 additions on the acid durability of a Li2O–ZnO–SiO2 glass and its glass-ceramic Ceram Int 2003; 29:463–69.

[5] Gao T, Aro HT, Ylänen H, Vuorio E. Silica-based bioactive glasses modulate expression of bone morphogenetic protein-2 mRNA in Saos-2 osteoblasts in. Biomaterials 2001; 22:1475-83

[6] Hench LL. Genetic design of bioactive glass J Eur Cer Soc 2009; 29:1257–65
[7] Verné E, Miola M, Vitale Brovarone C, Cannas M, Gatti S, Fucale G, Maina G, Massé A, Di Nunzio S. Surface Silver-doping of Biocompatible Glass to Induce Antibacterial Properties. Part I: Massive Glass – J Mat Sc Mat Med 2009; 20(3): 733-40, .
[8] Miola M, Ferraris S, Di Nunzio S, Robotti PF, Bianchi G, Fucale G, Maina G, Cannas M, Gatti S, Massè A, Vitale Brovarone C, Vernè E. Surface silver-doping of biocompatible glasses to induce antibacterial properties. Part II: plasma sprayed glass-coatings J Mater Sci: Mater Med 2009; 20:741–49
[9] Vernè E, Ferraris S, Miola M, Fucale G, Maina G, Martinasso G, Canuto RA, Di Nunzio S, Vitale-Brovarone C. Synthesis and characterisation of bioactive and antibacterial glass–ceramic Part 1 – Microstructure, properties and biological behaviour. Adv Appl Ceram 2008;107:234-44
[10] Vernè E, Ferraris S, Miola M, Fucale G, Maina G, Robotti P, Bianchi G, Martinasso G, Canuto RA. Vitale-Brovarone C. Synthesis and characterisation of bioactive and antibacterial glass-ceramic Part 2 – plasma spray coatings on metallic substrates. Adv Appl Ceram 2008; 107:245-53
[11] Vitale-Brovarone C, Miola M, Balagna C, Verne E. 3D-glass–ceramic scaffolds with antibacterial properties for bone grafting. Chem Eng J 2008; 137: 129–36
[12] Balagna C, Vitale-Brovarone C, Miola M, Verne E, Canuto RA, Saracino S, Muzio G, Fucale G, Maina G. Biocompatibility and antibacterial effect of silver doped 3D-glass-ceramic scaffolds for bone brafting. J Biomater Appl 2010, doi: 10.1177/0885328209356603
[13] Di Nunzio S, Verne E. Process for the production of silver-containing prosthetic devices. WO2006/058906 A1.
[14] Xu Y, Cheng J, Zheng W, Gao D. Study on the preparation and properties of silver-doped borosilicate antibacterial glass. J Non-Cryst Solids 2008; 354:1341–45
[15] Ahmed I, Ready D, Wilson M, Knowles JC. Antimicrobial effect of silver-doped phosphate based glasses. J Biomed Mater Res A. 2006; 79: 618-26
[16] Kawashita M, Tsuneyama S, Miyaji F, Kokubo T, Kozuka H, Yamamoto K. Antibacterial silver-containing silica glass prepared by sol-gel method. Biomaterials 2000; 21:393-98
[17] Bellantone M, Williams HD, Hench LL. Broad-spectrum bactericidal activity of Ag_2O-doped bioactive glass. Antimicrob Agent Chemother 2002;.46:1940-45.
[18] Balamurugan A, Balossier G, Kannan S, Michel J, Rebelo AHS, Ferriera JMF Development and *in vitro* characterization of sol–gel derived CaO–P2O5–SiO2–ZnO bioglass. Acta Biomater 2007; 3: 255–62
[19] Aina V, Petardi A, Bergandi L, Malavasi G, Menabue L, Morterra C, Ghigo D. Cytotoxicity of zinc-containing bioactive glasses in contact with human osteoblasts. Chem-Biol Interact 2007; 167:
[20] Atmaca S, Gül K, Çiçek R. The Effect of Zinc on Microbial growth. Journal of medical sciences 1998; 28:595-97.
[21] Valappil SP, Ready D, Abou Neel EA, Pickup DM, Chrzanowski W, O'Dell LA, Newport R J, Smith ME, Wilson M, Knowles JC. Antimicrobial gallium-doped phosphate-based glasses. Adv Funct Mater 2008; 18: 732–41
[22] Valappil SP, Ready D, Abou Neel EA, Pickup DM, O'Dell LA, Chrzanowski W, Pratten J, Newport RJ, Smith ME, Wilson M, Knowles JC. Controlled delivery of antimicrobial gallium ions from phosphate-based glasses. Acta Biomater 2009; 5:1198–1210

[23] Mulligan AM, Wilson M, Knowles JC. The effect of increasing copper content in phosphate-based glasses on biofilms of *Streptococcus sanguis*. Biomaterials 2003;24: 1797-1807.

[24] Abou Neela EA, Ahmeda I, Prattenb J, Nazhata SN, Knowles JC. Characterisation of antibacterial copper releasing degradable phosphate glass fibres. Biomaterials 2005; 26:2247–54

[25] Xiao SJ, Kenausis G, Textor M, Biochemical modification of titanium surfaces, in: Brunette DM, Tengvall P, Textor M, Thomsen P, Titanium in Medicine, Springler-Verlag, Berlin Heidelberg New York; 2001; pp. 417-453

[26] De Jonge LT, Leeuwenburgh SCG, Wolke JGC, Jansen JA, Organic-inorganic surface modifications for titanium implant surfaces. Pharm Res 2008; 25:2357-2369

[27] Spriano S, Ferraris S, Bianchi CL, Cassinelli C, Torricelli P, Fini M, Rimondini L, Giardino R, Bioactive titanium surfaces, in: Titanium alloys: preparation, properties and applications, ISBN: 978-1-60876-151-7 Editor: Pedro N. Sanchez 2010 Nova Science Publishers, Inc.

[28] Luginbuehl V, Meinel L, Merkle H P, Gander B. Localized delivery of growth factors for bone repair. Eur J Pharm Biopharm 2004; 58: 197-208

[29] Lee S-H, Shin H. Matrices and scaffold for delivery of bioactive molecules in bone and cartilage tissue engineering. Adv Drug Deliv Rev 2007; 59: 339-359

[30] Chen QZ, Rezwan K, Armitage D, Nazhat SN, Boccaccini AR. The surface functionalization of 45S5 Bioglass®-based glass-ceramic scaffolds and its impact on bioactivity. J Mater Sci Mater Med 2006; 17: 979-87

[31] Chen QZ, Rezwan K, Françon V, Armitage D, Nazhat SN, Jones FH, Boccaccini AR. Surface functionalization of Bioglass®-derived porous scaffolds. Acta Biomater 2007; 3: 551-62

[32] Chen X, Guo C, Zhao N. Preparation and characterization of the sol-gel-nano-bioactive glasses modified by the coupling agent gamma aminopropyltriethoxysilane. Appl Surf Sci 2008; 255:466-68

[33] Sun J, Li Y, Li L, Zhao W, Li L, Gao J, Ruan M, Shi J. Functionalization and bioactivity *in vitro* of mesoporous bioactive glasses. J Non-Cryst Solids 2008; 354: 3799-3805

[34] Hattar S, Asselin A, Greenspan D, Oboeuf M, Berdal A, Sautier J.M. Potential of biomimetic surfaces to promote *in vitro* osteoblast-like cell differentiation. Biomaterials 2005; 26:839–48

[35] Leach JK, Kaigler D, Wang Z, Krebsbach PH, Mooney DJ. Coating of VEGF-releasing scaffolds with bioactive glass for angiogenesis and bone regeneration. Biomaterials 2006; 27: 3249–55

[36] Vernè E; Ferraris S.; Vitale Brovarone C; Spriano S; Bianchi C.L; Naldoni A; Morra M; Cassinelli C, Alkaline phosphatase grafting on bioactive glasses and glass-ceramics. Acta Biomater 2010; 6:229–40

[37] Verne' E, Ferraris S, Vitale Brovarone C, Spriano S, Bianchi CL, Morra M, Cassinelli C. Surface functionalization of biomaterials with alkaline phosphatase. Key Eng Mater 2008; 361-363: 593-596

[38] Nandi SK, Mukeherjee P, Roy S, Kundu B, De DK, Basu D. Local antibiotic delivery systems for the treatment of osteomyelitis – A review. Mater Sci Eng C 2009; 29:2478-85

[39] Diefenbeck M, Mückley T, Hofmann GO. Prophylaxis and treatment of implant-related infections by local application of antibiotics. Injury Int J Care Injured 2006; 37: S95—S104

[40] Cui Q, Mihalko WM, Shields JS, Ries M, Saleh KJ. Antibiotic-impregnated cement spacers for the treatment of infection associated with total hip or knee arthroplasty. J Bone Joint Surg Am. 2007; 89: 871-82

[41] Jiranek WA, Hanssen AD, Greenwald AS. Antibiotic-loaded bone cement for infection prophylaxis in total Joint replacement. JBJS Am 2006; 88:2487-2500,

[42] Arcos D, Ragel CV, Vallet-Regi M. Bioactivity in glass/PMMA composites used as drug delivery systems. Biomaterials 2001; 22:701-8

[43] Kawanabe K, Okada Y, Matsusue Y, Iida H, Nakamura T. Treatment of osteomyelitis with antibiotic-soaked porous glass-ceramic. JBJS 1998; 80B: 527-30

[44] Nandi SK, Kundu B, Mukherjee P, Mandal TK, Datta S, De DK, Basu D. *In vitro* and *in vivo* release of cefuroxime axetil from bioactive glass as implantable delivery system in experimental osteomyelitis. Ceram Int 2009; 35:3207-16

[45] Zhang X, Jia W, Gu YF, Xiao W, Liu X, Wang DP, Zhang CQ, Huang WH, Rahaman MN, Day DE, Zhou N. Teicopain-loaded borate bioactive glass implants for treating chronic bone infection in a rabbit osteomyelitis mode.l Biomaterials 2010; 31:5865-74207–18

[46] Xie Z, Liu X, Jia W, Zhang C, Huangb W, Wang J. Treatment of osteomyelitis and repair of bone defect by degradable bioactive borate glass releasing vancomycin. J Control Release 2009; 139:118-26

[47] S.lubbe A, Alexiou C, Bergemann C. Clinical applications of magnetic drug targeting. J Surg Res 2001; 95:200-6

[48] Patil YB, Toti US, Khdair A, Mac L, Panyamc J. Single-step surface functionalization of polymeric nanoparticles for targeted drug delivery. Biomaterials 2009; 30:859–66

[49] Nie Y, Zhang Z, Li L, Luo K, Ding H, Gu Z. Synthesis, characterization and transfection of a novel folatetargeted multipolymeric nanoparticles for gene delivery. J Mater Sci: Mater Med 2009; 20:1849–57

[50] Itokazu M, Esaki M, Yamamoto K, Tanemori T, Kasai T. Local drug delivery system using ceramics: vacuum method for impregnating a chemotherapeutic agent into a porous hydroxyapatite block. J Mater Sc. Mater Med 1999; 10:249-52

[51] Vernè E, Miola M, Ferraris S, Bianchi CL, Naldoni A, Maina G, Bretcanu O. Surface activation of a ferrimagnetic glass-ceramic for antineoplastic drugs grafting, Adv Biomater DOI: 10.1002/adbi.200900082

[52] Bretcanu O, Verne′ E, Cöisson M, Tiberto P, Allia P. Temperature effect on the magnetic properties of the coprecipitation derived ferrimagnetic glass-ceramics. J Magn Magn Mater 2006; 300:412–17

[53] Bretcanu O, Verne′ E, Cöisson M, Tiberto P, Allia P. Magnetic properties of the ferrimagnetic glass-ceramics for hyperthermia. J Magn Magn Mater 2006; 305:529–33.

INDEX

A

B

Lia Rimondini, Claudia Bianchi and Enrica Vernè (Eds)

R

S

Z

Made in the USA
Las Vegas, NV
03 November 2026

57425088R00252